Optimization of Energy Systems

Optimization of Energy Systems

Ibrahim Dincer and Marc A. Rosen
University of Ontario Institute of Technology, Canada

Pouria Ahmadi
Simon Fraser University, Canada

Registered Office
John Wiley & Sons Ltd, The Atrium, Southern Gate, Chichester, West Sussex, PO19 8SQ, UK

Editorial Office
The Atrium, Southern Gate, Chichester, West Sussex, PO19 8SQ, UK

For details of our global editorial offices, customer services, and more information about Wiley products visit us at www.wiley.com.

Library of Congress Cataloging-in-Publication Data applied for

ISBN: 9781118894439

Cover Design: Wiley
Cover Images: (Background) © nadla/Gettyimages
Inset Image: Courtesy of Author

Set in 10/12pt WarnockPro by SPi Global, Chennai, India
Printed and bound in Malaysia by Vivar Printing Sdn Bhd

10 9 8 7 6 5 4 3 2 1

Contents

Acknowledgements

Completion of this book on *Optimization of Energy Systems* was challenging, and could not have been completed without the support of others. In particular, the contributions of some graduate students and colleagues are gratefully acknowledged, including Parisa Heidarnejad, Hasan Barzegaravval, Saeed Javan, Saeed Rohani and Shoaib Khanmohammadi as are materials from Dr. Tahir Ratlamwala and Dr. Hassan Hajabdollahi.

Ibrahim Dincer warmly registers his appreciations to his wife, Gulsen Dincer and his children Meliha, Miray, Ibrahim Eren, Zeynep and Ibrahim Emir Dincer for their support and motivation as well as his parents Fatma and Hasan Dincer, who passed away many years ago, for their dedication and unconditional love and support.

Marc A. Rosen thanks his wife, Margot, his children, Allison and Cassandra, and his parents, Sharon and Lou, for their inspiration, love and support.

Finally, Pouria Ahmadi would like to thank his parents, Morad and Batoul, for their understanding and encouragement throughout his education.

Preface

Optimization is a significant tool in engineering for determining the best, or optimal, value for the decision variable(s) of a system. For various reasons, it is important to optimize processes so that a chosen quantity, known as the objective function, is maximized or minimized. For example, the output, profit, productivity, product quality, and so on, may be maximized, or the cost per item, investment, energy input, and so on, may be minimized. The success and growth of industries today is strongly based on their ability to optimize systems and processes, as well as their designs. With the advent in recent years of new materials, such as composites and ceramics, and new manufacturing processes, several traditional industries (e.g., steel) have faced significant challenges and, in some cases, diminished in size, while many new fields have emerged. It is important to exploit new techniques for product improvement and cost reduction in traditional and new industries. Even in expanding areas, such as consumer electronics, the prosperity of a company is closely connected to its ability to apply optimization to new and existing process and system designs. Consequently, engineering design, which has always been important, has become increasingly coupled with optimization.

Energy engineering is a field where optimization plays a particularly important role. Engineers involved in thermal engineering, for instance, are required to answer such questions as

- What processes or equipment should be selected for a system, and how should the parts be arranged for the best outcome?
- What are the best characteristics for the components (e.g., size, capacity, cost)?
- What are the best process parameters (e.g., temperature, pressure, flow rate, and composition) of each stream interacting with the system?

In order to answer such questions, engineers are required to formulate an appropriate optimization problem. Proper formulation is usually the most important and sometimes the most difficult step in optimization. To formulate an optimization problem, there are numerous elements that need to be defined, including system boundaries, optimization criteria, decision variables, and objective functions. In order to have an optimized system that can reduce the cost and environmental impact and at a same time increase the efficiency of the system, optimization is useful.

This book is a research-oriented textbook. It provides comprehensive coverage of fundamentals and main concepts, and can be used for system design, analysis, assessment, optimization, and hence improvement. The book includes practical features in a usable format often not included in other solely academic textbooks. The book can used by senior undergraduate and graduate students in mainstream engineering fields (such as mechanical and electrical engineering) and as well as specialized engineering programs on energy systems.

This book consists of twelve chapters. Chapter 1 addresses general concepts, fundamental principles and basic aspects of thermodynamics, energy, entropy, and exergy. These topics are covered in a broad manner, so as to furnish the reader with the background information necessary for subsequent chapters. Chapter 2 describes several modeling techniques and optimization methods and the formulation of an optimization problem. Objective functions and how to select them for enhanced sustainability, optimization constraints for energy systems, and optimization algorithms are explained in detail in this chapter. Chapter 3 focuses on modeling and optimization of thermal components. Chapter 4 covers the modeling and optimization of various types of heat exchanger as well as sensitivity analyses of the optimization results. Chapter 5 provides necessary information for modeling and optimization of refrigeration systems. Chapter 6 describes the modeling and optimization of heat pump systems. Chapter 7 focuses on fuel cell system modeling, analysis, assessment, and optimization. Chapter 8 covers a range of renewable energy systems and their modeling, analysis, and optimization. Chapter 9 focuses on conventional power generating systems. Chapter 10 addresses the modeling and optimization of cogeneration and trigeneration systems. Chapter 11 delves into the modeling, analysis, and optimization of integrated multigeneration systems for the production of multiple useful outputs.

Incorporated throughout are many illustrative examples and case studies, which provide the reader with a substantial learning experience, especially in areas of practical application. Complete references are included to point the curious reader in the right direction. Information on topics not covered fully in the text can, therefore, be easily found. We hope this book brings a new dimension to energy system modeling and optimization practice and helps the community implement better solutions for a better future.

Ibrahim Dincer, Marc A. Rosen, and Pouria Ahmadi

1

Thermodynamic Fundamentals

1.1 Introduction

Energy plays a critical role in driving almost all practical processes and is essential to sustain life. Energy exists in several forms, for example, light, heat, and electricity. Energy systems are widespread and used in diverse industries such as power generation, petrochemical processing, refrigeration, hydrogen production, chemical processing, and manufacturing. Interest is growing in producing superior energy products at minimal cost, while satisfying concerns regarding environmental impact, safety, and other issues. It is no longer adequate to develop a system that simply performs a desired task. For various reasons, it is often important to optimize processes so that a chosen quantity, known as the objective function, is maximized or minimized. For example, the output, profit, productivity, product quality, and so on, may be maximized, or the cost per item, financial investment, energy input, and so on, may be minimized. The success and growth of industries today is strongly based on their ability to optimize designs and systems.

When an engineer undertakes the analysis of an energy system and/or its application, she or he should deal with several basic factors first. These depend on the type of the problem being studied, and often involve such disciplines as thermodynamics, fluid mechanics, and heat transfer. Consequently, it is helpful to introduce several fundamental definitions and concepts before moving on to detailed energy systems applications, especially for readers who lack a background in thermodynamics, fluid mechanics, or heat transfer.

This chapter provides such a review, and is intended to give novice and practicing energy systems engineers a strong understanding of fundamentals, including physical phenomena, basic laws and principles, and governing relations, as well as a solid grounding in practical aspects. This introductory chapter covers relevant fundamentals involved in the optimization of energy systems. We begin the chapter with a summary of fundamental definitions and physical quantities, with their units, dimensions, and interrelations. We then consider introductory aspects of thermodynamics, with a particular focus on energy, exergy, and heat transfer.

1.2 Thermodynamics

Energy is needed for almost every activity. In simple terms, energy is usually thought of as the ability to carry out useful tasks like producing work and heating. Energy is

Optimization of Energy Systems, First Edition. Ibrahim Dincer, Marc A. Rosen, and Pouria Ahmadi.
© 2017 John Wiley & Sons Ltd. Published 2017 by John Wiley & Sons Ltd.

contained in the fuel we use, the food we eat, and the places we live. Energy enables such outcomes as transportation, fresh water, and thermal comfort in buildings.

Energy use has drawbacks too. It can be dangerous if not used cautiously and often leads to pollution and environmental damage.

Energy can be converted from one form to another, but cannot be created or destroyed. Work and heat are two categories of energy in transit. Thermodynamics plays a key role in the analysis of processes, systems, and devices in which energy transfers and transformations occur. The implications of thermodynamics are far reaching and applications span the range of human enterprise. Nature allows the conversion of work completely into heat, but not the reverse. Additionally, converting heat into work requires a device, which is often complex (e.g., an engine).

Although energy can be transformed into different forms, the principle of conservation of energy states that the total energy of a system can only change if energy is transferred into or out of the system. This implies that it is impossible to create or destroy energy. The total energy of a system can this be calculated by adding all forms of energy in the system. Examples of energy transfer and transformation include generating or making use of electric energy, performing chemical reactions and lifting an object. Lifting against gravity performs work on the object and stores gravitational potential energy; if the object falls, gravity does work on the object, which transforms its potential energy into kinetic energy associated with its speed.

The name "thermodynamics" stems from the Greek words *therme* (heat) and *dynamis* (power), which is descriptive of efforts to convert heat into power [1]. The discipline of thermodynamics is based primarily on two fundamental natural laws, known as the first and second laws. The first law of thermodynamics is simply an expression of the conservation of energy principle. It states that energy, as a thermodynamic quantity, is neither created nor destroyed during a process. The second law of thermodynamics states that energy has quality as well as quantity, and that processes naturally occur in the direction of decreasing quality of energy [2].

1.3   The First Law of Thermodynamics

The first law of thermodynamics (FLT) embodies the principle of energy conservation, which states that, although energy can change form, it can be neither created nor destroyed. The FLT defines internal energy as a state function and provides a formal statement of the conservation of energy [2]. However, the first law provides no information about the direction in which processes can spontaneously occur, that is, reversibility aspects of thermodynamic processes. For example, the FLT cannot indicate how cells can perform work while existing in an isothermal environment. The FLT provides no information about the inability of any thermodynamic process to convert heat fully into mechanical work, or any insight into why mixtures cannot spontaneously separate or unmix themselves. A principle to explain these phenomena and to characterize the availability of energy is required. Such a principle is embodied in the second law of thermodynamics (SLT), which we explain later in this chapter.

A simple example of energy conversion is the process in which a body that has some potential energy at some elevation falls, and part of the potential energy is converted into kinetic energy. Experimental data show that the decrease in potential energy is equal to the increase in kinetic energy if air resistance is negligible. This simple example demonstrates the conservation of energy principle. In order to analyze energy systems, we need to use energy balance equations, which express the balance of the energy entering and leaving a system and the energy change in the system. That is, the net change in the total energy of the system during a process is equal to the difference between the total energy entering and the total energy leaving the system during that process, or

$$\begin{pmatrix} \text{Total energy} \\ \text{entering system} \end{pmatrix} - \begin{pmatrix} \text{Total energy} \\ \text{leaving system} \end{pmatrix} = \begin{pmatrix} \text{Change in total} \\ \text{energy of system} \end{pmatrix}$$

This relation can also be written as

$$E_{\text{in}} - E_{\text{out}} = \Delta E_{\text{system}} \tag{1.1}$$

The energy E may include internal energy U, kinetic energy (KE) and potential energy (PE) terms as follows:

$$E = U + KE + PE \tag{1.2}$$

For a change of state from state 1 to state 2 with a constant gravitational acceleration (g), Equation 1.2 can be used to show the following:

$$E_2 - E_1 = (U_2 - U_1) + \tfrac{1}{2}m(V_2^2 - V_1^2) + mg(Z_2 - Z_1) \tag{1.3}$$

where m denotes the fixed amount of mass in the system, V the velocity, and Z the elevation. In order to apply the FLT to an energy system, we need to know some further concepts, which are described in the following sections.

1.3.1 Thermodynamic System

A thermodynamic system is a region or device or combination of devices that contains a certain quantity of matter. It is important to carefully define the system under consideration during an analysis and its boundaries. Three important types of systems can be defined:

- **Closed system**. Such a system is defined as one across the boundaries of which no material crosses. It therefore contains a fixed quantity of matter. Sometimes this is also called a control mass.
- **Open system**. This is defined as a system in which material (mass) is allowed to cross the boundaries. The term open system is sometimes referred to as a control volume.
- **Isolated system**. This is a closed system that is independent and unaffected by the surroundings. No mass, heat, or work crosses its boundary.

1.3.2 Process

A process is a physical or chemical change in the properties of matter or the conversion of energy from one form to another. In some processes, one property remains constant. The prefix "iso" is employed to describe such as process, for example isothermal (constant temperature), isobaric (constant pressure), and isochoric (constant volume).

1.3.3 Cycle

A cycle is a series of thermodynamic processes in which the end point conditions or properties of the matter are identical to the initial conditions.

1.3.4 Heat

Heat is the thermal form of energy, and heat transfer takes place when a temperature difference exists within a medium or between different media. The definitive experiment that showed heat to be a form of energy, convertible into other forms, was carried out by Scottish physicist James Joule. Heat transfer always requires a difference in temperature, and higher temperature differences provide higher heat transfer rates. The units for heat are joules or kilojoules in the International System (SI) and the foot pound force or British thermal unit (Btu) in the English system. In terms of sign conventions in thermodynamic calculations, a common one states that heat transfer to a system is considered positive, while heat transfer from a system is negative. If there is no heat transfer involved in a process, it is called adiabatic.

1.3.5 Work

Work is the energy that is transferred by a difference in pressure or force of any kind, and is subdivided into shaft work and flow work. Shaft work is the mechanical energy used to drive a mechanism such as a pump, compressor, or turbine. Flow work is the energy transferred into a system by a fluid flowing into, or out of, it. Both forms are usually expressed in kilojoules. Work can also be expressed on a unit mass basis (e.g., in kJ/kg). By one common convention, work done by a system is usually considered positive and work done on a system (work input) is considered negative. The SI unit for power or rate of work is joules per second, which is a Watt (W).

1.3.6 Thermodynamic Property

A thermodynamic property is a physical characteristic of a substance, often used to describe its state. Any two properties usually define the state or condition of a substance, and all other properties can be derived from these. Some examples of properties are temperature, pressure, enthalpy, and entropy. Thermodynamic properties can be classified as intensive (independent of the size or scale, e.g., pressure, temperature and density) and extensive properties (dependent on size or scale, e.g., mass and total volume). Extensive properties become intensive properties when expressed on a per unit mass basis, such as is the case for specific volume.

Property diagrams of substances can be presented in graphical form and present the main properties listed in property tables, for example refrigerant or steam tables. In analyzing an energy system, the thermodynamic properties should be defined so as to permit the simulation (and hence optimization) of the system. In this book, Engineering Equation Solver (EES) and Refprop software are utilized to calculate thermodynamic properties.

1.3.6.1 Specific Internal Energy

Internal energy represents a molecular state type of energy. Specific internal energy is a measure per unit mass of the energy of a simple system in equilibrium, and can be expressed as the function $c_v dT$. For many thermodynamic processes in closed systems, the only significant energy changes are internal energy changes, and the work done by the system in the absence of friction is the work of pressure-volume expansion, such as in a piston-cylinder mechanism.

The specific internal energy of a mixture of liquid and vapor can be written as

$$u = (1 - x)u_{\text{liq}} + xu_{\text{vap}} = u_{liq} + xu_{\text{liq,vap}} \tag{1.4}$$

where $u_{\text{liq,vap}} = u_{\text{vap}} - u_{\text{liq}}$.

1.3.6.2 Specific Enthalpy

Specific enthalpy is another measure of the energy per unit mass of a substance. Specific enthalpy, usually expressed in kJ/kg or Btu/lb, is normally expressed as a function of $c_p dT$. Since enthalpy is a state function, it is necessary to measure it relative to some reference state. The usual practice is to determine the reference values, which are called the standard enthalpy of formation (or the heat of formation), particularly in combustion thermodynamics.

The specific enthalpy of a mixture of liquid and vapor components can be written as

$$h = (1 - x)h_{\text{liq}} + xh_{\text{vap}} = h_{\text{liq}} + xh_{\text{liq,vap}} \tag{1.5}$$

where $h_{\text{liq,vap}} = h_{\text{vap}} - h_{\text{liq}}$.

1.3.6.3 Specific Entropy

Entropy is the ratio of the heat added to a substance to the absolute temperature at which it is added, and is a measure of the molecular disorder of a substance at a given state.

The specific entropy of a mixture of liquid and vapor components can be written as

$$s = (1 - x)s_{\text{liq}} + xs_{\text{vap}} = s_{\text{liq}} + xs_{\text{liq,vap}} \tag{1.6}$$

where $s_{\text{liq,vap}} = s_{\text{vap}} - s_{\text{liq}}$.

1.3.7 Thermodynamic Tables

Thermodynamic tables were first published in 1936 as steam tables by Keenan and Keyes; in 1969 and 1978 these were revised and republished. Thermodynamic tables are available for many substances, ranging from water to refrigerants, and are commonly employed in process design calculations. Some thermodynamic tables include steam and vapor tables. In this book, we usually imply these types of tables when we refer to thermodynamic tables. Such tables normally have distinct phases (parts); for example, four different parts for water tables include those for saturated water, superheated water vapor, compressed liquid water, and saturated solid-saturated vapor water. Similarly, two distinct parts for R-134a include saturated and superheated tables. Most tables are tabulated according to values of temperature and pressure, and then list values of various other thermodynamic parameters such as specific volume, internal energy, enthalpy, and entropy. Often when we have values for two independent variables, we may obtain other data from the respective table. In learning how to use these tables, it is important to specify the state using any two independent parameters (unless more are needed). In some design calculations, if we do not have the exact values of the parameters, we use interpolation to find the necessary values.

Beyond thermodynamic tables, much attention has recently been paid to computerized tables for design calculations and other purposes. Although computerized tables can eliminate data reading problems, they may not provide a good understanding of the concepts involved and a good comprehension of the subject. Hence, in thermodynamics courses, it is important for the students to know how to obtain thermodynamic data from the appropriate thermodynamic tables.

The Handbook of Thermodynamic Tables [3] is one of the most valuable data resources, with data for numerous solid, liquid, and gaseous substances.

1.3.8 Engineering Equation Solver (EES)

EES is a software package that solves a system of linear or nonlinear algebraic or differential equations numerically. It consists of a large library of built in thermodynamic properties as well as mathematical functions. Unlike other available software packages, EES does not solve engineering problems explicitly; rather it solves the equations, and it is the user's responsibility to apply relevant physical laws, relations, and understanding. EES saves the user considerable time and effort by solving the mathematical equations, and has the capability to connect to other professional software such as Matlab.

EES is one of the most suitable software packages for energy systems analyses and thermodynamic properties. The software is straightforward to use, as shown in Figure 1.1, where the thermodynamic modeling of a vapor compression refrigeration system and its components is presented along with the results of a parametric study conducted by varying the temperature at point 1. This software comes with several examples and has been used to model and analyze complex systems such as advanced power plants, combined heat and power systems, desalination plants, hydrogen production plants, and renewable energy-based systems.

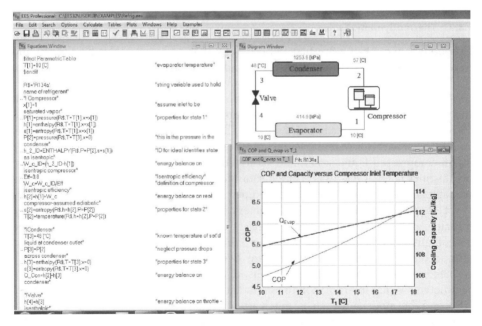

Figure 1.1 Sample screen of Engineering Equation Solver (EES) for thermodynamic analyses of energy systems.

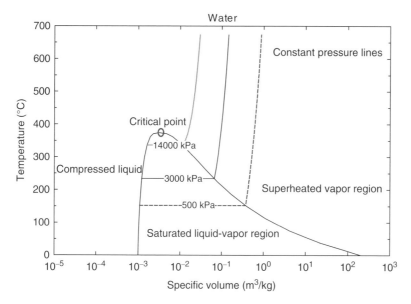

Figure 1.2 Temperature-specific volume diagram for phase changes of water.

The professional version of the software can perform single objective optimization but not multi-objective optimization. However, several methods can be implemented to connect EES with Matlab software to conduct multi-objective optimization, as is discussed in subsequent chapters.

In EES, we can easily create thermodynamic diagrams (e.g., T-V, T-S, P-V, P-h, and h-S). For example, Figure 1.2 shows a temperature-volume diagram for water at several pressures. The state of a system or substance is defined as the condition of the system or substance characterized by certain observable macroscopic values of its properties such as temperature and pressure.

The term "state" is often used interchangeably with the term "phase," for example solid phase or gaseous phase of a substance. Each of the properties of a substance at a given state has only one definite value, regardless of how the substance reached the state. For example, when sufficient heat is added or removed under certain conditions, most substances undergo a state change. The temperature remains constant until the state change is complete. As shown in Figure 1.2, there are three regions on the T-v diagram for water: compressed liquid, saturated two-phase region (which is also known as the wet region) and superheated vapor region. As the pressure increases, the saturated line continues to shrink, as shown in Figure 1.2, and eventually it becomes a point when the pressure reaches 22.06 MPa for water. This point is called the critical point and is defined as the point which the saturated liquid and saturated vapor are identical. This is indicated in Figure 1.2. EES can also generate temperature-specific entropy (T-s) diagrams for various refrigerants, as shown in Figure 1.3. The software contains a large fluids database. Table 1.1 lists the properties of the various refrigerants shown in Figure 1.3. In this table, $T_b{}^a$ is the normal boiling point, while $P_{cr}{}^b$ and $T_{cr}{}^a$ are the critical pressure and critical temperature, respectively.

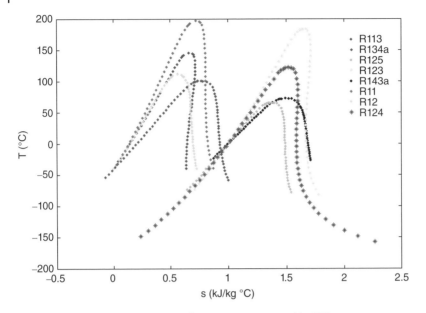

Figure 1.3 *T-s* diagrams of selected refrigerants, as generated by EES.

Table 1.1 Thermophysical properties of selected working fluids.

Substance	Molecular mass (kg/kmol)	$T_b{}^a$ (K)	$P_{cr}{}^b$ (MPa)	$T_{cr}{}^a$ (K)
R123	152.93	300.97	3.66	456.83
R134a	102.03	247.08	4.059	374.21
R124	136.48	261.22	3.62	395.43
R11	137.37	296.86	4.40	471.11
R12	120.91	243.4	4.13	385.12
R143a	84.04	161.34	3.76	345.86
R113	187.38	320.74	3.39	487.21
R125	120.02	172.52	3.61	339.17
R141b	116.95	305.2	4.46	479.96

Example 1.1: Apply EES to the vapor compression refrigeration cycle using R134a as a working fluid, as shown in Figure 1.4. The working fluid mass flow rate is 0.1 kg/s. Assume that there are no pressure drops across the condenser and that the compressor isentropic efficiency is 0.8.

Part 1) Show the cycle on a *P-h* diagram with all temperatures at each state point.
Part 2) Calculate the coefficient of performance (COP), compressor specific work and cooling load when the evaporator and condenser temperatures are 10°C and 45°C, respectively.
Part 3) Plot the effect of evaporator temperature on COP and cooling load of the cycle when the evaporator temperature varies from 10°C to 18°C, and discuss the results.

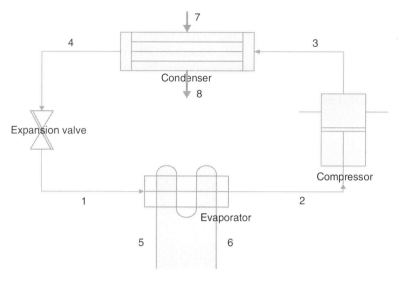

Figure 1.4 Schematic of vapor compression refrigeration cycle.

Solution: A refrigeration system operating on a vapor compression cycle is considered here.

Assumptions: 1: Steady operating conditions exist. 2: Kinetic and potential energy changes are negligible.

Analysis:

1) The working fluid leaving the evaporator is a saturated vapor with a quality $x = 1$. Therefore this point lies on the saturated vapor line in a *P-h* diagram, as shown in Figure 1.5. We know that the pressure at point 1, P_1, is the saturation pressure

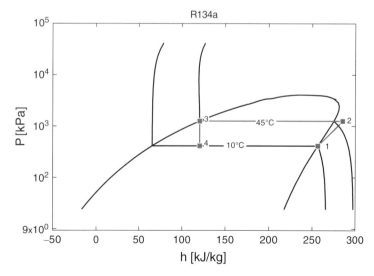

Figure 1.5 *P-h* diagram of vapor compression refrigeration cycle.

at the evaporator temperature $(T_{EVP} = 10°C)$. Using EES and the thermodynamic properties for R134a, this pressure is found to be 538 kPa. Since there is no pressure drop across the evaporator, $P_1 = P_4$. Similarly, the pressure at point 3 is equal to the saturation pressure at the condenser temperature. This information yields that $P_3 = 1161$ kPa. Connecting these points forms the P-h diagram.

2) In order to calculate the COP and cooling load, the thermodynamic properties at all points need to be determined. Using EES we find:

$$x_1 = 1$$
$$P_1 = \text{pressure } (R134a, \ T = T_1, \ x = x_1) = 415 \text{ kPa}$$
$$h_1 = \text{specific enthalpy } (R134a, \ T = T_1, \ x = x_1) = 256.2 \text{ kJ kg}^{-1}$$
$$s_1 = \text{specific entropy } (R134a, \ T = T_1, \ x = x_1) = 0.926 \text{ kJ(kg.K)}^{-1}$$
$$P_2 = \text{pressure } (R134a, \ T = T_3, \ x = 0) = 1161 \text{ kPa}$$
$$h_{2,ideal} = \text{specific enthalpy } (R134a, \ P = P_2, \ s_2 = s_1) = 277.5 \text{ kJ kg}^{-1}$$
$$w_{comp,ideal} = (h_{2,ideal} - h_1) = 21.31 \text{ kJ kg}^{-1}$$
$$\eta_{comp} = \frac{w_{comp,ideal}}{w_{comp,act}} \rightarrow w_{comp,act} = \frac{(h_{2,ideal} - h_1)}{\eta_{comp}} = 26.64 \text{ kJ kg}^{-1}$$
$$h_2 = h_1 + w_{comp,act} = 282.8 \text{ kJ kg}^{-1}$$
$$s_2 = \text{specific entropy } (R134a, \ h = h_2, \ P = P_2) = 0.926 \text{ kJ(kg.K)}^{-1}$$
$$T_2 = \text{temperature } (R134a, \ h = h_2, \ P = P_2) = 53°C$$

Since the condenser temperature is given (45°C) and $P_3 = P_2$ because there is no pressure drop across the condenser, we have:

$$h_3 = \text{specific enthalpy } (R134a, \ T = T_3, \ x = 0) = 115.8 \text{ kJ kg}^{-1}$$
$$s_3 = \text{specific entropy } (R134a, \ T = T_3, \ x = 0) = 0.418 \text{ kJ(kg.K)}^{-1}$$

The process in the throttle valve is isenthalpic, so $h_4 = h_3$. Since there is no pressure drop across the evaporator, $P_4 = P_1$. Point 4 is in the saturation region and the quality at point 4 is defined as follows:

$$x_4 = \text{quality } (R134a, \ h = h_4, \ P = P_4) = 0.26$$
$$s_4 = \text{specific entropy } (R134a, \ h = h_4, \ P = P_4) = 0.43 \text{ kJ(kg.K)}^{-1}$$
$$T_4 = \text{temperature } (R134a, \ h = h_4, \ P = P_4) = 10°C$$

Now that the thermodynamic properties at all points are calculated, we can use the energy balance equation for a control volume around the evaporator and compressor in order to determine the cooling load and compressor work rate as follows:

$$\dot{Q}_{Evp} = \dot{m}(h_1 - h_4) = 14.04 \text{ kW}$$
$$\dot{W}_{comp} = \dot{m}w_{comp,act} = 2.66 \text{ kW}$$

The coefficient of performance (COP) of the refrigerator is expressed as

$$\text{COP} = \frac{\text{Cooling load}}{\text{Work input rate}} = \frac{\dot{Q}_{Evp}}{\dot{W}_{comp}} = \frac{14.04 \text{ kW}}{2.66 \text{ kW}} = 5.27$$

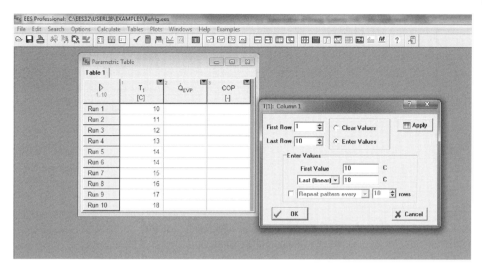

Figure 1.6 Screen shot of the parametric analysis of the system using EES.

3) To assess the effect of evaporator temperature on the COP and the cooling load, we use parametric tables in EES. To do this, we need to remove the evaporator temperature (T_1) in EES and go to the Table option in EES and select New Parametric Table and add T_1, COP and $\dot{Q}_{Evp}$. Next, we need to enter the range for evaporator temperature as shown in Figure 1.6. After entering all inputs, we press the green button to run and obtain the results (see Figure 1.7).

Figure 1.7 shows the variation of evaporator temperature on both cooling load and system COP. An increase in this temperature while other design parameters are fixed results in an increase the specific enthalpy at point 1, which eventually leads to an increase in the evaporator cooling load. Similarly, this increase results in an increase in the COP of the system according to the COP definition (COP = (Cooling load produced)/(Work input)). It is clear that the higher the evaporator temperature, the higher the cooling load and the COP of the system.

Example 1.2: Consider the adiabatic combustion of methane at 25°C with a stoichiometric amount of air at 25°C, as shown in Figure 1.8. Plot the variation of adiabatic combustion temperature with percentage of excess air.

Solution: An adiabatic reactor is considered in which combustion takes place with a stoichiometric amount of air at 25°C.

The combustion reaction for methane with stoichiometric air at 25°C and X% excess air is:

$$CH_4 + 2\left(1 + \frac{X}{100}\right)(O_2 + 3.76\,N_2) \leftrightarrow CO_2 + 2H_2O$$
$$+ 3.76\left(2 + \frac{2X}{100}\right)N_2 + \frac{2X}{100}O_2$$

Denoting the adiabatic combustion temperature T_f, we can write the energy balance for a control volume around the reactor:

$$\sum N_p H_p = \sum N_R H_R$$

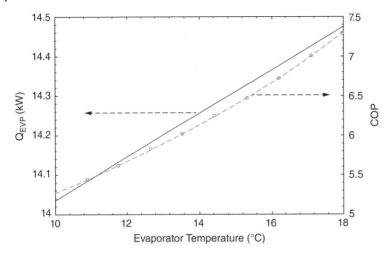

Figure 1.7 Effect of varying evaporator temperature on cooling load and COP of the system.

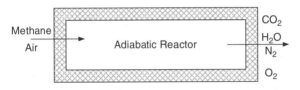

Figure 1.8 Schematic of adiabatic reactor.

where N_P and N_R are the stoichiometric coefficients for products and reactants and H_R and H_p are the enthalpies of the products and reactants, calculated as:

$$H_R = \text{enthalpy } (CH_4, T = 298) + 2\left(1 + \frac{X}{100}\right) \text{entahlpy } (O_2, T = 298 \text{ kJ kg}^{-1})$$
$$+ 3.76\left(2 + \frac{2X}{100}\right) \text{enthalpy } (N_2, T = 298 \text{ kJ kg}^{-1})$$
$$H_p = \text{enthalpy}(CO_2\, T = T_f) + 2\,\text{entahlpy } (H_2O, T = T_f)$$
$$+ 3.76\left(2 + \frac{2X}{100}\right) \text{enthalpy } (N_2, T = T_f) + \frac{2X}{100}\text{enthalpy } (O_2, T = T_f)$$

Then, we solve the energy balance equation ($\sum N_p H_p = \sum N_R H_R$) for T_f at any given excess air percentage X% (see Figure 1.9).

1.4 The Second Law of Thermodynamics

As mentioned earlier, the first law is the energy conservation principle. The second law of thermodynamics (SLT) is instrumental in determining inefficiencies of practical thermodynamic systems, and indicates that it is impossible to achieve 100% efficiency (in terms of reversible conversion) in energy conversion processes. Two primary statements of the second law follow:

- **Kelvin/Planck statement**: It is impossible to construct a device, operating in a cycle (e.g., a heat engine), that accomplishes only the extraction of heat from some source and its complete conversion to work. This statement describes the impossibility of having a heat engine with a thermal efficiency of 100%.

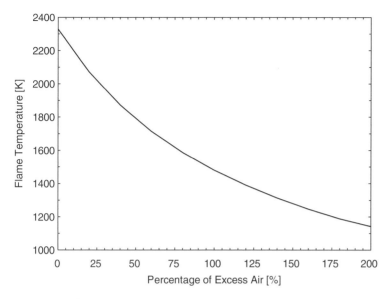

Figure 1.9 Effect of varying excess air on combustion flame temperature.

- **Clausius statement**: It is impossible to construct a device, operating in a cycle (e.g., refrigerator or heat pump), that transfers heat from a low temperature (cooler) region to a high temperature (hotter) region, of itself.

A simple way to illustrate the implications of both the first and second laws is a desktop game (known as "Newton's Cradle") that consists of several pendulums (with metal balls at the ends), one in contact with the other. When you raise the first of the balls, you give energy to the system in the form of potential energy. Releasing this ball allows it to gain kinetic energy at the expense of the potential energy. When this ball hits the second ball, a small elastic deformation transforms the kinetic energy, again as a form of potential energy. The energy is transferred from one ball to the other. The last ball again gains kinetic energy, allowing it to rise. The cycle continues, but every time the balls rise to a slightly lower level, until all motion finally stops. The first law concerns why the balls keep moving, while the second law explains why they do not do it forever. In this game the energy is lost in sound and heat, as the motion declines.

The second law also states that the entropy in the universe always increases. As mentioned before, entropy is a measure of degree of disorder, and every process happening in the universe increases the entropy of the universe to a higher level. The entropy of the state of a system is proportional to (depends on) its probability, which provides us with an opportunity to define the second law in a broader manner as "the entropy of a system increases in any heat transfer or conversion of energy within a closed system." That is why all energy transfers or conversions are irreversible. From the entropy perspective, the basis of the second law is the statement that the sum of the entropy changes of a system plus that of its surroundings must always be positive. Recently, much effort has been invested in minimizing the entropy generation (irreversibilities) in thermodynamic systems and applications.

Dincer and Rosen [2] indicate that exergy, which is based mainly on the SLT, can help with four tasks: design, analysis, performance assessment, and improvement.

Consequently, the second law is the linkage between entropy and the usefulness of energy. Second law analysis has found applications in a wide variety of disciplines, such as chemistry, economics, ecology, environment, and sociology, some of which are far removed from engineering thermodynamic applications.

1.5 Reversibility and Irreversibility

Reversibility and irreversibility are two important concepts in assessing thermodynamic processes and systems. Reversibility is defined by the statement that only for a reversible process can both a system and its surroundings be returned to their initial states, without additional external energy input. Such a process is only theoretical.

Irreversibility during a process describes the destruction of useful energy or its availability. Without new inputs, both a system and its surroundings cannot be returned to their initial states due to the irreversibilities that occur, for example friction, heat transfer or rejection, electrical, and mechanical effects. For instance, an actual system provides an amount of work that is less than the ideal reversible work, with the difference between these two values determining the irreversibility of that system. In real applications, there are always such differences, and therefore real processes and cycles are always irreversible.

1.6 Exergy

Exergy (also called availability) is defined as the maximum amount of work that can be produced by a stream of matter or energy (heat, work, etc.) as it comes to equilibrium with a reference environment. Exergy is a measure of the potential of a flow or system to cause change, as a consequence of not being in complete stable equilibrium relative to a reference environment. For exergy analysis, the state of the reference environment, or the reference state, must be specified completely. This is commonly done by specifying the temperature, pressure, and chemical composition of the reference environment. Exergy is not subject to a conservation law. Rather, exergy is consumed or destroyed due to irreversibilities in any process. Table 1.2 compares energy and exergy from the point of view of thermodynamics.

As pointed out by Dincer and Rosen [2], exergy is a measure of usefulness or quality. It is also a measure of the potential of a flow or system to cause change, and therefore can be seen as a type of measure of the potential of a substance to impact on the environment.

Exergy analysis is useful for improving the efficiency of energy-resource use, for it quantifies the locations, types, and magnitudes of wastes and losses. In general, more meaningful efficiencies are evaluated with exergy analysis than with energy analysis, since exergy efficiencies are always a measure of the approach to the ideal. Therefore, exergy analysis identifies accurately the margin available to design more efficient energy systems by reducing inefficiencies. Many engineers and scientists suggest that thermodynamic performance is best evaluated using exergy analysis because it provides more insights and is more useful in efficiency improvement efforts than energy analysis.

From the point of view of energy and exergy efficiencies, it is important to note that if a fossil fuel-based energy source is used for a low temperature thermal application like

Table 1.2 Comparison between energy and exergy.

Energy	Exergy
• Dependent on parameters of matter or energy flow only, and independent of environment parameters. • Has values different from zero (and is equal to mc² in accordance with Einstein's equation). • Guided by the first law of thermodynamics for all processes. • Limited by the second law of thermodynamics for all processes (including reversible ones).	• Dependent both on the parameters of matter or energy flow and on environment parameters. • Can equal zero (in dead state, by virtue of being in equilibrium with the environment). • Guided by the first and second laws of thermodynamics for reversible processes (where it is conserved) and irreversible processes (where it is destroyed partly or completely). • Not limited for reversible processes owing to the second law of thermodynamics.

Source: Dincer 2012. Reproduced with permission of Elsevier.

space heating or cooling, there would be a great difference between the corresponding energy and exergy efficiencies, perhaps with values of 50–70% for the energy efficiency and 5% for the exergy efficiency [2]. One may ask why and, to address that question, we point out the following:

- High quality (e.g., capable of high temperature heating) energy sources such as fossil fuels are often used for relatively low-quality (e.g., low temperature heating) processes like water and space heating or cooling.
- Exergy efficiency permits a better matching of energy sources and uses, leading to high quality energy being reserved for high quality tasks and not employed for low-quality end uses.

1.6.1 Exergy Associated with Kinetic and Potential Energy

Kinetic energy is a form of mechanical energy, and therefore can be entirely converted into work. Thus, the exergy of the kinetic energy of a system is equal to the kinetic energy itself regardless of the temperature and pressure of the environment, and is

$$ex_{KE} = \frac{V^2}{2} (\text{kJ kg}^{-1})$$ (1.7)

where V is the velocity of the system in relation to the environment.

Potential energy is also a form of mechanical energy, and can also be entirely converted into work. The exergy of the potential energy of a system is equal to the potential energy itself and does not depend on pressure and temperature [2, 4]. That is,

$$ex_{KE} = gZ$$ (1.8)

where g is the gravitational acceleration and Z is the elevation of the system.

In summary, the exergy values associated with kinetic and potential energy are equal for each and they are entirely available to do work. Kinetic and potential exergy have the same values as the corresponding energy terms. Kinetic exergy is particularly relevant where speeds are significant, as in a turbine, whereas potential exergy is especially relevant for electrical or hydraulic systems. In many practical cases involving industrial processes, kinetic and potential exergy terms can be neglected.

1.6.2 Physical Exergy

Physical exergy represents the maximum amount of work that can be obtained from a system as its pressure and temperature are changed to the pressure and temperature of the reference environment. The specific physical exergy ex_{ph} is determined with specific enthalpy and entropy values of the stream at a given temperature and pressure, and the reference environmental state temperature (T_0) and pressure (P_0), as follows [2, 5]:

$$ex_{ph} = [h(T, P) - h(T_0, P_0)] - T_0[s(T, P) - s(T_0, P_0)] \tag{1.9}$$

where $h(T, P)$ and $s(T, P)$ are the specific enthalpy and specific entropy of the stream at a given temperature and pressure, respectively. Values for specific enthalpy and specific entropy can be found either from thermodynamic tables or EES.

1.6.3 Chemical Exergy

Chemical exergy represents the maximum work that can be obtained when a substance is brought from the reference environment state to the dead state by a process including heat transfer and exchange of substances only with the reference environment. The maximum work is attained when the process is reversible. Alternatively, chemical exergy can also be viewed as the exergy of a substance that is at the reference environment state.

Chemical exergy is also equivalent to the minimum amount of work necessary to produce a substance at the reference environment state from the constituents of the reference environment. Chemical exergy has two main parts, reactive exergy resulting from the chemical reactions necessary to produce species which do not exist as stable components in the reference environment, and concentration exergy resulting from the difference between the chemical concentration of a species in a system and its chemical concentration in the reference environment. The concentration part is related to the exergy of purifying or diluting a substance, such as separating oxygen from air.

To determine a substance's chemical exergy, we need to define a reference environment in terms of its temperature T_0, pressure P_0, and chemical composition. In some reference environment models, substances present in the atmosphere, the hydrosphere, and upper part of the crust of the earth, at P_0 and T_0, form the basis of the reference environment. In other models, these substances are allowed to react with each other hypothetically and allowed to reach a stable state with a minimum Gibbs energy, at sea level, at rest without other force fields [2, 5, 6].

Once a reference environment is defined for an exergy analysis, the exergy of any substance at pressure P and temperature T can be evaluated relative to the reference environment. Note that it is not possible to obtain work by allowing substances in the reference environment to interact with each other.

A stream of matter also carries chemical exergy, conceptually determined as discussed above for a quantity of a substance.

1.6.3.1 Standard Chemical Exergy

In thermodynamics, the standard molar chemical exergy of a constituent i is defined as consisting of the molar free enthalpy Δg_f^0 for the formation of the compound in the standard state from its constituent elements and the stoichiometric sum of the standard chemical exergy values of the elements in their stable state at the temperature T_0 and pressure P_0.

Standard Chemical Exergy for Components of Air In the natural environment, there are many substances that, like nitrogen in the atmosphere, cannot react toward a more stable configuration to produce a new material. They can be considered as part of the reference environment. Transformations, including chemical and nuclear reactions, cannot convert these components into more stable components [6]. So, we cannot extract useful work from these substances and a specific exergy value of 0 kJ mol^{-1} can be assigned to them.

Often, these substances include the normal composition of air (including gases such as N_2, O_2, CO_2, H_2O, Ar, He, Ne), at $T_0 = 298.15$ K and $P_0 = 100$ kPa. The partial pressure P_i and molar fraction of each of these substances in air at a given relative humidity are given in Table 1.3.

The standard chemical exergy at P_0 for air can be written as

$$\overline{ex}_{ch}^0 = RT_0 \ln \left(\frac{P_0}{P_i} \right) \tag{1.10}$$

Standard chemical values for the main constituents of air are listed in Table 1.4. Note that exergy values for elements in their stable condition at $T_0 = 298.15$ K and $P_0 = 101.325$ kPa are called standard chemical exergies, and these are used in the calculation of chemical exergy for various substances.

1.6.3.2 Chemical Exergy of Gas Mixtures

The chemical exergy of a mixture of N gases, in which all are constituents of the environment, can be obtained similarly. In this case, we hypothesize N chambers. If each gas has a molar fraction x_k and enters the chamber at T_0 and with a partial pressure $x_k P_0$, then each gas exits at the same temperature and a partial pressure x_k^e. Summing for all constituents, the chemical exergy per mole of the mixture can be calculated as follows:

$$\overline{ex}_{ch} = \sum x_k \overline{ex}_{ch}^k + RT_0 \sum x_k \ln(x_k) \tag{1.11}$$

1.6.3.3 Chemical Exergy of Humid Air

The state of the local atmosphere is determined by its intensive parameters: T_0, P_0, and its composition. For humid air, the gas composition for all species other than vapor can

Table 1.3 Partial pressures and molar fractions of various constituents of air.

Constituent	P_i (kPa)	Molar fraction (%)
N_2	75.78	75.67
O_2	20.39	20.34
CO_2	0.00335	0.03
H_2O	2.2	3.03
He	0.00048	0.00052
Ne	0.00177	0.0018
Ar	0.906	0.92
Kr	0.000097	0.000076

Source: [7].

Table 1.4 Standard chemical exergy values at P_0 and T_0 of various constituents of air.

Constituent	$\overline{ex}_{ch}^0$ (kJ/mol)	Constituent	$\overline{ex}_{ch}^0$ (kJ/mol)
N_2	0.72	He	30.37
O_2	3.97	Ne	27.19
CO_2	19.87	Ar	11.69
H_2O	9.49	Kr	34.36

Source: [2, 6].

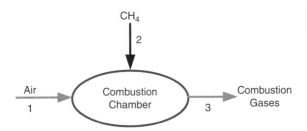

Figure 1.10 Combustion chamber used in gas turbine cycle.

be considered fixed. So, the composition variation is linked to the moisture content. To calculate the molar fraction of water vapor, we use:

$$x_{H_2O}^e = \varphi \frac{P_{gas,0}}{P_0} \tag{1.12}$$

Here φ is relative humidity (RH) and $P_{gas,0}$ is the saturation pressure at T_0. In the case when the molar fraction of dry air is assumed to be constant, the local atmospheric molar fraction of gases other than water is expressed as $x_i^e = (1 - x_{H_2O}^e)x_i^{dry}$, in which the molar fraction of constituent i corresponds to the atmospheric partial pressure of the species.

1.6.3.4 Chemical Exergy of Liquid Water and Ice

The molar chemical exergy of liquid water is expressed as [7]:

$$\overline{ex}_{ch}^w = v(P_0 - P_{gas,0}) - RT_0 \ln \varphi \tag{1.13}$$

Here v is the molar fraction of compressed water, $P_{gas,0}$ is the saturation pressure at T_0 and φ is relative humidity. This equation is valid for solid water (ice) as well [7].

Example 1.3: The combustion chamber is a major component of both gas turbine and combined cycle plants. A combustion chamber for a gas turbine engine is shown in Figure 1.10. Compressed air enters at point 1, and methane (CH_4) fuel is injected into the combustion chamber at point 2 at a mass flow rate of 1.5 kg s^{-1}. Combustion gases exit at point 3. In this example, air enters the combustion chamber at $T_1 = 550$ K and $P_1 = 10$ bar and the combustion gases exit at $T_3 = 1400$ K. The molar composition of air is taken to be 0.7748 N_2, 0.2059 O_2, 0.0003 CO_2 and 0.019 H_2O.

The following assumptions are invoked:

- The combustion chamber operates at steady state.
- The air and combustion gases can both be treated as ideal gas mixtures.

- Combustion is complete.
- Changes in potential and kinetic energies are negligible.

1) The heat loss from the combustion chamber is 2% of the lower heating value of the fuel. Determine the fuel-to-air ratio, and the molar fractions of the combustion gases.
2) Find the specific physical and chemical exergy values at point 3.

Solution: A combustion chamber is considered here.

Part 1:

To solve this problem, we first define the molar fuel-air ratio $\overline{\lambda}$ as:

$$\frac{\dot{n}_F}{\dot{n}_a} = \overline{\lambda}$$

So,

$$\frac{\dot{n}_P}{\dot{n}_a} = \frac{\dot{n}_a + \dot{n}_F}{\dot{n}_a} = \overline{\lambda} + 1$$

On a per mole of air basis, the combustion equation occurring in the combustion chamber can be written as:

$$\overline{\lambda}\, CH_4 + [0.7748\, N_2 + 0.2059\, O_2 + 0.0003\, CO_2 + 0.019\, H_2O]$$
$$\rightarrow [\overline{\lambda} + 1][x_{N_2}\, N_2 + x_{O_2}\, O_2 + x_{CO_2}\, CO_2 + x_{H_2O}\, H_2O]$$

To find the molar fraction of combustion gases, carbon, hydrogen, oxygen and nitrogen balances are written:

$$x_{N_2} = \frac{0.7748}{\overline{\lambda} + 1}, \quad x_{O_2} = \frac{0.2059 - 2\overline{\lambda}}{\overline{\lambda} + 1}$$

$$x_{CO_2} = \frac{0.0003 + \overline{\lambda}}{\overline{\lambda} + 1}, \quad x_{H_2O} = \frac{0.019 + 2\overline{\lambda}}{\overline{\lambda} + 1}$$

The molar breakdown of the combustion gases is known once $\overline{\lambda}$ is determined. An energy rate balance is used to determine the fuel-air ratio as follows:

$$\dot{Q}_{CV} - \dot{W}_{CV} + \dot{n}_F \overline{h}_F + \dot{n}_a \overline{h}_a - \dot{n}_P \overline{h}_P = 0$$

Since heat loss from the combustion chamber is 2% of the lower heating value of the fuel,

$$\dot{Q}_{CV} = -0.02\, \dot{n}_F\, \overline{LHV} = \dot{n}_a(-0.02\overline{\lambda} \times \overline{LHV})$$

Combining the above two equations yields

$$-0.02\overline{\lambda}\, \overline{LHV} + \overline{h}_a + \overline{\lambda}\, \overline{h}_F - (1 + \overline{\lambda})\overline{h}_P = 0$$

Employing ideal gas mixture principles to calculate the enthalpies of the air and the combustion gases, where $T_1 = 550$ K and $T_3 = 1400$ K, we obtain:

$$\overline{h}_a = [0.7748\, \overline{h}_{N_2} + 0.2059\, \overline{h}_{O_2} + 0.0003\, \overline{h}_{CO_2} + 0.019\, \overline{h}_{H_2O}]_{\text{at } T_1}$$

$$\overline{h}_P = \frac{1}{\overline{\lambda} + 1}[0.7748\, \overline{h}_{N_2} + (0.2059 - 2\overline{\lambda})\overline{h}_{O_2} + (0.0003 + \overline{\lambda})\overline{h}_{CO_2}$$
$$+ (0.019 + \overline{\lambda})\overline{h}_{H_2O}]_{\text{at } T_3}$$

Then, solving the energy rate balance for $\bar{\lambda}$ yields:

$$\bar{\lambda} = \frac{0.7748\Delta\bar{h}_{N_2} + 0.2059\Delta\bar{h}_{O_2} + 0.0003\Delta\bar{h}_{CO_2} + 0.019\Delta\bar{h}_{H_2O}}{\bar{h}_F - 0.02\overline{LHV} - (-2\bar{h}_{O_2} + \bar{h}_{CO_2} + 2\bar{h}_{H_2O})(T_4)}$$

Using specific enthalpy values from thermodynamic tables [4] and considering

$$\bar{h}_F = -74{,}872 \text{ kJ/kmol}, \quad \overline{LHV} = 802{,}361 \text{ kJ/kmol}$$

we find $\bar{\lambda} = 0.0393$. Using this value and the combustion gas molar fractions determined earlier, the following molar breakdown of the combustion products is obtained:

Component	N_2	CO_2	H_2O	O_2
Molar breakdown (%)	74.55	3.81	9.39	12.24

Part 2:

To calculate the specific physical exergy at point 3, the specific physical exergy expression described earlier is employed:

$$ex_{ph,3} = (h_3 - h_0) - T_0(s_3 - s_0) = C_p(T_3 - T_0) - T_0\left[C_p \ln\left(\frac{T_3}{T_0}\right) - R\ln\left(\frac{P_3}{P_0}\right)\right]$$

Since the pressure drop through the combustion chamber is treated as negligible $P_3 = P_1 = 10$ bar. For the combustion gases, we consider a fixed specific heat at constant pressure as $C_p = 1.14$ kJ/kg K. Substituting these values into the specific physical exergy expression yields $ex_{ph,3} = 933.1$ kJ/kg.

To determine the specific chemical exergy at point 3, Equation 3.15 is used:

$$
\begin{aligned}
ex_{ch,3} &= \sum x_k ex_{ch}^k + RT_0 \sum x_k \ln x_k \\
&= (0.1224 \times 3970 + 0.0381 \times 19\,870 + 0.0939 \times 9490 + 0.745 \times 720) \\
&\quad + (8.314 \times 298.15 \\
&\quad \times [0.1224 \ln(0.1224) + 0.0381 \ln(0.0381) + 0.0939 \ln(0.0939) \\
&\quad + 0.745 \ln(0.745)]) \\
&= 632.1 \frac{\text{kJ}}{\text{kmol}} = 22.95 \frac{\text{kJ}}{\text{kg}}
\end{aligned}
$$

When a mixture including gaseous combustion products containing water vapor is cooled at a constant pressure, the dew point temperature, which is the saturation temperature corresponding the partial pressure of water vapor, leads to the formation of liquid water. Thus, cooling such a mixture at constant pressure below the dew point temperature can result in condensation. For example, if the combustion gas mixture were cooled to 25°C at a pressure of 1 atm, some condensation would occur. In this case we can model the results at 25°C as a gas phase containing saturated water vapor in equilibrium with a liquid water phase. To find the dew point temperature, we first calculate the partial pressure of water vapor. According to Part 1 of this problem, the molar fraction of H_2O in the combustion gases is 0.0939, so the partial pressure of water vapor is $P_v = x_v P_0 = (0.0939)(1.013 \text{ bar}) = 0.0951$ bar. The corresponding saturation temperature at this pressure is 44.81°C; the reference environment temperature is therefore below the dew point, which leads to the formation of liquid water. On the basis of 1 kmol of combustion products, the gas phase at 25°C consists of 0.9061 kmol of dry products (0.7455 N_2, 0.3810 CO_2, 0.1224 O_2) plus n_v kmol of water vapor. The partial pressure of water vapor is equal to the saturation pressure at 25°C, which is 0.0317 bar. The amount of water vapor is calculated as $P_v = x_v P_0 = n_v/(0.9061 + n_v) \times P_0$, where n_v is the amount of water vapor. Hence, 0.0317 bar = $(n_v \times 1.013 \text{ bar})/(0.9061 + n_v)$ which yields n_v = 0.02927 kmol. Thus, the molar fractional composition of the combustion products at 25°C and 1 atm is 0.7455 N_2, 0.3810 CO_2, 0.1224 O_2, 0.02927 H_2O (g), and 0.06583 H_2O (l). At point 3 in the present analysis, 0.06583 kmol of liquid water is present on the basis of 1 kmol of mixture, following the method outlined by Bejan *et al.* [5]. Therefore, the specific chemical exergy at point 3 after modification is:

$$ex_{ch,3} = (1 - 0.06583) \times 22.95 + 0.06583 \times \frac{900}{18} = 24.32 \text{ kJ/kg}$$

Here, the first term is the percentage of dry combustion gases multiplied by the specific chemical exergy obtained from part 2, and the second term is the product of the molar fraction of liquid water due to the condensation and specific chemical exergy of liquid water.

Finally, the specific exergy at point 3 can be determined as:

$$ex_3 = ex_{ch,3} + ex_{ph,3} = 24.32 + 933.1 = 957.42 \text{ kJ/kg}$$

1.6.3.5 Chemical Exergy for Absorption Chillers

For the absorption cooling system, because a water and LiBr solution is not ideal, the following expression is used for the molar chemical exergy calculation:

$$\overline{ex}_{ch} = (1/\overline{M}_{sol}) \left[\sum_{i=1}^{n} y_i \overline{ex}_{ch}^k + \overline{R} T_0 \sum_{i=1}^{n} y_i \ln(a_i) \right] \tag{1.14}$$

Extending this equation for a LiBr-water solution we obtain:

$$\overline{ex}_{ch} = (1/\overline{M}_{sol}) \left[\begin{array}{c} y_{H_2O} \overline{ex}_{H_2O}^0 + y_{LiBr} \overline{ex}_{LiBr}^0 + \\ \overline{R} T_0 (y_{H_2O} \ln(a_{H_2O}) + y_{LiBr} \ln(a_{LiBr})) \end{array} \right] \tag{1.15}$$

Here, a_{H_2O} is the water activity defined as the vapor pressure of water in the mixture divided by the vapor pressure of pure water, and a_{LiBr} is LiBr activity defined as the vapor pressure of LiBr in the mixture divided by the vapor pressure of LiBr. This equation consists of two parts, standard chemical exergy of the pure species and exergy due to the dissolution process, defined as follows:

$$\overline{ex}_{ch}^0 = \frac{1}{\overline{M}_{sol}} (y_{H_2O} \overline{ex}_{H_2O}^0 + y_{LiBr} \overline{ex}_{LiBr}^0) \tag{1.16}$$

$$\overline{ex}_{ch}^{dis} = \frac{RT_0}{\overline{M}_{sol}} [y_{H_2O} \ln(a_{H_2O}) + y_{LiBr} \ln(a_{LiBr})] \tag{1.17}$$

where y_i is the molar fraction defined as

$$y_{H_2O} = \frac{(1 - x_{1w})\overline{M}_{LiBr}}{(1 - x_{1w})\overline{M}_{LiBr} + x_{1w}\overline{M}_{H_2O}} \tag{1.18}$$

$$y_{LiBr} = 1 - y_{H_2O} \tag{1.19}$$

Table 1.5 Standard molar chemical exergy values for selected substances at $T_0 = 298.15$ K and $P_0 = 1$ atm.

Element	$\overline{ex}^0_{ch}$ (kJ/mol)	Element	$\overline{ex}^0_{ch}$ (kJ/mol)
Ag (s)	70.2	Kr (g)	34.36
Al (s)	888.4	Li (s)	393.0
Ar (s)	11.69	Mg (s)	633.8
As (s)	494.6	Mn (s_α)	482.3
Au (s)	15.4	Mo (s)	730.3
B (s)	628.8	N_2 (g)	0.72
Ba (s)	747.4	Na (s)	336.6
Bi (s)	274.5	Ne (g)	27.19
Br_2 (l)	101.2	Ni (s)	232.7
C (s, graphite)	410.26	O_2 (g)	3.97
Ca (s)	712.4	P (s, red)	863.6
Cd (s_α)	293.2	Pb (s)	232.8
Cl_2 (g)	123.6	Rb (s)	388.6
Co (s_α)	265.0	S (s, rhombic)	609.6
Cr (s)	544.3	Sb (s)	435.8
Cs (s)	404.4	Se (s, black)	346.5
Cu (s)	134.2	Si (s)	854.6
D_2 (g)	263.8	Sn (s, white)	544.8
F_2 (g)	466.3	Sn (s)	730.2
Fe (s_α)	376.4	Ti (s)	906.9
H_2 (g)	236.1	U (s)	1190.7
He (g)	30.37	V (s)	721.1
Hg (l)	115.9	W (s)	827.5
I_2 (s)	174.7	Xe (g)	40.33
K (s)	366.6	Zn (s)	339.2

Source: [2, 8].

Here, x_{1w} is defined as

$$x_{1w} = \frac{x_{LiBr}}{100} \tag{1.20}$$

where x_{LiBr} is the LiBr-water solution concentration in percent, and $\overline{M}_{LiBr}$ and $\overline{M}_{H_2O}$ are 86.85 kg/kmol and 18.02 kg/kmol, respectively.

To calculate the chemical exergy for components not listed in Table 1.5, we may refer to reactions for which the standard chemical exergy of constituents are given. In this case, we can calculate the chemical exergy for the new constituent. Since the standard chemical exergy of LiBr is not listed in Table 1.5, the following reaction is used to calculate the molar chemical exergy of LiBr [8]:

$$\overline{ex}^0_{ch} = \overline{g}_f^0 + \sum_{i=1}^{n} \overline{ex}^0_{ch,i} \tag{1.21}$$

$$\text{Li} + \frac{1}{2}\text{Br}_2 \rightarrow \text{LiBr} \tag{1.22}$$

$$\overline{ex}_{ch,LiBr}^0 = \overline{g}_{f,LiBr}^0 + \overline{ex}_{ch,Li}^0 + \frac{1}{2}\overline{ex}_{ch,Br_2}^0 \tag{1.23}$$

Here, $\overline{g}_{f,LiBr}^0 = -324 \frac{kJ}{mol}$ [9].

Figure 1.11 shows the variation of chemical exergy as a function of LiBr mass basis concentration based on Equations 1.16 and 1.17. As shown in this figure, an increase in LiBr concentration results in an increase in the total chemical exergy of the LiBr–water solution.

Therefore, based on the LiBr concentration, the total chemical exergy at each point in a single effect absorption chiller can be straightforwardly calculated.

1.6.4 Exergy Balance Equation

By combining the conservation principle for energy and non-conservation principle for entropy (i.e., the second law of thermodynamics), the exergy balance equation can be obtained as follows:

$$\sum \text{Exergy Input} = \sum \text{Exergy Output} + \text{Exergy Destruction} \tag{1.24}$$

In terms of symbols, the following exergy rate balance can be written:

$$\dot{Ex}_Q + \sum_i \dot{m}_i ex_i = \sum_e \dot{m}_e ex_e + \dot{Ex}_W + \dot{Ex}_D \tag{1.25}$$

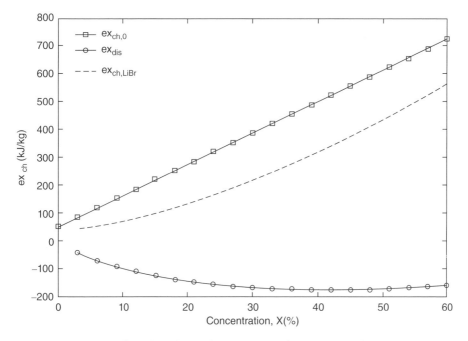

Figure 1.11 Variation of standard chemical exergy ($ex_{ch,0}$), chemical exergy due to dissolution (ex_{dis}) and total specific chemical exergy as a function of LiBr mass basis concentration at $T_0 = 25°C$.

where subscripts i and e denote the control volume inlet and outlet, respectively, $\dot{E}x_D$ is the exergy destruction rate and other terms are defined as follows:

$$\dot{E}x_Q = \left(1 - \frac{T_0}{T_i}\right)\dot{Q}_i \tag{1.26}$$

$$\dot{E}x_w = \dot{W} \tag{1.27}$$

$$ex = ex_{ph} + ex_{ch} \tag{1.28}$$

Here, $\dot{E}x_Q$ is the exergy rate of heat transfer crossing the boundary of the control volume at absolute temperature T, the subscript 0 refers to the reference environment conditions and $\dot{E}x_W$ is the exergy rate associated with shaft work. Table 1.6 lists exergy destruction rate expressions for some standard components.

1.6.5 Exergy Efficiency

Efficiencies are often evaluated as ratios of energy quantities, and are often used to assess and compare various systems. Power plants, heaters, refrigerators, and thermal storages, for example, are often compared based on energy efficiencies or energy-based measures of merit. However, energy efficiencies are often misleading in that they do not always provide a measure of how nearly the performance of a system approaches ideality. Further, the thermodynamic losses that occur in a system (i.e., those factors that cause performance to deviate from ideality) are often not accurately identified and assessed using energy analysis. The results of energy analysis can indicate the main inefficiencies to be within the wrong sections of a system, and a state of technological efficiency different than actually exists. Exergy efficiency computes the efficiency of a process taking the second law of thermodynamics into account. From the second law of thermodynamics, it can be demonstrated that no real system can ever achieve 100% efficiency. When calculating the energy efficiency of a system, no indication is provided of how the system compares to a thermodynamically perfect one operating under the same conditions. Exergy efficiency is a better measure and shows how a system works compared to a perfect one. The exergy efficiency is usually defined as the product exergy output divided by the exergy input.

Exergy efficiencies often give more illuminating insights into process performance than energy efficiencies because (1) they weigh energy flows according to their exergy contents, and (2) they separate inefficiencies into those associated with effluent losses and those due to irreversibilities. In general, exergy efficiencies provide a measure of potential for improvement.

1.6.6 Procedure for Energy and Exergy Analyses

A simple procedure for performing energy and exergy analyses involves the following steps:

- Subdivide the process under consideration into as many sections as desired, depending on the depth of detail and understanding desired from the analysis.
- Perform conventional mass and energy balances on the process, and determine all basic quantities (e.g., work, heat) and properties (e.g., temperature, pressure).
- Based on the nature of the process, the acceptable degree of analysis complexity and accuracy and the questions for which answers are sought, select a reference environment model.

Table 1.6 Expressions for exergy destruction rates for some selected components.

Component	Illustration	Exergy destruction rate expression
Air compressor		$\dot{Ex}_{D,AC} = \dot{Ex}_1 - \dot{Ex}_2 + \dot{W}_{AC}$
Turbine		$\dot{Ex}_{D,T} = \dot{Ex}_1 - \dot{Ex}_2 - \dot{W}_T$
Combustion chamber		$\dot{Ex}_{D,CC} = \dot{Ex}_1 + \dot{Ex}_f - \dot{Ex}_2$ $\dot{Ex}_f = \dot{m}_f \times \Phi \times LHV$ $\Phi_{CH_4} = 1.06$ $\Phi_{H_2} = 0.985$
Air preheater		$\dot{Ex}_{D,PH} = \dot{Ex}_2 + \dot{Ex}_5 - \dot{Ex}_3 - \dot{Ex}_6$
Condenser		$\dot{Ex}_{D,Cond} = \dot{Ex}_1 - \dot{Ex}_2 - \dot{Ex}_Q$
Pump		$\dot{Ex}_{D,Pump} = \dot{Ex}_1 - \dot{Ex}_2 + \dot{W}$

(Continued)

Table 1.6 (Continued)

Component	Illustration	Exergy destruction rate expression
Ejector		$\dot{Ex}_{D,EJ} = \dot{Ex}_1 + \dot{Ex}_2 - \dot{Ex}_3$
Boiler		$\dot{Ex}_{D,\text{Boiler}} = \dot{Ex}_1 + \dot{Ex}_2 + \dot{Ex}_5 + \dot{Ex}_7 - \dot{Ex}_3 - \dot{Ex}_4 - \dot{Ex}_6 - \dot{Ex}_8$
Flat plate solar collector		$\dot{Ex}_{D,FPC} = \dot{Ex}_1 + \dot{Ex}_{\text{sun}} - \dot{Ex}_2$ $\dot{Ex}_{\text{sun}} = I_{\text{sun}} A_{FPC}$ $\times \left[1 - \frac{4}{3}\left(\frac{T_0}{T_{\text{sun}}}\right) + \frac{1}{3}\left(\frac{T_0}{T_{\text{sun}}}\right)^4 \right]$
Deaerator		$\dot{Ex}_{D,DEA} = \dot{Ex}_1 + \dot{Ex}_2 - \dot{Ex}_3 - \dot{Ex}_4$

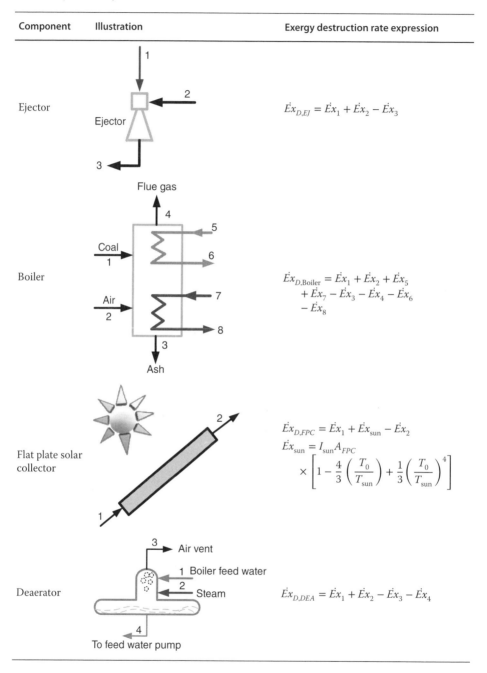

- Evaluate energy and exergy values relative to the selected reference environment model.
- Perform exergy balances including the determination of exergy consumptions.
- Select efficiency definitions, depending on the measures of merit desired, and evaluate values for the efficiencies.
- Interpret the results and draw appropriate conclusions and recommendations. These can relate to such issues as designs and design changes, retrofitted plant modifications, and so on.

1.7   Concluding Remarks

In this chapter, a summary is presented of the general introductory aspects of thermodynamics and related fundamental definitions and physical quantities, to help provide necessary background for understanding energy systems and applications, and their operations. The coverage of fundamentals provided here is particularly useful for the energy, exergy, and other analyses presented subsequently. EES was introduced as useful software for energy systems analyses. Some illustrative examples are presented and discussed. The FLT refers to energy conservation and treats all energy forms equally, thus not identifying losses of energy quality or work potential and possible improvements in the use of resources. For example, energy alone cannot identify the losses in an adiabatic throttling process. However, the SLT involves exergy and entropy concepts and considers irreversibilities and the consequent non-conservation of exergy and entropy.

References

1 Dincer, I. and Cengel, Y. A. (2001) Energy, entropy and exergy concepts and their roles in thermal engineering. *Entropy* **3**:116–149.
2 Dincer, I. and Rosen, M. A. (2012) *Exergy: Energy, Environment and Sustainable Development*. Second Edition, Elsevier.
3 Raéznjeviâc, K. (1995) *Handbook of Thermodynamic Tables*, Begell House, New York.
4 Cengel, Y. A., Boles, M. A. and Kanoğlu, M. (2011) *Thermodynamics: An Engineering Approach*, McGraw-Hill, New York.
5 Bejan, A., Tsatsaronis, G. and Moran, M. J. (1996) *Thermal Design and Optimization*. John Wiley & Sons, Inc., New York.
6 Kotas, T. J. (1986) Exergy method of thermal and chemical plant analysis. *Chemical Engineering Research & Design* **64**:212–229.
7 Szargut, J., Morris, D. R. and Steward, F. R. (1988) *Energy Analysis of Thermal, Chemical, and Metallurgical Processes*, Hemisphere, New York.
8 Kotas, T. J. (1985) *The Exergy Method of Thermal Plant Analysis*, Krieger, Malabar, Florida.
9 Palacios-Bereche, R., Gonzales, R. and Nebra, S. A. (2012) Exergy calculation of lithium bromide–water solution and its application in the exergetic evaluation of absorption refrigeration systems LiBr-H2O. *International Journal of Energy Research* **36**:166–181.

Study Questions/Problems

1   Define the following forms of energy and explain their differences: internal energy, thermal energy, heat, sensible energy, latent energy, chemical energy, nuclear energy, flow energy, flow work and enthalpy.

2   What is specific heat? Define two commonly used specific heats. Is specific heat a function of temperature?

3   What is the difference between an adiabatic system and an isolated system?

4   Define the critical point and explain the difference between it and the triple point.

5   Consider example 1.1, and perform an exergy analysis, including calculation of the exergy destruction rate for each component and the exergetic COP of the refrigeration system. Also, plot the variation of reference environment temperature and evaporator temperature with exergetic COP and total exergy destruction, and compare the results.

6   Consider example 1.3, for a reference state of $T_0 = 298.15$ K and $P_0 = 101.325$ kPa. Determine the exergy destruction rate and exergy efficiency of the combustion chamber and compare the results with information in the literature.

7   Define the terms energy, exergy, entropy, and enthalpy.

8   What is a second law efficiency? How does it differ from a first law efficiency?

9   What is the relationship between entropy generation and irreversibility?

10 When we enter a pool of water at 22°C, it often feels more than cool—it often feels cold. When we walk into a 22°C room, it is comfortable. Why is there such a difference between these cases, even though our body is surrounded by the same temperature?

11 How does an exergy analysis help further the goal of more efficient energy-resource use? What are the advantages of using exergy analysis?

12 In example 1.3, what is the effect of varying P_1 and T_1 on the specific chemical exergy at point 3 and the total exergy destruction rate of the combustion chamber?

13 On a hot day of summer, occupants return to their well-sealed house and find the house is at 32°C. They turn on the air conditioner, which cools the entire house to 21°C in 15 min. If the COP of the air conditioning system is 2.5, determine the power drawn by the air conditioner. Assume the entire mass within the house is equivalent to 700 kg of air for which $C_v = 0.72$ kJ/kg.K and $C_p = 1$ kJ/kg.K.

14 Air is compressed steadily by a 10 kW compressor from 100 kPa and 15°C to 600 kPa and 170°C at a rate of 3 kg/min (see Figure 1.12). Neglecting changes in kinetic and potential energies, determine (a) the increase in the exergy of the air, and (b) the rate of exergy destruction during this process. Assume the surroundings to be at 15°C.

Figure 1.12 Schematic of an air compressor.

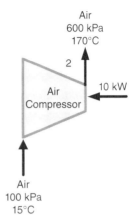

Air
600 kPa
170°C

2

10 kW

Air
Compressor

Air
100 kPa
15°C

15 Biomass is a biological material formed from living or recently living organisms, and is often viewed as a renewable source of energy. As an energy source, biomass can either be used directly, or converted into other energy products such as biofuels. Calculate the chemical exergy of the biomass "pine sawdust" using the following composition and other data:

Biomass type	Pine sawdust
Moisture content in biomass (by weight)	10%
Elemental analysis (dry basis by weight)	
Carbon (C)	50.54%
Hydrogen (H)	7.08%
Oxygen (O)	41.11%
Sulfur (S)	0.57%

Also, determine the molar composition of the product gases when pine sawdust is combusted according to the following reaction:

$$C_x H_y O_z + \omega H_2O + \gamma(O_2 + 3.76 N_2) \rightarrow a\,CO_2 + b\,H_2O + c\,N_2$$

16 Consider the simple gas turbine power plant shown in Figure 1.13. Air at ambient conditions enters the air compressor at point 1 and exits after compression at point 2. The hot air enters the combustion chamber (CC) into which fuel is injected, and hot combustion gases exit (point 3) and

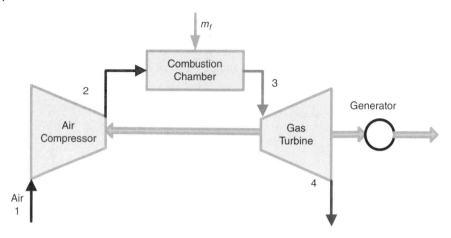

Figure 1.13 Schematic of a gas turbine power plant.

pass through a gas turbine to produce shaft power. The hot gas expands in the gas turbine to point 4. Natural gas (with a volumetric composition of $CH_4 = 96.57\%$, $C_2H_6 = 2.63\%$, $C_3H_8 = 0.1\%$, $C_4H_{10} = 0.7\%$) is injected into the combustion chamber with a mass flow rate of 2 kg/s. The compressor pressure ratio is 7, the gas turbine inlet temperature is 1400 K and the compressor isentropic efficiency and gas turbine isentropic efficiency are 0.83 and 0.87, respectively. If the net output power of this power plant is 15 MW, and there is no pressure drop in the combustion chamber and heat loss in the compressor and gas turbine:

- Determine the fuel-to-air ratio, and the molar fractions of the combustion gases.
- Find the specific physical and chemical exergy values at all points.
- For a reference state of $T_0 = 298.15$ K and $P_0 = 101.325$ kPa, determine the exergy destruction rate and exergy efficiency of the all components as well as the overall cycle.
- If the compressor pressure ratio varies from 6 to 12 while other parameters are fixed, plot the variation of system exergy efficiency with compressor pressure ratio for various gas turbine isentropic efficiencies.
- If the gas turbine inlet temperature (GTIT) varies from 1100 K to 1500 K, plot the variation of system exergy efficiency and total exergy destruction with GTIT.

17 Consider a house with a floor space area of 180 m^2 and average height of 2.7 m. The initial temperature of the house is uniform at 10°C. To maintain thermal comfort inside the house, an electric heater is turned on to raise the temperature to an average value of 22°C.
   a) Determine the amount of energy transferred to the air assuming no air escapes during the heating process.
   b) Determine the amount of energy transferred to the air assuming some air escapes through cracks as the heated air expands at constant pressure.
   c) Determine the cost of heat for either case when the cost of electricity in the area is $0.075/kWh.

18 Consider a solar powered steam power plant with feedwater heating, as shown in Figure 1.14. The conditions of the steam generated in the boiler are 8600 kPa and

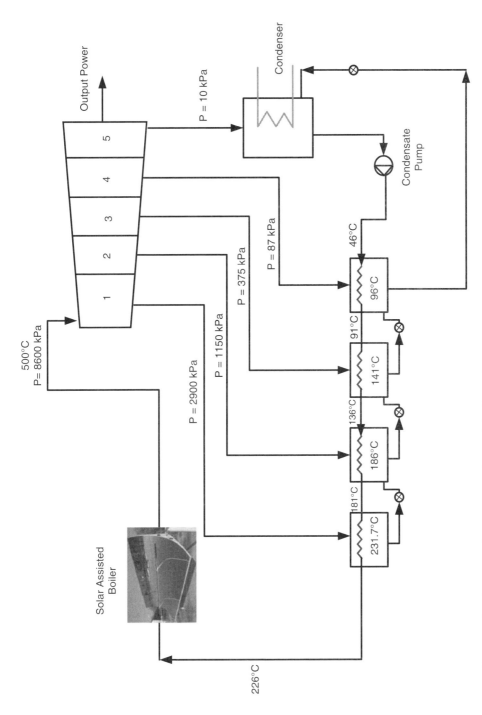

Figure 1.14 Schematic of a solar powered steam power plant with feedwater heating.

500°C. The exhaust pressure of the turbine, 10 kPa, is constant. The saturation temperature of the exhaust steam is therefore 45.83°C. Allowing for slight subcooling of the condensate, we fix the temperature of the liquid water from the condenser at 45°C. The feedwater pump, which operates under exactly the conditions of the pump, causes a temperature rise of about 1°C, making the temperature of the feedwater entering the series of heaters equal to 46°C. The saturation temperature of steam at the boiler pressure of 8,600 kPa is 300.06°C, and the temperature to which the feedwater can be raised in the heaters is certainly less. This temperature is a design variable, which is ultimately fixed by economic considerations. However, a value must be chosen before any thermodynamic calculations can be made. We have therefore arbitrarily specified a temperature of 226°C for the feedwater stream entering the boiler. We have also specified that all four feedwater heaters accomplish the same temperature rise. Thus, the total temperature rise of $226-46 = 180$°C is divided into four 45°C increments. This establishes all intermediate feedwater temperatures at the values shown in the figure. The steam supplied to a given feedwater heater must be at a pressure high enough that its saturation temperature is above that of the feedwater stream leaving the heater. We have here presumed a minimum temperature difference for heat transfer of no less than 5°C, and have chosen extraction steam pressures such that the T_{sat} values shown in the feedwater heaters are at least 5°C greater than the exit temperatures of the feedwater streams. The condensate from each feedwater heater is flashed through a throttle valve to the heater at the next lower pressure, and the collected condensate in the final heater of the series is flashed into the condenser. Thus, all condensate returns from the condenser to the boiler by way of the feedwater heaters.

- Utilize the following solution procedure:
  - Write a problem statement
  - Show the cycle on a *T-s* diagram with respect to saturation lines
  - Make appropriate assumptions and approximations
  - Identify relevant physical laws
  - List all relevant properties
  - Write all balance equations (mass, energy, entropy, exergy)
  - Do calculations through the EES software
  - Determine which type of solar collector will be appropriate for this power plant and include solar data in the calculations
  - Conduct a performance evaluation in terms of energy and exergy efficiencies (and exergy destructions)
  - Discuss the results, providing reasoning and verification
  - Discuss performance, environmental, and sustainability issues
  - Make recommendations for better technical performance, environmental performance and sustainability.
- Do a parametric study to show how system performance is affected by varying the operating and environment conditions (e.g., *T*, *P*, *m*).

2

Modeling and Optimization

2.1 Introduction

The need to provide more efficient and cost effective energy systems has become of increasing importance. Greater global competition and the desire for better and more efficient processes have resulted in the need for better design practices. During the past decade, interest has grown in producing higher quality products at minimal cost, while satisfying increasing concerns regarding environmental impact, safety, and other factors, rather than developing a system that only performs a desired task. Energy use is directly linked to well-being, living standards, and prosperity, and meeting the growing demand for energy in a safe and environmentally responsible manner is an important challenge. A key driver of energy demand is the human desire to sustain and improve ourselves, our families, and our communities. There are around seven billion people on earth, and population growth will likely lead to an increase in energy demand, which depends on an adequacy of energy resources. In addition, increasing population and economic development in many countries may significantly affect the environment, because energy generation processes emit pollutants, many of which are harmful to ecosystems. The importance of energy in daily life makes it an important priority for the optimization of such systems.

It is often desired to optimize processes by maximizing or minimizing an objective function. For example, the economic profit or product quality can be maximized, while unit product cost per item and energy input can be minimized. Business success today is strongly based on an ability to optimize processes and systems. With the advent in recent years of new materials, such as composites and ceramics, and new manufacturing processes, several traditional industries (e.g., steel processing) have faced significant challenges and, in some cases, diminished in size, while new fields have emerged. It is important to exploit new techniques for product improvement and cost reduction in both traditional and new industries. Even in an expanding area such as consumer electronics, the prosperity of a company is closely connected to its ability to apply optimization to new and existing process and system designs. Consequently, engineering design, which has always been important, has become increasingly coupled with optimization.

Optimization is a significant tool in engineering for determining the best, or optimal, value for a decision variable of a system. Energy engineering is a field where optimization

Optimization of Energy Systems, First Edition. Ibrahim Dincer, Marc A. Rosen, and Pouria Ahmadi.
© 2017 John Wiley & Sons Ltd. Published 2017 by John Wiley & Sons Ltd.

plays a particularly critical role. Engineers involved in this area are required to answer many questions. Some examples of these are:

- What materials should be used?
- What types of energy should be used in a given application?
- What processes or equipment should be selected for a system, and how should the parts be arranged for the best outcome?
- What are the best characteristics for the components (e.g., size, capacity, cost)?
- What are the best process parameters (e.g., temperature, pressure, flow rate, and composition) of each stream interacting in a system?
- How can energy use be reduced/saved?
- How should wastes be disposed of?

In order to answer such questions, engineers are required to formulate an appropriate optimization problem. Proper formulation is usually the most important and sometimes the most difficult step in optimization. To formulate an optimization problem, numerous elements need to be defined, including system boundaries, optimization criteria, decision variables, and objective functions.

In this chapter, we describe several modeling techniques and optimization methods and the formulation of an optimization problem. We cover objective functions and how to select them for enhanced sustainability, optimization constraints for energy systems, and optimization algorithms.

2.2 Modeling

Modeling is used to obtain information about how something behaves without actually testing it in real life. For instance, if we want to design a heat exchanger and we are not sure about what size would lead to low cost and high efficiency, we could develop a model of the heat exchanger and apply it in conjunction with computer simulation to determine the efficiency for various heat exchanger sizes. In modeling a heat exchanger, we obtain useful insights that inform decisions we may make for the heat exchanger, without actually building the device.

The use of modeling and simulation in engineering is well recognized and accepted. To ensure that the results of a simulation are realistic and representative of the physical application, engineering practitioners need to understand the assumptions, concepts, and implementation constraints of this field. Otherwise simulation results may not be applicable in design and construction of a product. Modeling and simulation provide several benefits for engineers, of which some of the main ones are:

- Modeling is an attractive method since it is generally less expensive and safer than performing experiments with a prototype of the final product or actual system.
- Modeling can often be even more realistic than conventional experiments, as it can permit environment parameters found in the operational application field of a product or process to be easily varied. In addition, modeling can be used to perform parametric studies in order determine the effects of varying selected design parameters on performance, thereby enhancing understanding of a product or process.
- Modeling can often be conducted faster than experimentation and testing. This allows models to be used for efficient analyses of alternative options, especially when the necessary data to initialize a simulation is obtainable from operational data.

The study of physical phenomena in engineering involves two major steps. In the first, all variables that affect phenomena are identified and reasonable assumptions and approximations are made for each, appropriate physical laws and principles are invoked, and the problem is formulated mathematically. In the second step, the problem is solved using an appropriate approach, and the results are presented. Reasonably accurate results to meaningful practical problems can often be obtained with relatively little effort using suitable and realistic mathematical models. The development of models requires a good knowledge of the natural phenomena involved and relevant laws. An engineer experienced in the modeling of practical problems is usually able to decide between an accurate but complex model and a simple but relatively inaccurate model. The appropriate choice normally depends on the situation. It is usually the simplest model that provides sufficient results for the problem under consideration. For example, we can model and analyze the heat losses from a building in order to select the correct size for a heater in the building; we determine the heat losses under extreme conditions and select a furnace that can provide sufficient energy to offset those losses. The development of an accurate and complex model is usually not overly difficult.

At the minimum, a model should reflect the essential features of the corresponding physical problem. Many real problems can be analyzed with simple models. Note that the results obtained from an analysis are only as accurate as the assumptions made to simplify the analysis. A solution that is not consistent with the observed nature of the problem is too crude and cannot be used reasonably. Figure 2.1 illustrates how a system can be modeled in order to determine outputs. From this figure, it can be observed that an accurate model depends on several factors such as reasonable assumptions

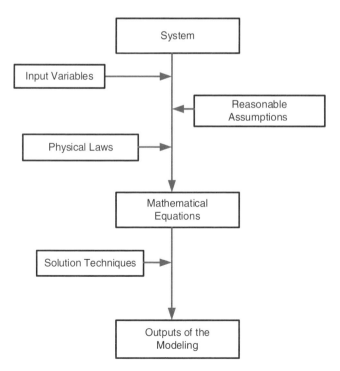

Figure 2.1 Flowchart of the mathematical modeling of a system.

and applying relevant laws that can simulate the system well. Modeling is a key tool in energy engineering as it can predict the behavior of an energy system, and form a foundation for optimization. If modeling is inaccurate, optimization results become unrealistic and unusable. Thus, modeling needs to be carefully carried out before optimization is performed.

Validation of models is normally required. This becomes increasingly important for cases of advanced energy systems that are complicated and unknown. This objective can be achieved by using selected inputs as the real case for subsystems of the advanced energy system, and then comparing the outputs to ensure the validity and correctness of the model.

There are two approaches in modeling: descriptive and predictive. Most people are familiar with descriptive modeling, which is used to describe and explain technical and other phenomena. Examples of descriptive models in engineering include models for heat exchangers and heat pumps, which are useful for explaining how such devices work. These models can be three-dimensional mock ups, made of plastic or metal, which shows the internal mechanisms. Such descriptive models are often used to improve understanding of physical principles.

Predictive models are used to predict the performance of a system and consequently are of great importance in engineering design and optimization. For instance, the equations governing the temperature distribution inside a wall with a given thickness and constant thermal conductivity represent a predictive model because they permit the temperature variation with time and position inside the wall to be obtained. Modeling is particularly important in energy system engineering and thermal design because of the complex nature of the transport phenomena. The complexity results in part from the variations of a system and its parts with time and location, the boundary conditions applicable to a given problem, heat and mass transfer effects, and complexities in material properties. Numerous types of predictive models have been developed to represent a wide range of energy systems. Each model has its own characteristics and usually is appropriate for specific applications.

The emphasis of this chapter is on predictive models for thermal systems subject to various operating conditions and design parameters. Such models can be used for optimization activities, as covered in subsequent chapters. Since thermodynamic modeling plays a significant role in thermal systems analysis, we present below models for selected general components of such systems. These components are widely employed in traditional and advanced energy systems, such as power plants, petrochemical plants, cogeneration and trigeneration plants, hydrogen production systems, and thermal energy storage systems.

2.2.1 Air compressors

An air compressor is a device that uses electrical power (usually from an electric motor, a diesel engine, or a gasoline engine) to compress air by pressurizing it. The air can be stored and/or released on demand for various applications. Numerous types of air compressor exist, including positive-displacement or negative-displacement types. We can model an air compressor using energy balances and other relations, and such models can be used for optimizing air compressors considering thermodynamic and other factors. Figure 2.2 shows a thermodynamic model of an air compressor (AC) as a control volume.

Figure 2.2 Model of an air compressor.

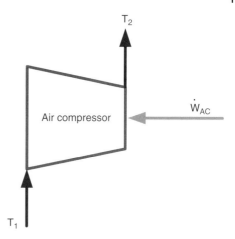

For the air compressor, we can write expressions for the outlet temperature T_2 and the work rate requirement $\dot{W}_{AC}$, respectively, as follows:

$$T_2 = T_1 \left\{ 1 + \frac{1}{\eta_{AC}} \left[r_{AC}^{\frac{\gamma_a - 1}{\gamma_a}} - 1 \right] \right\} \tag{2.1}$$

$$\dot{W}_{AC} = \dot{m}_a C_{pa} (T_2 - T_1) \tag{2.2}$$

Here, $\gamma_a = c_p / c_v$. Also, $\dot{m}_a$ denotes air mass flow rate, η_{AC} compressor isentropic efficiency, and c_{pa} specific heat at constant pressure of air, which can be expressed a function of temperature as follows [1]:

$$c_{pa} = 1.048 - \frac{3.837 \times T}{10^4} + \frac{9.453 \times T^2}{10^7} - \frac{5.490 \times T^3}{10^{10}} + \frac{7.929 \times T^4}{10^{14}} \tag{2.3}$$

The temperature T_1 is the air temperature entering the compressor (which is often at ambient conditions). Note that usually $T_2 > T_1$ since compression warms the air.

2.2.2 Gas Turbines

A turbine is a rotary mechanical device that extracts energy from a pressurized flow and converts it to useful work. A typical turbine has a rotor assembly: a moving part consisting of a shaft or drum with blades attached. Turbines are widely used in power generation, petrochemical processing, liquefaction, and other applications. There are various turbine types, with varying work outputs and operating on specific cycles. Thus we have gas turbines, steam turbines, wind turbines, and hydraulic turbines.

Here, we consider a gas turbine, which is commonly used in gas turbine power plants and cogeneration facilities. A thermodynamic model of a gas turbine is shown in Figure 2.3. To model the gas turbine, we typically need to be able to calculate its outlet temperature. The gas turbine outlet temperature can be written a function of the gas turbine's isentropic efficiency η_{GT}, inlet temperature T_3 and pressure ratio P_3/P_4, as follows:

$$T_4 = T_3 \left(1 - \eta_{GT} \left(1 - \left(\frac{P_3}{P_4} \right)^{\frac{1 - \gamma_g}{\gamma_g}} \right) \right) \tag{2.4}$$

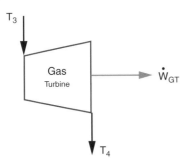

T_3

$\dot{W}_{GT}$

Gas
Turbine

T_4

Figure 2.3 Model of a gas turbine.

The gas turbine output power can be expressed as

$$\dot{W}_{GT} = \dot{m}_g C_{pg}(T_3 - T_4) \tag{2.5}$$

Here, $\dot{m}_g$ is the gas turbine mass flow rate, which is calculated as follows:

$$\dot{m}_g = \dot{m}_f + \dot{m}_a \tag{2.6}$$

The net output power of a gas turbine cycle can be expressed as

$$\dot{W}_{net} = \dot{W}_{GT} - \dot{W}_{AC} \tag{2.7}$$

The specific heat C_{pg} is taken to be a function of temperature as follows [1]:

$$C_{pg}(T) = 0.991 + \left(\frac{6.997T}{10^5}\right) + \left(\frac{2.712T^2}{10^7}\right) - \left(\frac{1.2244T^3}{10^{10}}\right) \tag{2.8}$$

These equations allow us to predict the effect of gas turbine inlet pressure, isentropic efficiency, and expansion ratio on gas turbine outlet power.

2.2.3 Pumps

A pump is a device that pressurizes and/or moves liquids, by mechanical action. Pumps can be classified into three major groups according to the method they use to move the fluid: direct lift, displacement, and gravity. Pumps are commonly driven by electricity, although they can also be linked with and driven by such devices as engines and wind turbines. Pumps come in many sizes, ranging from microscopic, for use in medical applications, to large, for power generation and industrial processes. Pumps are commonly employed in thermodynamic cycles where pressure increases of liquids are required, for example, Rankine cycles, absorption refrigeration systems, cogeneration systems, cryogenic energy storage systems, desalination units, and other energy systems. In large power plants, the main pump, called the boiler feed pump, consumes about 2–4% of the net electrical output of the plant.

Figure 2.4 shows a thermodynamic model of a typical pump with fluid inlets and outlets. Writing the energy rate balance for the control volume around the pump gives

$$\dot{W} = \dot{m}v(P_2 - P_1) \tag{2.9}$$

where v denotes the specific volume of the fluid and $\dot{m}$ the fluid mass flow rate through the pump. The outlet enthalpy of a pump is in the subcooled region and can be expressed as follows:

$$h_2 = h_1 + \frac{\dot{W}}{\dot{m}} \tag{2.10}$$

Figure 2.4 Model of a pump.

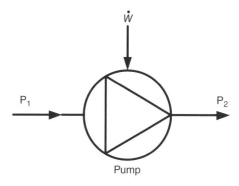

Figure 2.5 Model of a heat exchanger.

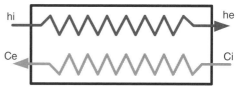

Substituting Equation 2.9 into Equation 2.10 yields:

$$h_2 = h_1 + v(P_2 - P_1) \tag{2.11}$$

Hence, the outlet enthalpy of the pump is a function of inlet enthalpy and pressure difference across the pump. The outlet pressure and outlet enthalpy of a pump are often defined, allowing the outlet temperature to be calculated straightforwardly using thermodynamic tables or other sources of property data.

2.2.4 Closed Heat Exchanger

Figure 2.5 shows a model of a closed heat exchanger. Considering the heat exchanger as a control volume and treating it as adiabatic, an energy rate balance can be written as:

$$\dot{m}_{hi}h_{hi} + \dot{m}_{ci}h_{ci} = \dot{m}_{he}h_{he} + \dot{m}_{ce}h_{ce} \tag{2.12}$$

Here, it is assumed that heat loss from the heat exchanger is negligible, although a term to account for heat loss can be added to Equation 2.12 straightforwardly. Also, kinetic and potential energy are neglected. The enthalpy change is generally determined using thermodynamic tables.

A condenser is an example of a heat exchanger commonly used in power plants. In the condenser of a steam power plant, expanded steam from the turbine outlet is condensed and conveyed to pumps as a saturated liquid.

In a closed heat exchanger, the amount of heat removed from the hot fluid can be expressed as a function of its enthalpy drop as:

$$\Delta Q = \dot{m}_{hi}(h_{hi} - h_{he}) \tag{2.13}$$

This is equal to the heat gained by the cold fluid:

$$\Delta Q = \dot{m}_{ci}(h_{ce} - h_{ci}) \tag{2.14}$$

When a substance flowing through a heat exchanger can be treated as ideal gas, the following equation can be used for the specific enthalpy difference:

$$h_i - h_e = \int_i^e C_p dT \tag{2.15}$$

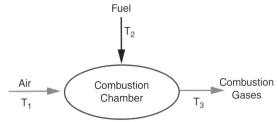

Fuel

T_2

Air
T_1

Combustion Chamber

Combustion Gases

T_3

Figure 2.6 Model of a heat exchanger.

2.2.5 Combustion Chamber (CC)

A combustion chamber is a device in which fuel is burned. Combustion chambers are widely used in engines, power plants and cogeneration plants. Figure 2.6 presents a model of a general combustion chamber. The outlet properties of the combustion chamber are dependent on the air mass flow rate, the fuel flow rate and lower heating value (LHV_f) and the combustion chamber efficiency η_{cc}. These terms are related as follows:

$$\dot{m}_a h_1 + \dot{m}_f LHV_f = \dot{m}_g h_3 + (1 - \eta_{cc}) \dot{m}_f LHV_f \tag{2.16}$$

The combustion chamber outlet pressure is determined by considering the pressure drop across the combustion chamber ΔP_{cc} as follows:

$$\frac{P_3}{P_2} = 1 - \Delta P_{CC} \tag{2.17}$$

The combustion reaction occurring and its species coefficients can be expressed as follows:

$$\frac{\dot{n}_f}{\dot{n}_a} = \bar{\lambda} \tag{2.18}$$

where λ is the fuel-to-air ratio on a molar basis. Hence,

$$\frac{\dot{n}_p}{\dot{n}_a} = \frac{\dot{n}_a + \dot{n}_f}{\dot{n}_a} = \bar{\lambda} + 1$$

Consider as an example a gas turbine power plant. Then, the fuel in Figure 2.6 is natural gas, which can be modeled as pure methane (CH_4). Typically, the combustion chamber heat loss is about 2% of the energy input with fuel.

On a per mole of air basis, the combustion equation occurring in the combustion chamber can be written as

$$\bar{\lambda} \, CH_4 + [0.7748 \, N_2 + 0.2059 \, O_2 + 0.0003 \, CO_2 + 0.019 \, H_2O]$$
$$\rightarrow [\bar{\lambda} + 1][x_{N_2} \, N_2 + x_{O_2} \, O_2 + x_{CO_2} \, CO_2 + x_{H_2O} \, H_2O] \tag{2.19}$$

To find the molar fractions of combustion gases, carbon, hydrogen, oxygen, and nitrogen balances are written as follows:

$$x_{N_2} = \frac{0.7748}{\bar{\lambda} + 1}, \qquad x_{O_2} = \frac{0.2059 - 2\bar{\lambda}}{\bar{\lambda} + 1}$$

$$x_{CO_2} = \frac{0.0003 + \bar{\lambda}}{\bar{\lambda} + 1}, \qquad x_{H_2O} = \frac{0.019 + 2\bar{\lambda}}{\bar{\lambda} + 1}$$

The molar breakdown of the combustion gases can be found once $\overline{\lambda}$ is determined. An energy rate balance is used to determine the fuel-air ratio $\overline{\lambda}$ as follows:

$$\dot{Q}_{CV} - \dot{W}_{CV} + \dot{n}_f \overline{h}_f + \dot{n}_a \overline{h}_a - \dot{n}_p \overline{h}_p = 0 \tag{2.20}$$

Since the heat loss from the combustion chamber is 2% of lower heating value of the fuel,

$$\dot{Q}_{CV} = -0.02\, n_f\, \overline{LHV} = \dot{n}_a(-0.02\overline{\lambda} \times \overline{LHV}) \tag{2.21}$$

Combining the above two equations yields

$$-0.02\overline{h}\,\overline{LHV} + \overline{h}_a + \overline{\lambda}\overline{h}_f - (1 + \overline{\lambda})\overline{h}_p = 0$$

Employing ideal gas mixture principles to calculate the molar enthalpies of the air and combustion gases, we obtain:

$$\overline{h}_a = \left[0.7748\,\overline{h}_{N_2} + 0.2059\,\overline{h}_{O_2} + 0.0003\,\overline{h}_{CO_2} + 0.019\,\overline{h}_{H_2O}\right]_{at\ T_1}$$

$$\overline{h}_p = \frac{1}{\overline{\lambda} + 1}$$
$$\left[0.7748\,\overline{h}_{N_2} + (0.2059 - 2\overline{\lambda})\overline{h}_{O_2} + (0.0003 + \overline{\lambda})\overline{h}_{CO_2} + (0.019 + \overline{\lambda})\,\overline{h}_{H_2O}\right]_{at\ T_3}$$

Then, solving the energy rate balance for $\overline{\lambda}$ yields:

$$\overline{\lambda} = \frac{0.7748\,\Delta\overline{h}_{N_2} + 0.2059\,\Delta\overline{h}_{O_2} + 0.0003\,\Delta\overline{h}_{CO_2} + 0.019\,\Delta\overline{h}_{H_2O}}{\overline{h}_f - 0.02\overline{LHV} - (-2\overline{h}_{O_2} + \overline{h}_{CO_2} + 2\overline{h}_{H_2O})(T_4)} \tag{2.22}$$

This equation allows the molar fractions of the combustion gases exiting the combustion chamber to be determined.

2.2.6 Ejector

An ejector, which acts as a kind of pump, uses the Venturi effect of a converging–diverging nozzle to convert the mechanical energy (pressure) of a motive fluid to kinetic energy (velocity), creating a low pressure zone that draws in and entrains a suction fluid. After passing through the throat of the injector, the mixed fluid expands and the velocity is reduced, recompressing the mixed fluids by converting velocity back to pressure. The motive fluid may be a liquid, steam or any other gas.

The process occurring in an ejector (Figure 2.7) is assumed to be steady state, one dimensional, and adiabatic, and no work is done during the process. The velocities at the inlet and outlet of the ejector can be considered negligible [2]. For simplicity, the effect of losses in the nozzle, mixing section, and diffuser are accounted for by the efficiency for each section of the ejector. In this study, the primary motive flow enters the ejector at point 1, and the suction flow exits the evaporator at point 2.

The process in the ejector includes expansion of the high pressure prime motive flow through the nozzle, mixing with the low pressure secondary flow in the mixing section at constant pressure, and pressurization in the diffuser where the kinetic energy of the flow is converted to pressure head. The flow exits the ejector at point 3. An important parameter for the secondary flow is the entrainment ratio, defined as

$$\omega = \frac{\dot{m}_2}{\dot{m}_1} \tag{2.23}$$

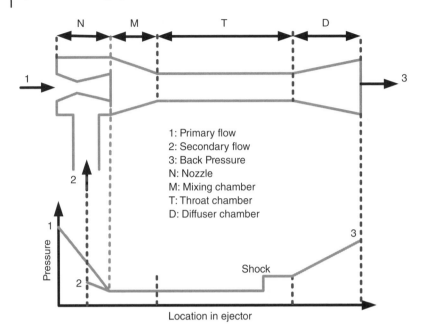

Figure 2.7 Model of ejector and pressure profile through it. *Source*: Adapted from Wang 2009.

In the nozzle section in Figure 2.7, the inlet velocity of the primary flow $V_{pf,n1}$ is negligible, so the exit enthalpy and velocity of the primary flow can be expressed as:

$$V_{pf,n_2} = \sqrt{2\eta_{noz}(h_{pf,n_1} - h_{pf,n_2,s})} \tag{2.24}$$

where h_{pf,n_1} is the specific enthalpy at point 30, $h_{pf,n_2,s}$ is the specific enthalpy of the primary flow exiting after isentropic expansion, and η_{noz} is the nozzle efficiency.

A momentum conservation equation for the mixing chamber area gives:

$$\dot{m}_1 V_{pf,n_2} + \dot{m}_2 V_{sf,n_2} = (\dot{m}_1 + \dot{m}_2)V_{mf,m,s} \tag{2.25}$$

Neglecting the secondary flow velocity V_{sf,n_2} compared to the primary flow velocity V_{pf,n_2}, the exit velocity of the mixed flow $V_{mf,m,s}$ is expressible as

$$V_{mf,m,s} = \frac{V_{pf,n_2}}{1 + \omega} \tag{2.26}$$

The mixing chamber efficiency can be expressed as:

$$\eta_{mix} = \frac{V_{mf,m}^2}{V_{mf,m,s}^2} \tag{2.27}$$

Therefore, the actual velocity of the mixed flow is

$$V_{mf,m} = \frac{V_{pf,n2}\sqrt{\eta_{mix}}}{1 + \omega} \tag{2.28}$$

An energy equation for the mixing chamber gives:

$$\dot{m}_1\left(h_{pf,n_2} + \frac{V_{pf,n_2}^2}{2}\right) + \dot{m}_2\left(h_{sf,n_2} + \frac{V_{sf,n_2}^2}{2}\right) = \dot{m}_3\left(h_{mf,m} + \frac{V_{mf,m}^2}{2}\right) \tag{2.29}$$

By simplifying this equation and using Equations. (2.23) and (2.28), the specific enthalpy of the mixed flow is obtained:

$$h_{mf,m} = \frac{h_{pf,n_1} + \omega h_{sf,n_2}}{1 + \omega} - \frac{V_{mf,m}^2}{2} \tag{2.30}$$

In the diffuser section, the mixed flow converts kinetic energy to a pressure increase. Assuming the exit velocity of the mixed flow to be negligible and considering the diffuser efficiency, the actual exit enthalpy of the mixed flow can be calculated as

$$h_3 = h_{mf,m} + (h_{mf,d,s} - h_{mf,m})/\eta_{\text{dif}} \tag{2.31}$$

where $h_{mf,d,s}$ is the ideal exit specific enthalpy of the mixed flow for isentropic compression, and η_{dif} is the diffuser efficiency.

Using these equations, the entrainment ratio is expressed as [2]:

$$\omega = \sqrt{\eta_{\text{noz}}\eta_{\text{mix}}\eta_{\text{dif}}\frac{h_1 - h_a}{h_2 - h_b}} - 1 \tag{2.32}$$

where η_{noz}, η_{mix} and η_{dif} are the nozzle, mixing chamber, and diffuser efficiencies, respectively.

2.2.7 Flat Plate Solar Collector

Flat plate solar collectors are being used for domestic hot water heating and for space heating, applications where the required temperature is low. Many models of flat plate collectors are available. Water enters the solar collector and is heated, and the rate of heat gain by the working fluid can be written as

$$\dot{Q}_u = \dot{m}C_p(T_{\text{out}} - T_{\text{in}}) \tag{2.33}$$

where T_{out}, T_{in}, C_p and $\dot{m}$ are the water outlet temperature, inlet temperature, specific heat at constant pressure and mass flow rate. The Hottel-Whillier equation for the heat gained by a flat plate collector considering heat losses from the collector is used [3]:

$$\dot{Q}_u = A_p F_R[(\tau\alpha)I - Q_L] \tag{2.34}$$

where $(\tau\alpha)$ is the optical efficiency, I is solar radiation intensity, T_0 is the ambient temperature and the F_R is heat removal factor, expressible as:

$$F_R = \frac{\dot{m}C_p}{U_l A_P}\left[1 - e^{\left\{-\frac{F'U_lA_p}{\dot{m}C_p}\right\}}\right] \tag{2.35}$$

Here, F' is the collector efficiency factor and U_l is the overall collector loss coefficient. Also, Q_L in Equation 2.34 can be written as:

$$Q_L = U_l(T_{\text{in}} - T_0) \tag{2.36}$$

The energy efficiency of the flat plate solar collector is expressed as:

$$\eta = \frac{\dot{Q}_u}{IA_P} \tag{2.37}$$

2.2.8 Solar Photovoltaic Thermal (PV/T) System

Photovoltaic/thermal solar collectors, also known as hybrid PV/T or PVT systems, convert solar radiation into thermal and electrical energy. As a PV panel temperature increases, its efficiency drops; we can therefore improve the efficiency of a PV/T collector by cooling it. Such systems combine a photovoltaic cell, which converts the electromagnetic radiation of the sun into electricity, with a solar thermal collector, which captures the remaining solar input energy to provide heating. By harvesting both electricity and heat, this device achieves a higher exergy output and is more efficient than a solar photovoltaic (PV) or solar thermal collector alone.

To model a PV/T system, we calculate the electrical power produced by the PV module as follows [4]:

$$\dot{W} = \eta_c \dot{I} \beta_c \tau_g A \tag{2.38}$$

where η_c is the solar cell efficiency, β_c is the packing factor of solar cell and τ_g is the transitivity of the solar panel glass and A is the solar collector area (in m^2). The rate of useful thermal energy obtained from the PV/T air collector can thus be expressed as follows:

$$\dot{Q}_{solar} = \frac{\dot{m}_{air} C p_{air}}{U_L} \{(h_{P2Z} Z \dot{I}) - U_L(T_{air,in} - T_0)\} \times 1 - \exp\left(\frac{-bU_L L}{\dot{m}_{air} C p_{air}}\right) \tag{2.39}$$

where

$$Z = \alpha_b \tau_g^2 (1 - \beta_c) + h_{P1G} \tau_g \beta_c (\alpha_c - \eta_c) \tag{2.40}$$

Here α_c denotes the absorptivity of the solar cell and α_b the absorptivity of a black surface. In Equation 2.39, U_L is the overall heat transfer coefficient between the solar cell and the ambient environment through the top and back surfaces of the insulation. The air outlet temperature of the PV/T panel is calculated based on its energy balance:

$$T_{air,out} = \left[T_0 + \frac{h_{P2G} Z \dot{I}}{U_L}\right] \left\{1 - \frac{1 - \exp\left(\frac{-bU_L L}{\dot{m}_{air} C p_{air}}\right)}{\frac{bU_L L}{\dot{m}_{air} C p_{air}}}\right\} + T_{air,in} \left[\frac{1 - \exp\left(\frac{-bU_L L}{\dot{m}_{air} C p_{air}}\right)}{\frac{bU_L L}{\dot{m}_{air} C p_{air}}}\right] \tag{2.41}$$

The thermal efficiency of the PV/T collector is expressed as

$$\eta_{th} = \frac{\dot{Q}}{\dot{I} b L} \tag{2.42}$$

Here, $\dot{I}$ is the solar intensity and b and L are the width and length of the PV/T panel.

2.2.9 Solar Photovoltaic Panel

Solar photovoltaic (PV) technology converts renewable energy directly into electricity. PV systems are used in many buildings, including almost all net-zero energy buildings. In urban and suburban areas, PV modules are often mounted on roofs of houses and non-residential buildings.

The modeling of a PV system is described in this section. The performance of a solar cell is normally evaluated under standard conditions for an average solar spectrum with an air mass coefficient (AM) of 1.5, which corresponds to a solar zenith angle Z of 48.2^0,

Figure 2.8 Equivalent circuit of a PV cell.

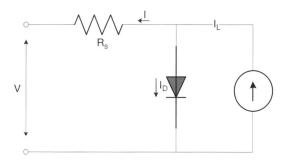

and an irradiance normalized to $1000\,\text{W/m}^2$. A primary electric model of a PV cell contains a current source and a diode (see Figure 2.8). Applying basic circuit laws allows the cell terminal voltage to be determined as

$$I = I_L - I_D \tag{2.43}$$

Here, the light current I_L depends on the solar irradiance G and the cell temperature T_c, and is calculated according to design reference conditions:

$$I_L = \left(\frac{G}{G_{\text{ref}}}\right)(I_{L,\text{ref}} + k_t(T_{\text{cell}} - T_{\text{ref}})) \tag{2.44}$$

where the values of G_{ref}, $I_{L,\text{ref}}$, T_{ref}, and k_t are given by manufacturers. The cell temperature is a function of wind speed, solar irradiance, and ambient temperature, and can be determined (in °C) with the following correlation:

$$T_{\text{cell}} = 0.943T_0 + 0.028G - 1.528V_{\text{wind}} + 4.3 \tag{2.45}$$

which was developed by Chenni *et al.* [5] using experimental data for six solar cell technologies. Chenni *et al.* state that the dependencies of the cell temperature on ambient temperature, solar irradiance, and wind speed are fairly independent of location.

The diode current I_D is given as a function of the reverse saturation current I_0 and the cell terminal voltage and current. The diode current is expressed by the Shockley equation as

$$I_D = I_0 \left[\exp(\frac{q(V + IR_s)}{\gamma k T_c}) - 1\right] \tag{2.46}$$

where V denotes the terminal voltage (V), I the saturation current (A), γ the shape factor, R_s the series resistance, q the electron charge, and k the Boltzmann constant. The reverse saturation current is defined as:

$$I_0 = DT_{\text{cell}}^3 \exp\left(\frac{-q\varepsilon_G}{AkT_{\text{cell}}}\right) \tag{2.47}$$

where D is the diode diffusion factor, and ε_G is the material bandgap energy, which is 1.12 eV for Si and 1.35 eV for Ga. The reverse saturation current is calculated using Equation 2.47 at two temperatures in order to eliminate D. Therefore, I_0 can be expressed as follows:

$$I_0 = I_{0,\text{Ref}} \left(\frac{T_{\text{cell}}}{T_{\text{ref}}}\right)^3 \exp\left[\left(\frac{q\varepsilon_G}{Ak}\right)\left(\frac{1}{T_{\text{ref}}} - \frac{1}{T_{\text{cell}}}\right)\right] \tag{2.48}$$

Further details of the derivations are presented by Chenni *et al.* [5]. Therefore, the I-V characteristics of the PV modules can generally be expressed as [5]:

$$I = \left(\frac{G}{G_{ref}}\right)(I_{L,ref} + k_t(T_{cell} - T_{ref})) - I_0\left[\exp\left(\frac{q(V + IR_s)}{\gamma k T_{cell}}\right) - 1\right] \tag{2.49}$$

where q and k are the electron charge and the Boltzmann constant, respectively, R_s is the series resistance, and γ is the shape factor, which is a function of the completion factor and the number of cells in the module, and which is defined as:

$$\gamma = AN_{CS}N_S \tag{2.50}$$

Here, N_{CS} and N_S are the number of cells connected in series and the number of modules connected in series respectively. Such data are usually available in PV module catalogs. The power output of the PV cell is the product of its terminal current and voltage. Equation 2.49 has a nonlinear characteristic and the point at which the maximum current and voltage occurs is called the maximum power point:

$$P_{mp} = I_{mp}V_{mp} \tag{2.51}$$

The following implicit expression can be used to calculate the maximum current of the PV cell [5]:

$$I_{mp} + \frac{(I_{mp} - I_L - I_0)\left[\ln\left(\frac{I_L - I_{mp}}{I_0} + 1\right) - \frac{qI_{mp}R_s}{k\gamma T_{ref}}\right]}{1 + (I_L - I_{mp} + I_0)\frac{qR_s}{k\gamma T_{ref}}} = 0 \tag{2.52}$$

This expression can be solved using the Newton–Raphson method. At the maximum power point, the first derivative of power with respect to voltage is zero. Rearranging the consequent equations results in an explicit expression for the maximum voltage as a function of the maximum terminal current:

$$V_{mp} = \frac{kT_{ref}}{q}\ln\left(\frac{I_L - I_{mp}}{I_0} + 1\right) - I_{mp}R_s \tag{2.53}$$

The control system for the PV system is designed so that the system operates at the maximum power point. The new values of maximum power point voltage and maximum power point current are obtained from the *I–V* characteristic curve and Equation 2.53 simultaneously at new climatic and operating conditions (Figure 2.9).

The electrical efficiency of a PV module can be defined as the ratio of actual electrical output to input solar energy incident on the PV surface [5]:

$$\eta_{el} = \frac{V_{MP}I_{MP}}{S} \tag{2.54}$$

To model PV panels in this book, silicon photovoltaic modules manufactured by Sun Power Corporation with the specifications in Table 2.1 are selected. To validate the simulation code developed in Matlab software, the outputs of various simulation codes are compared with data provided in the literature and manufacturer's catalogs. Figure 2.9 compares results from the simulation code and reference [5]. The results of the PV simulation solar panel's *I-V* characteristic curves for the standard test condition irradiance of $1000\,W/m^2$ and a reference temperature of $25\,°C$ are shown in Figure 2.9. The maximum power point at the standard test condition is observed to vary by only 1.1% from the manufacturer's reported values.

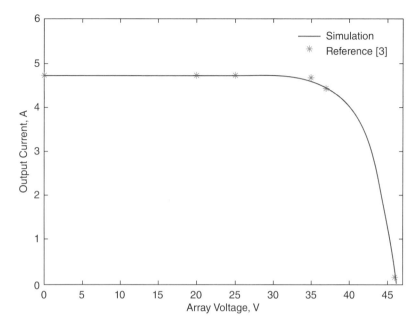

Figure 2.9 Validation of PV simulation results with experimental data.

Table 2.1 PV module specifications/parameters.

Parameter	Value
Short circuit current, I_{SC} (A)	5.75
Open circuit voltage V_{OC} (V)	47.7
Maximum point current, I_{MP} (A)	5.45
Maximum point voltage, V_{MP} (A)	41
Array area, A (m^2)	1.24
Number of cells in modules, N_{CS}	72
Number of cells in modules, N_S	7
Temperature coefficient of short circuit (A/°C)	3.5×10^{-3}
Temperature coefficient of open circuit (A/°C)	-1.36×10^{-1}
Series resistance (Ω)	0.041
Reference total irradiance (W/m^2)	1000
Reference light current (A)	6.81
Reference temperature (°C)	25
Wind speed (m/s)	5

2.3 Optimization

From a mathematical point of view, optimization is the process of maximizing or minimizing a function subject to several constraints, for a number of variables, for each of which a range exists [5]. Put more simply and practically, optimization involves finding the best possible configuration for a given problem subject to reasonable constraints.

When an optimization problem involves only one objective function, the task of finding the optimum solution is called single objective optimization. Single objective optimization thus considers the solution to the problem with respect to just one criterion. Single objective optimization has been applied for decades to a wide range of applications.

The need to consider more than one objective function and the importance of doing so led to the advent of multi-objective optimization. In management disciplines, such optimization problems are commonly known as multiple criterion decision-making. Most real world optimization problems inherently involve multiple objective functions. The principles and intent of optimization cannot be reasonably applied with only one objective function when other objectives are also important.

Some important optimization concepts and terms are described and defined in the next four subsections.

2.3.1 System Boundaries

The first step in any optimization problem is to define the system boundaries. All subsystems that affect system performance should be included. When the system is overly complex, it is often desirable to divide it into smaller subsystems. In this case, it is often reasonable to carry out optimization on each subsystem independently, that is, sub-optimization of the subsystems is performed.

2.3.2 Objective Functions and System Criteria

The next step in an optimization problem is to define the system criteria, which are sometimes called objective functions. An objective function is based on the desire or purpose of the decision maker, and it can be either maximized or minimized. Optimization criteria can vary widely. For instance, optimization criteria can be based on economic aims (e.g., total capital investment, total annual levelized costs, cost of exergy destruction, cost of environmental impact), efficiency aims (e.g., energy, exergy, other), other technological goals (production rate, production time, total weight), environmental impact objectives (reduced pollutant emissions), and other objectives.

Note that we can consider more than one objective function when solving for the optimal solution for an optimization problem, via multi-objective optimization.

2.3.3 Decision Variables

Another essential step in formulating an optimization problem is the selection of independent decision variables that adequately characterize the possible design options. To select decision variables, it is important to (a) include all important variables that can affect the performance and cost effectiveness of the system, (b) not include variables of minor importance, and (c) distinguish among independent variables whose values are amenable to change. In a given optimization problem, only decision variables are changing. Variables whose values are calculated from the independent variables using mathematical models are dependent variables.

2.3.4 Constraints

The constraints in a given design problem arise due to limitations on the ranges of the physical variables, basic conservation principles which must be satisfied, and other restrictions. Restrictions on variables may arise due to limitations on the space, equipment, and materials that are employed. That is, we may restrict the physical dimensions of a system, the temperatures (high and/or low) that components can attain, maximum allowable pressure, material flow rate and force generated, and so on. Also, minimum values of the temperature may be specified for thermoforming of a plastic and for ignition to occur in an engine. Thus, both minimum and maximum values of design variables may be involved in constraints.

Many constraints in thermal systems arise because of conservation laws, particularly those related to mass, momentum, and energy. For instance, under steady state conditions, mass inflow to a system must equal mass outflow. This condition gives rise to an equation that must be satisfied by the relevant design variables, thus restricting the values that may be employed in the search for an optimum. Similarly, energy balance considerations are important in thermal systems and may limit the range of temperatures, heat fluxes, dimensions, and so on, that may be used.

Several such constraints are often satisfied during modeling and simulation because the governing equations are based on conservation principles. In this way, the objective function being optimized already considers these constraints. In such cases, only the additional limitations that define the boundaries of the design domain remain to be considered.

2.3.5 Optimization Methods

2.3.5.1 Classical Optimization
Classical optimization techniques are useful for finding the optimum solution or unconstrained maximum or minimum of continuous and differentiable functions. Some specifications for classical optimization can be selected based on this understanding, as described below:

- These are analytical methods that make use of differential calculus in locating the optimum solution.
- Classical methods have limited scope in practical applications as these often involve objective functions that are not continuous and/or differentiable.
- These methods assume that the function is differentiable twice with respect to the design variables and that the derivatives are continuous.
- Three main types of problems can be handled by classical optimization techniques:
  - Single variable functions.
  - Multivariable functions with no constraints.
  - Multivariable functions with both equality and inequality constraints. In problems with equality constraints, the Lagrange multiplier method can be used. If the problem has inequality constraints, the Kuhn-Tucker conditions can be used to identify the optimum solution.

2.3.5.2 Numerical Optimization Methods
Several categories of this optimization technique exist:

- **Linear programming**: Applies to the case in which an objective function f is linear and the set A, where A is the design variable space, is specified using only linear equalities and inequalities.
- **Integer programming**: Applies to linear programs in which some or all variables are constrained to take on integer values.

- **Quadratic programming**: Allows the objective function to have quadratic terms, while the set *A* must be specified with linear equalities and inequalities.
- **Nonlinear programming**: Applies to the general case in which the objective function or the constraints or both contain nonlinear parts.
- **Stochastic programming**: Applies to the case in which some of the constraints depend on random variables.
- **Dynamic programming**: Applies to the case in which the optimization strategy is based on dividing the problem into smaller sub-problems.
- **Combinatorial optimization**: Concerns problems where the set of feasible solutions is discrete or can be reduced to a discrete one.
- **Evolutionary algorithm**: Involves numerical methods based on random search.

2.3.5.3 Evolutionary Algorithms

An evolutionary algorithm utilizes techniques inspired by biological evaluation, reproduction, mutation, recombination, and selection. The candidate solutions to the optimization problem play the role of individuals in a population, and a fitness function determines the environment within which the solutions "live." Evolutionary algorithm methods include *genetic algorithms* (GAs), *artificial neural networks* (ANNs), and *fuzzy logic*. These approaches are discussed further below. Each of the approaches is available in toolboxes developed by Math Works and can thus be used straightforwardly with Matlab software.

Genetic Algorithm　A genetic algorithm is a search method used for obtaining an optimal solution. The method is based on evolutionary techniques that are similar to processes in evolutionary biology, including inheritance, learning, selection, and mutation. The process starts with a population of candidate solutions called individuals, and progresses through generations, with the fitness of each individual being evaluated. Fitness is defined based on the objective function. Then multiple individuals are selected from the current generation based on fitness and modified to form a new population. This new population is used in the next iteration and the algorithm progresses toward the desired optimal point (Goldberg, 1989; Schaffer, 1985).

In a genetic algorithm, a population of candidate solutions—called individuals, creatures, or phenotypes—to an optimization problem evolves toward better solutions. Each candidate solution has a set of properties that can be mutated and altered; traditionally, solutions are represented in binary as strings of 0 s and 1 s, but other encodings are also possible. The evolution usually starts from a population of randomly generated individuals, and is an iterative process, with the population in each iteration called a generation. In each generation, the fitness of each individual in the population is evaluated; the fitness is usually the value of the objective function in the optimization problem to be solved. The more fit individuals are stochastically selected from the current population, and each individual's genome is modified to form a new generation. The new generation of candidate solutions is then used in the next iteration of the algorithm. Commonly, the algorithm terminates when either a maximum number of generations has been produced, or a satisfactory fitness level has been reached for the population.

Some advantages of using genetic algorithms for optimization are:

- Genetic algorithms can solve any optimization problem that can be described with the chromosome encoding.

- Genetic algorithms can provide multiple solutions for problems.
- Since the genetic algorithm execution technique is not dependent on the error surface, they can be utilized to solve multi-dimensional, non-differential, non-continuous, and even non-parametrical problems.
- Structural genetic algorithms provide the possibility of solving solution structures and solution parameter problems simultaneously.
- Genetic algorithms are easy to understand and require little knowledge of mathematics.
- Genetic algorithms are easily transferred to existing simulations and models.

Artificial Neural Network Artificial neural networks (ANNs) are interconnected groups of processing elements, called artificial neurons, similar to those in the central nervous system of the body. The approach is thus analogous to some elements of neuroscience. The characteristics of the processing elements and their interconnections determine the processing of information and the modeling of simple and complex processes. Functions are performed in parallel and the networks have both non-adaptive and adaptive elements, which change with the inputs and outputs and the problem. The ANN approach leads to nonlinear, distributed, parallel, local processing and adaptive representations of systems.

2.4 Multi-objective Optimization

Optimal conditions are generally strongly dependent on the chosen objective function. However, several aspects of performance are often important in practical applications. In thermal and energy systems design, efficiency (energy and/or exergy), production rate, output, quality, and heat transfer rate are common quantities that are to be maximized, while cost, input, environmental impact, and pressure are quantities to be minimized. Any of these can be chosen as the objective function for a problem, but it is usually more meaningful and useful to consider more than one objective function.

Users of simple optimization are able to determine the minimum and maximum of a single variable function, and can utilize first and second derivative techniques to find the optimal value of a given function. At the advanced level, users of optimization are able to find an optimum value of multivariable functions. In addition, they are can solve multivariable optimization problems with constraints. Constrained optimization is an important subject in practice science since most real world problems contain constraints.

Multi-objective optimization has been extensively used and studied. There exist many algorithms and application case studies involving multi-objective optimization.

One of the common approaches for dealing with multiple objective functions is to combine them into a single objective function that is then minimized or maximized. For example, in the design of heat exchangers and cooling systems for electronic equipment, it is desirable to maximize the heat transfer rate. However, this often comes at the cost of increased fluid flow rates and corresponding frictional pressure losses. A multi-objective optimization problem has objective functions that are either minimized or maximized. As with single objective optimization, multi-objective optimization involves several constraints that any feasible solution including the optimal solution must satisfy.

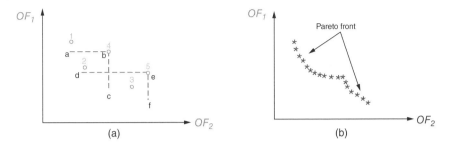

Figure 2.10 Multi-objective optimization with two objective functions OF_1 and OF_2 that are to be minimized, showing (a) dominant designs and (b) the Pareto frontier.

A multi-objective optimization problem can be formulated as:

$$\text{Minimize/maximize} \quad f_n(x) \qquad n = 1, 2, \ldots N$$
$$\text{Subject to} \qquad g_j(x) > 0 \qquad j = 1, 2, \ldots J$$
$$h_k(x) = 0 \qquad k = 1, 2, \ldots K$$
$$x_i^{(L)} \leq x_i \leq x_i^{(U)} \qquad i = 1, 2, \ldots n$$

A solution of this problem is x, which is a vector of n decision variable or design parameters. The last set of constraints here is called the variable bounds, which restrict the searching bound. Any solution of the decision variables should lie within a lower bound $(x_i^{(L)})$ and upper bound $(x_i^{(U)})$. To illustrate, we consider two objective functions, OF_1 and OF_2. We assume that these are to be minimized (although maximization can be similarly handled since it is equivalent to minimization of the negative of the function). Figure 2.10a shows values for the two objective functions at five design points. Design 2 is clearly seen in this figure to be preferable to design 4 because both objective functions are smaller for design 2 compared to design 4. Similarly, design 3 is preferable to design 5. In addition, designs 1, 2, and 3 are not dominated, by any other designs.

The set of non-dominated designs is introduced as the Pareto frontier, representing the best collection of design points. This is shown in Figure 2.10b. Note that any point on the Pareto frontier can be considered as an optimal design condition. The selection of a specific design from the set of points constituting the Pareto frontier is at the discretion of the decision maker, typically an engineer or designer.

2.4.1 Sample Applications of Multi-objective Optimization

Multi-objective optimization has a broad range of applications in the real world, from aircraft design to petrochemical plant design. Here, we briefly discuss some of the main practical applications of multi-objective optimization.

2.4.1.1 Economics

In economics, many problems involve several objectives along with some constraints that affect what combinations of the objectives are attainable. For example, consumer demand for various goods is determined by the process of maximization of the utilities derived from those goods, subject to a constraint based on how much income is available to spend on them and their prices. These constraints allow more of one good to be purchased only at the sacrifice of consuming less of another good; therefore, the various objectives are in conflict with each other. A common method to analyze such a problem is to use a graph of indifference curves, representing preferences, and a budget constraint, representing the trade-offs with which the consumer is faced.

2.4.1.2 Finance

In finance, a common problem is to choose a portfolio when there are two conflicting objectives. The desire is for the expected value of portfolio returns to be as high as possible. This problem is often represented by a graph in which the efficient frontier shows the best combinations of risk and expected return that are available, and in which indifference curves show the investor's preferences for various risk-expected return combinations.

2.4.1.3 Engineering

Most problems in engineering are multi-objective problems subject to several reasonable linear and nonlinear constraints. In mechanical engineering, for instance, optimization plays a significant role. In designing a heat exchanger, both heat exchanger cost and efficiency are simultaneously significant, in that the higher the efficiency of the heat exchanger, the higher the cost of the system. Therefore, the solution of this multi-objective optimization should be selected in the priority of objective functions. Another example is the optimization of power generation units where cost, efficiency, and environmental impacts are important and selection of either affects the other. Even if there is not an exact solution for such a problem, the final solution depends on the decision maker to design and build the system. However, in most cases the preference is for lower cost and, where global warming is viewed as important, lower environmental impacts.

2.4.2 Illustrative Example: Air Compressor Optimization

To enhance understanding of the application of optimization, an example involving an air compressor, a relatively simple device, is illustrated in detail. Objective functions, constraints, and multi-objective optimization are involved.

An air compressor uses electrical power to compress air, and numerous air compression methods exist, including positive-displacement and negative-displacement types. To optimize an air compressor using thermodynamics and other factors, the device is first modeled using energy balances and other relations. Figure 2.11 shows an air compressor (AC) as a control volume for thermodynamic modeling.

2.4.2.1 Thermodynamic and Economic Modeling and Analysis

Modeling We can write expressions for the air compressor outlet temperature T_2 and work rate requirement $\dot{W}_{AC}$, respectively, as follows:

$$T_2 = T_1 \left\{ 1 + \frac{1}{\eta_{AC}} \left[r_{AC}^{\frac{\gamma_a - 1}{\gamma_a}} - 1 \right] \right\} \tag{2.55}$$

$$\dot{W}_{AC} = \dot{m}_a C_{pa}(T_2 - T_1) \tag{2.56}$$

where $\dot{m}_a$ is air mass flow rate, η_{AC} is the compressor isentropic efficiency, $\gamma_a = c_p/c_v$ and c_{pa} is the specific heat at constant pressure of air, which can be expressed a function of temperature as follows:

$$c_{pa} = 1.048 - \frac{3.837 \times T}{10^4} + \frac{9.453 \times T^2}{10^7} - \frac{5.490 \times T^3}{10^{10}} + \frac{7.929 \times T^4}{10^{14}} \tag{2.57}$$

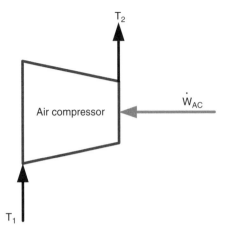

T$_2$

Figure 2.11 Model of an air compressor.

$\dot{W}_{AC}$

Air compressor

T$_1$

Also, T_1 is the ambient air temperature entering the air compressor and T_2 is the hot air temperature exiting.

Exergy Analysis As exergy analysis is discussed extensively in other chapters, we focus here on the exergy destruction and efficiency of the air compressor, which can exist in isolation or as a component of a larger system. An exergy analysis for the air compressor follows, but it is noted that further details about exergy analyses of compressors, applied within gas turbine based power generation, are provided elsewhere [6–8].

The specific exergy of the air entering the compressor can be written as

$$ex_1 = (h_1 - h_0) - T_0(s_1 - s_0) \tag{2.58}$$

where

$$(h_1 - h_0) = C_p(T_1 - T_0) \tag{2.59}$$

$$(s_1 - s_0) = C_p \ln\left(\frac{T_1}{T_0}\right) - R\ln\left(\frac{P_1}{P_0}\right) \tag{2.60}$$

and where T_0 and P_0 denote the reference environment temperature and pressure. These values are taken here to be the ambient temperature and pressure, which are 20 °C and 1 atm, respectively.

The compression ratio for the air compressor and the specific exergy of the outlet air can be expressed as

$$R_{AC} = \frac{P_2}{P_1} \tag{2.61}$$

$$ex_2 = (h_2 - h_0) - T_0(s_2 - s_0) = C_p(T_2 - T_0) - T_0\left(C_p \ln\left(\frac{T_2}{T_0}\right) - R\ln\left(\frac{P_2}{P_0}\right)\right) \tag{2.62}$$

An exergy balance for the air compressor can be written as

$$\dot{m}_1 ex_1 + \dot{W}_{AC} = \dot{m}_2 ex_2 + \dot{E}x_D \tag{2.63}$$

and the exergy efficiency as:

$$\psi_{AC} = \frac{\dot{E}x_2 - \dot{E}x_1}{\dot{W}_{AC}} \tag{2.64}$$

Exergoeconomic Analysis   An exergoeconomic analysis of the air compressor is described. Further details on exergoeconomic analysis, cost balances, and exergoeconomic factors are discussed earlier in this book and elsewhere [8–10]. To determine the cost of exergy destruction for the air compressor, a cost rate balance can be utilized. The cost rate balance for this component can be written as follows:

$$c_1 \dot{E}x_1 + c_w \dot{W}_{AC} + \dot{Z}_{AC} = c_2 \dot{E}x_2 \tag{2.65}$$

where c_1, c_2 and c_w are the unit costs of the inlet air, outlet air, and work. Here, the inlet air is taken to be free, so its unit cost is zero, that is,

$$c_1 = 0 \tag{2.66}$$

Also, $\dot{Z}_{AC}$ is the purchase cost rate of air compressor, which can be expressed as follows:

$$\dot{Z}_{AC} = \frac{Z_{AC} \times CRF \times \varphi}{N \times 3600} \tag{2.67}$$

where Z_{AC} is the purchase cost of the air compressor and CRF is the capital recovery factor, which is dependent on the interest rate i and equipment lifetime n, and is determined as

$$CRF = \frac{i(1+i)^n}{(1+i)^n - 1} \tag{2.68}$$

Also, N denotes the annual number of operation hours for the unit, and φ is the maintenance factor, which is often 1.06 [10].

The purchase cost of the air compressor can be approximated as follows [10]:

$$Z_{AC} = c_{11} \dot{m}_a \frac{1}{c_{12} - \eta_{AC}} R_{AC} \ln(R_{AC}) \tag{2.69}$$

where

$$c_{11} = 44.71 \ \$/(\text{kg s}^{-1}) \tag{2.70}$$

$$c_{12} = 0.95 \tag{2.71}$$

For each optimization problem, defining objective functions is of great importance. In this example, multiple objective functions can be considered through multi-objective optimization. When high efficiency is desired, it is reasonable to include the exergy efficiency of the compressor as an objective function. As shown in Equation 2.64, the compressor exergy efficiency is a function of the compressor pressure ratio and isentropic efficiency. Another objective function can be the compressor cost, as expressed in Equation 2.69, which is a function of the compressor pressure ratio, the air mass flow rate through the compressor, and the compressor isentropic efficiency.

Considering these two objective functions (OFs), we can write:

$$\text{OF I} = \psi_{AC} = \frac{\dot{E}x_2 - \dot{E}x_1}{\dot{W}_{AC}} \tag{2.72}$$

$$\text{OF II} = \dot{Z}_{AC} = \frac{\left(c_{11} \dot{m}_a \frac{1}{c_{12} - \eta_{AC}} R_{AC} \ln R_{AC}\right) CRF \times \varphi}{N \times 3600} \tag{2.73}$$

2.4.2.2   Decision Variables

The main design parameters for optimization of the air compressor are the compressor pressure ratio and the compressor isentropic efficiency. Thus, to perform a multi-objective optimization they are considered as our decision variables.

Table 2.2 Physical constraints for optimization of an air compressor.

Constraint	Reason
$R_{AC} < 22$	Material temperature limit
$\eta_{AC} < 0.92$	Commercial availability

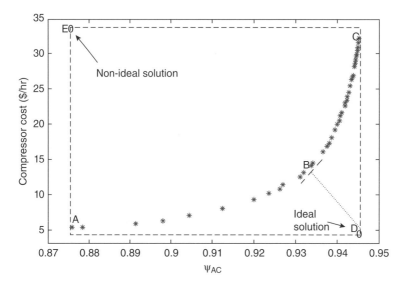

Figure 2.12 Pareto frontier for the optimization of an air compressor, highlighting the best trade-off among values for the objective functions.

2.4.2.3 Constraints

To formulate a meaningful optimization problem, constraints often exist that must be satisfied while performing the optimization, often to ensure solutions are reasonable and realistic. Here, two constraints are considered, as described in Table 2.2.

2.4.2.4 Multi-objective Optimization

To determine the best among the optimal design parameters for an air compressor, a modified version of a genetic algorithm developed with Matlab software is used. Figure 2.12 shows the Pareto frontier for multi-objective optimization of an air compressor where the pressure ratio and isentropic efficiency of the compressor are the two main design variables. The range of values shown is limited by the problem constraints.

It can be seen in Figure 2.12 that the total compressor cost increases moderately as the compressor exergy efficiency increases up to about 92%. Increasing the total exergy efficiency further increases the cost significantly. It is seen in Figure 2.12 that the maximum exergy efficiency exists at design point *C* (94.44%), while the compressor cost rate is the greatest at this point (33.1 $/hr). On the other hand, the minimum value for the compressor cost rate occurs at design point *A* and is about 5.12 $/hr. Design point *C* is the optimal situation when exergy efficiency is the sole objective function, while design point *A* leads the optimum design when total cost rate of product is the sole objective

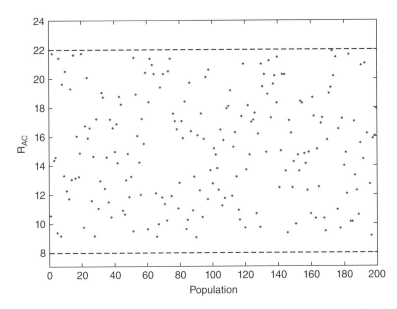

Figure 2.13 Scatter distribution of compressor isentropic efficiency and its allowable range with population in Pareto frontier.

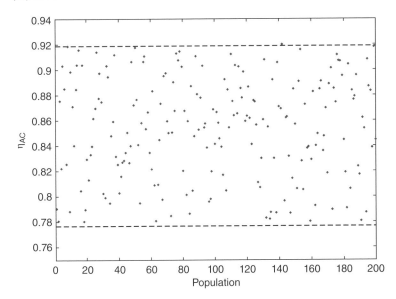

Figure 2.14 Scatter distribution of compressor pressure ratio and its allowable range with population in Pareto frontier.

function. Point D is the ideal solution of the multi-objective optimization because both objective functions are at their optimal values, that is, at higher exergy efficiency and lower total cost rate. Since this point is not located on the Pareto frontier, point B could be selected as one of the best solutions because it is close to the ideal solution.

The variations of compressor pressure ratio and compressor isentropic efficiency are illustrated in Figures 2.13 and 2.14, respectively, where scatter distributions for the

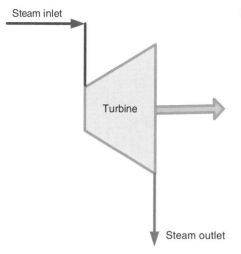

Steam inlet

Figure 2.15 Model of a steam turbine.

Turbine

Steam outlet

populations in the Pareto frontier are shown for each of these design parameters. The points in these figures are obtained from the developed Matlab code, and show how the design parameters change within their allowable ranges.

In a genetic algorithm, a population (called chromosomes or the genotype of the genome), which encode candidate solutions (called individuals, creatures, or phenotypes) to an optimization problem, evolves toward better solutions. The scatter distribution of design parameters are within the range exhibiting good selections of these two parameters for optimization purposes. It is noted in Figures 2.13 and 2.14 that the points are not just near the boundaries but they are scattered almost randomly within their ranges (8 to 22 for R_{AC}, and 0.78 to 0.92 for η_{AC}). In real optimization, the selection of decision variables is based on the scattered distribution of the decision variables, providing an efficient search for the best optimal solution of the objective function.

2.4.3 Ilustrative Example: Steam Turbine

Steam turbines convert the energy in steam to mechanical energy and are a key component of many power plants. Optimization is applied to a steam turbine, a relatively simple device, as an illustrative example. Objective functions, constraints, and multi-objective optimization are involved. Figure 2.15 shows a thermodynamic model of steam turbine. Considering this steam turbine as a control volume, the output power can be determined using an energy balance equation:

$$\dot{m}_{in}h_i = \dot{W}_T + \dot{m}_{out}h_{out} \tag{2.74}$$

Also, an expression for the isentropic efficiency of the steam turbine follows:

$$\eta_{ORC \cdot T} = \frac{\dot{W}_{act}}{\dot{W}_{is}} \tag{2.75}$$

To model the steam turbine, we consider what information is usually known. The given information for modeling purposes usually includes the temperature, pressure, and mass flow rate at the steam turbine inlet, as well as the turbine outlet pressure and isentropic efficiency. The steam turbine outlet pressure is roughly equivalent to the condenser pressure when the turbine exhausts to a condenser. With this information, we can calculate such quantities as outlet temperature and output power, and exergy flow and destruction rates.

For this application of multi-objective optimization, we need to define objective functions. One common objective function is the total cost rate of the turbine, which can be expressed as follows:

$$\dot{Z}_{ST} = \frac{Z_{ST} \times CRF \times \varphi}{N \times 3600} \tag{2.76}$$

where Z_{ST} is the purchase cost of the steam turbine and CRF is the capital recovery factor, which is dependent on the interest rate i and equipment lifetime n, and is determined as

$$CRF = \frac{i(1+i)^n}{(1+i)^n - 1} \tag{2.77}$$

Also, N denotes the annual number of operation hours for the unit, and φ is the maintenance factor, which is often 1.06 [10].

The purchase cost of the steam turbine can be approximated as follows [10]:

$$Z_{ST} = c_{11}(\dot{W}_{ST})^{0.7}\left(1 + \left(\frac{0.05}{1 - \eta_{ST}}\right)^3\right)(1 + 5\exp\left(\frac{T_{in} - 866}{10.42}\right)) \tag{2.78}$$

where

$$c_{11} = 3880 \ \$/(\text{kW}^{0.7}) \tag{2.79}$$

For any optimization problem, defining objective functions is of great importance. In this example, multiple objective functions are considered through multi-objective optimization. When high performance is desired, it is reasonable to include the steam turbine output power as an objective function. As shown in Equation 2.74, the output power is a function of the turbine inlet conditions, outlet pressure, and isentropic efficiency. Another objective function can be the turbine cost, as expressed in Equation 2.78, which is a function of the turbine output power, inlet temperature, and isentropic efficiency.

Considering these two objective functions (OFs), we can write:

$$\text{OF I} = \dot{W}_{ST} = \dot{m}_{in}h_{in} - \dot{m}_{in}h_{out} \tag{2.80}$$

$$\text{OF II} = \dot{Z}_{AC} = \frac{\left[c_{11}(\dot{W}_{ST})^{0.7}\left(1 + \left(\frac{0.05}{1-\eta_{ST}}\right)^3\right)\left(1 + 5\exp\left(\frac{T_{in}-866}{10.42}\right)\right)\right]CRF \times \varphi}{N \times 3600} \tag{2.81}$$

2.4.3.1 Decision Variables

The main design parameters for optimization of the steam turbine are the inlet temperature, outlet pressure, and isentropic efficiency. These parameters are considered our decision variables in the multi-objective optimization.

2.4.3.2 Constraints

Several constraints are applied here to ensure solutions are reasonable and realistic, as listed and explained in Table 2.3.

Table 2.3 Constraints for optimization of a steam turbine.

Constraint	Reason
$300^{\circ}\text{C} < T_{in} < 350^{\circ}\text{C}$	Material temperature limit
$8 \text{ kPa} < P_{cond} < 20 \text{ kPa}$	Output power limitation
$0.75 < \eta_{ST} < 0.85$	Commercial availability

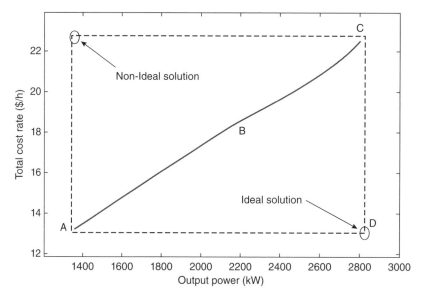

Figure 2.16 Pareto frontier for the optimization of a steam turbine, highlighting the best trade-off among values for the objective functions.

2.4.3.3 Multi-objective Optimization

To determine the best of the optimal design parameters for a steam turbine, a modified version of a genetic algorithm developed with Matlab software is used. Note that the steam turbine inlet mass flow rate and inlet pressure are taken to be 10 kg/s and 50 bar, respectively, based on previous studies [11]. Figure 2.16 shows the Pareto frontier for multi-objective optimization of the steam turbine as a function of the pressure ratio and isentropic efficiency, which are the two main design variables. The range of values shown is limited by the problem constraints.

It can be seen in Figure 2.16 that the total turbine cost increases linearly with turbine output power. It is shown in Figure 2.16 that the maximum turbine output power exists at design point C (2805 kW), while the turbine cost rate is the greatest at this point (22.48 $/hr). On the other hand, the minimum value for the turbine cost rate occurs at design point A and is about 13.12 $/hr. Design point C is the optimal situation when output power is the sole objective function, while design point A leads the optimum design when total cost rate of product is the sole objective function. Point D is the ideal solution of the multi-objective optimization because both objective functions are at their optimal values, that is, at higher output power and lower total cost rate. Since this point is not located on the Pareto frontier, point B could be selected as one of the best solutions because it is close to the ideal solution.

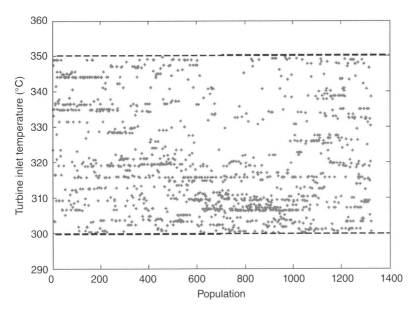

Figure 2.17 Scatter distribution of steam turbine inlet temperature with population in Pareto frontier.

The variations of steam turbine inlet temperature, isentropic efficiency, and outlet pressure are illustrated in Figures 2.17 through 2.19, respectively, where scatter distributions for the populations in the Pareto frontier are shown for each of these design parameters. The points in these figures are obtained from the developed Matlab code, and demonstrate how the design parameters change within their allowable ranges. In a genetic algorithm, a population which encodes candidate solutions of an optimization problem evolves toward better solutions. The scatter distributions of design parameters are within the ranges exhibiting good selections of these parameters for optimization purposes. Note that in Figures 2.17 and 2.19, the points are not just near the boundaries but are scattered almost randomly within their ranges. In real optimization, the selection of decision variables is based on the scattered distribution of the decision variables, providing an efficient search for the best optimal solution of one or more objective functions.

2.5 Concluding Remarks

In this chapter, general introductory aspects of thermodynamic modeling of common devices in energy systems are presented, along with their fundamental concepts and physical quantities, to provide a sound background for understanding applications of energy systems. The material covered is also useful for the energy, exergy, and other analyses presented subsequently. In addition, optimization techniques and the benefits of each are described. The coverage includes objective functions, constraints, and evolutionary algorithms. A focus is placed on multi-objective optimization, and validation and comparisons between base and optimized cases are discussed. To illustrate, an air compressor and a steam turbine are examined and multi-objective optimization is conducted. In order to enhance understanding, the variations of decision variables

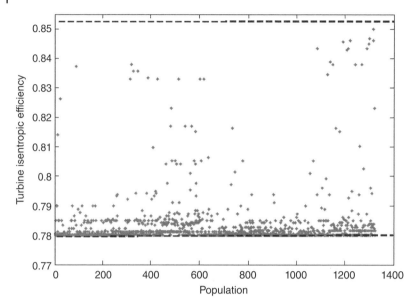

Figure 2.18 Scatter distribution of steam turbine isentropic efficiency with population in Pareto frontier.

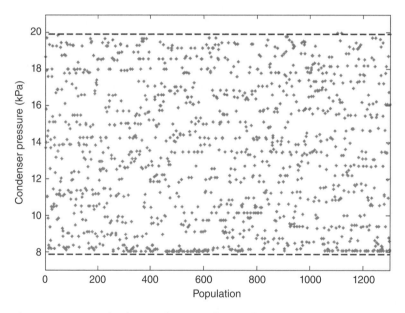

Figure 2.19 Scatter distribution of steam turbine outlet pressure with population in Pareto frontier.

are shown for the example optimizations to demonstrate how each decision variable changes when optimization is undertaken.

The material in this chapter presents knowledge needed for engineers to model a system and ensure each of its parts operate properly. In addition, the material can aid in selecting a suitable optimization technique based on the type of system considered, so as to better optimize the system. In the next chapter, other important devices and plants are modeled comprehensively and optimized.

References

1 Ahmadi, P., Dincer, I. and Rosen, M. A. (2013) Thermodynamic modeling and multi-objective evolutionary-based optimization of a new multigeneration energy system. *Energy Conversion and Management* **76**:282–300.

2 Wang, J., Dai, Y. and Sun, Z. (2009) A theoretical study on a novel combined power and ejector refrigeration cycle. *International Journal of Refrigeration* **32**:1186–1194.

3 Farahat, S., Sarhaddi, F. and Ajam, H. (2009) Exergetic optimization of flat plate solar collectors. *Renewable Energy* **34**:1169–1174.

4 Joshi, A., Tiwari, A., Tiwari, G., Dincer, I. and Reddy, B. (2009) Performance evaluation of a hybrid photovoltaic thermal (PV/T) (glass-to-glass) system. *International Journal of Thermal Sciences* **48**:154–164.

5 Chenni, R., Makhlouf, M., Kerbache, T. and Bouzid, A. (2007) A detailed modeling method for photovoltaic cells. *Energy* **32**:1724–1730.

6 Ahmadi, P., Dincer, I. and Rosen, M. A. (2013) Development and assessment of an integrated biomass-based multi-generation energy system. *Energy* **56**:155–166.

7 Ahmadi, P., Dincer, I. and Rosen, M. A. (2013) Performance assessment and optimization of a novel integrated multigeneration system for residential buildings. *Energy and Buildings* **67**:568–578.

8 Dincer, I. and Rosen, M. A. (2012) *Exergy: Energy, Environment and Sustainable Development*, Second Edition, Elsevier.

9 Kotas, T. J. (1995) *The exergy method of thermal plant analysis*. Krieger Melbourne, FL.

10 Bejan, A., Tsatsaronic, G. and Moran, M. J. (1996) *Thermal design and optimization*, John Wiley & Sons, Inc., New York.

11 Ahmadi, P., Dincer, I. and Rosen, M. A. (2011) Exergy, exergoeconomic and environmental analyses and evolutionary algorithm based multi-objective optimization of combined cycle power plants. *Energy* **36**:5886–5898.

Study Questions/Problems

1   Use Engineering Equation Solver (EES) to model the air compressor shown in Figure 2.2. Conduct a parametric study and discuss the effect of varying compressor pressure ratio and compressor isentropic efficiency on the compressor exergy efficiency and compressor work.

2   Consider a heat recovery steam generator that is used to recover the exhaust energy from flue gases exiting a gas turbine to produce saturated steam. How is the exergy efficiency defined for this device if it is considered as an objective function? What would the proper decision variable be for this device? Explain the effect of gas turbine outlet temperature on the selection of heat recovery steam generator (HRSG) decision variables.

3 In modeling a combustion chamber, how does the combustion reaction equation change if incomplete combustion is considered? How do the greenhouse gas emissions vary with an increase in combustion chamber air inlet temperature?

4 When an engineer or designer is deciding on the preferred optimal point on a Pareto frontier, what point should be selected and why?

5 What is the effect of interest rate i, as used in Equation 2.77, on the Pareto frontier? Draw a Pareto frontier for interest rates of 5%, 10%, and 15%.

6 Consider a gas turbine power plant that is composed of an air compressor, a combustion chamber, and a gas turbine. Model this power plant and define two main objective functions; apply a genetic algorithm in order to optimize the plant. Discuss and interpret the results.

3

Modeling and Optimization of Thermal Components

3.1 Introduction

Energy systems are often complex and operate based on various principles and physical laws. For an energy system, a set of coupled equations involving thermodynamics, heat and mass transfer, and fluid mechanics describe the system's behavior, and need be solved simultaneously, invoking reasonable assumptions as necessary, to provide predictive models. An energy system is in general composed of numerous thermal components, sometimes forming a system based on a cycle. It is often necessary to know the characteristics of each component (e.g., performance details, efficiencies) before they are integrated.

Consider a steam power plant as an example of how various thermal components can be connected to generate electricity. In the power plant, a boiler and superheater produce superheated steam, a steam turbine and generator generate electricity, a condenser rejects heat from the system and cools the steam working fluid so it condenses, several high and low pressure pumps increase the water pressure, and several fans and blowers take in air for combustion. These thermal components are integrated in a cycle and, in seeking good performance, the optimization of each component is significant. Possible problems are clearly evident when thermal components are integrated, like potential thermal losses.

This example illustrates one reason why exergy analysis has attracted increasing attention in recent years. Exergy analysis identifies the magnitudes of waste exergy emission and the locations of exergy destructions. The exergy efficiency of each thermal component in a system is an indicator that demonstrates how efficient the component is, accounting for irreversibilities. Exergy efficiency generally provides more illuminating insights into system and component performance than the corresponding energy efficiency, since the former weights energy flows according to their exergy contents and separates inefficiencies into those associated with effluent losses and those due to irreversibility. Furthermore, exergy efficiencies can provide a measure of the potential for improvement in energy systems. Note that improving the efficiency of a thermal component requires better designs and material, but that these usually come at the expense of higher capital cost. This type of situation is a good example of where multi-objective optimization plays an important role, for it can help identify the best compromise between the simultaneous desires for both greater efficiency and lower costs.

In this chapter, we model numerous thermal components that are commonly employed in advanced energy systems. An evolutionary algorithm-based optimization is applied to determine the optimal design parameters of each device. Since the

Optimization of Energy Systems, First Edition. Ibrahim Dincer, Marc A. Rosen, and Pouria Ahmadi.
© 2017 John Wiley & Sons Ltd. Published 2017 by John Wiley & Sons Ltd.

optimization of air compressors and turbines was covered in Chapter 2, we do not consider these two components here. The optimization depends on the objective functions considered, which can range from exergy efficiency to product cost and other parameters. In order to enhance understanding of the design criteria, sensitivity analyses are conducted to assess how a given objective function varies when changes are applied to selected design parameters. The chapter closes with some final remarks on optimal design of thermal components.

3.2   Air Compressor

As previously explained in Chapter 2, the compressor is an important component in various energy systems such as power plants, refrigeration systems, and petrochemical plants. The thermodynamic modeling of an air compressor is described in detail in section 2.4.2. Here, we include some brief results for multi-objective optimization. As indicated earlier, defining objective functions is of great importance for each optimization problem. In this case, multiple objective functions can be considered through multi-objective optimization. When high efficiency is desired, it is reasonable to include the exergy efficiency of the compressor as an objective function. As shown in Equation 3.1, the compressor exergy efficiency is a function of the compressor pressure ratio and isentropic efficiency. Another objective function can be the compressor cost, as expressed in Equation 3.2, which is a function of the compressor pressure ratio, the air mass flow rate through the compressor, and the compressor isentropic efficiency. All the formulas for these two objective functions are provided in Chapter 2.

Considering these two objective functions (OFs), we can write the following:

$$\text{OF I} = \Psi_{AC} = \frac{\dot{E}x_2 - \dot{E}x_1}{\dot{W}_{AC}} \tag{3.1}$$

$$\text{OF II} = \dot{Z}_{AC} = \frac{\left(c_{11} \dot{m}_a \frac{1}{c_{12} - \eta_{AC}} R_{AC} \ln R_{AC} \right) CRF \times \varphi}{N \times 3600} \tag{3.2}$$

Examining these two objective functions reveals that the main design parameters or decision variables for optimization of the air compressor are the compressor pressure ratio and the compressor isentropic efficiency. Thus, these are considered as the decision variables in performing a multi-objective optimization. The appropriate constraints for this optimization are given in Table 2.2 of Chapter 2. In order to determine the best design parameters, a genetic algorithm optimization method is applied and a Pareto curve is obtained as shown in Figure 3.1.

It can be seen in Figure 3.1 that the total compressor cost increases moderately as the compressor exergy efficiency increases up to about 92%. Increasing the total exergy efficiency further increases the cost significantly. It is seen in Figure 2.12 that the maximum exergy efficiency exists at design point C (94.5%), while the compressor cost rate is the greatest at this point (33.1 $/hr). But, the minimum value for the compressor cost rate occurs at design point A and is about 5.1 $/hr. Design point C is the optimal situation when exergy efficiency is the sole objective function, while design point A leads the optimum design when total cost rate of product is the sole objective function. Point D is the ideal solution of the multi-objective optimization because both objective functions are at their optimal values, that is, at higher exergy efficiency and lower total cost rate. More details about this optimization are given in Chapter 2.

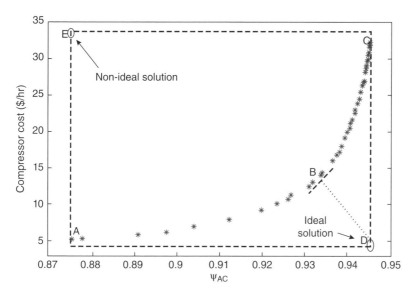

Figure 3.1 Pareto frontier for the optimization of an air compressor, highlighting the best trade-off among values for the objective functions.

3.3 Steam Turbine

Steam turbines convert the energy in steam to mechanical energy and are a key component of many power plants. Thermodynamic modeling and optimization of a steam turbine was previously covered in section 2.4.3. Here, we briefly discuss the optimization of a steam turbine. In this example multiple objective functions are considered through multi-objective optimization. When high performance is desired, it is reasonable to include the steam turbine output power as an objective function. As shown in Equation 2.74, the steam turbine output power is a function of the turbine inlet conditions, the outlet pressure and the isentropic efficiency. Another objective function can be the turbine cost, as expressed in Equation 2.78, which is a function of the turbine output power, inlet temperature, and isentropic efficiency.

Considering these two objective functions (OFs), we introduce the following:

$$\text{OF I} = \dot{W}_{ST} = \dot{m}_{in}h_{in} - \dot{m}_{in}h_{out} \tag{3.3}$$

$$\text{OF II} = \dot{Z}_{AC} = \frac{\left[c_{11}(\dot{W}_{ST})^{0.7} \left(1 + \left(\frac{0.05}{1-\eta_{ST}} \right)^3 \right) (1 + 5 \exp\left(\frac{T_{in}-866}{10.42} \right)) \right]}{N \times 3600} \tag{3.4}$$

The main design parameters for optimization of the steam turbine are the inlet temperature, outlet pressure, and isentropic efficiency. These parameters are considered as decision variables in the multi-objective optimization. The constraints for this optimization problem are given in Table 2.3 of Chapter 2. Figure 3.2 shows the Pareto frontier for multi-objective optimization of the steam turbine as a function of pressure ratio and isentropic efficiency of the steam turbine, which are the two main design variables. The range of values shown is limited by the problem constraints. It can be seen in Figure

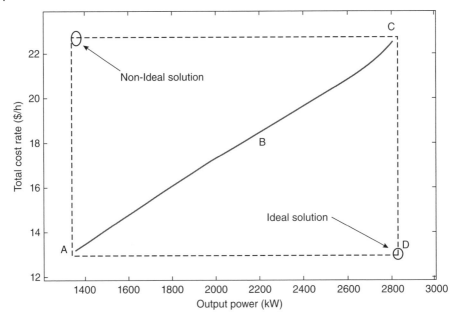

Figure 3.2 Pareto frontier for the optimization of a steam turbine, highlighting the best trade-off among values for the objective functions.

2.16 that the total turbine cost linearly increases with turbine output power. That figure also shows that the maximum turbine output power exists at design point C (2805 kW), while the turbine cost rate is the greatest at this point (22.48 $/hr). On the other hand, the minimum value for the turbine cost rate occurs at design point A and is about 13.12 $/hr. Design point C is the optimal situation when output power is the sole objective function, while design point A leads the optimum design when total cost rate of product is the sole objective function. Point D is the ideal solution of the multi-objective optimization because both objective functions are at their optimal values, that is, at higher output power and lower total cost rate.

Further details about the optimization of steam turbines are provided in Chapter 2.

3.4 Pump

A pump mechanically pressurizes and/or moves fluids. Pumps can be classified into three major groups according to the method they use to move the fluid: direct lift, displacement, and gravity pumps. Pumps operate via many energy sources, including manual operation, electricity, engines, or wind power and come in many sizes, from microscopic, for use in medical applications, to large industrial pumps for power generation in large power plants. A pump is a major part for most thermodynamic cycles where different pressure levels are required. Some samples are organic Rankine cycles, absorption refrigeration systems, cogeneration heat and power systems, and cryogenic energy storage and desalination units. In large power plants, the main pump is called the boiler feed pump, which consumes about 5 to 10% of the net output electricity of the plant. This shows the importance of this device in power generation systems. Since

Figure 3.3 Model of a pump.

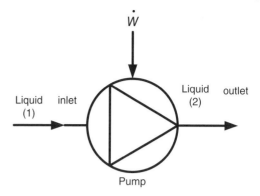

most energy systems include a pump, we are trying here to thermodynamically model this component. Figure 3.3 shows a model of a typical pump.

3.4.1 Modeling and Simulation of a Pump

As pointed out earlier, we can determine the rate that work is provided to the pump using an energy balance for the control volume around the pump:

$$\dot{W} = \dot{m}v(P_2 - P_1) \tag{3.5}$$

where v is the specific volume and $\dot{m}$ the mass flow rate of the fluid passing through the pump. The outlet specific enthalpy of the pump can be expressed as:

$$h_2 = h_1 + \frac{\dot{W}}{\dot{m}} \tag{3.6}$$

or, by combining Equations 3.5 and 3.6, as

$$h_2 = h_1 + v(P_2 - P_1) \tag{3.7}$$

The outlet specific enthalpy of the pump is dependent on inlet specific enthalpy and the pressure rise imparted by the pump.

3.4.2 Decision variables

The main design parameters or decision variables for multi-objective optimization of the pump are the pump inlet pressure (P_1), the pump outlet pressure (P_2) and the pump isentropic efficiency (η_{pump}).

3.4.3 Constraints

To formulate a meaningful optimization problem, there are three constraints that must be satisfied for the pump (see Table 3.1).

Table 3.1 Physical constraints for optimization of a pump.

Constraint	Reason
$40\,\text{bar} < P_{in} < 80\,\text{bar}$	Boiler feed water pressure limit
$9\,\text{kPa} < P_{in} < 20\,\text{kPa}$	Output power limitation
$0.65 < \eta_{pump} < 0.80$	Commercial availability

3.4.4 Multi-objective Optimization of a Pump

To perform an optimization, objective functions need to be defined for the pump. Here, two objective functions are considered: the exergy efficiency of the pump and its total cost rate. These are then written as

$$\text{OF I}: \Psi = \frac{\dot{E}x_2 - \dot{E}x_1}{\dot{W}_{\text{Pump}}} \tag{3.8}$$

$$\text{OF II}: \dot{Z}_{\text{Pump}} = \frac{Z_{\text{Pump}} \times CRF \times \varphi}{N \times 3600} \tag{3.9}$$

where Z_{Pump} denotes the initial cost of the pump and CRF is the capital recovery factor, which is dependent on the interest rate i and equipment lifetime n:

$$CRF = \frac{i(1+i)^n}{(1+i)^n - 1} \tag{3.10}$$

Also, N denotes the annual number of operation hours for the unit, and φ the maintenance factor, which is often 1.06 [7].

The initial cost of the pump can be approximated as follows [6, 8]:

$$Z_{ST} = c_{11}(\dot{W}_{\text{Pump}})^{0.7}\left(1 + \frac{2}{1 - \eta_{\text{pump}}}\right) \tag{3.11}$$

where

$$c_{11} = 705.48 \, \$/(\text{kW}^{0.7}) \tag{3.12}$$

and η_{pump} is the pump isentropic efficiency.

To determine the optimal design parameters for a pump, a modified version of a genetic algorithm developed with Matlab software is used. The pump inlet mass flow rate is taken to be 10 kg/s based on previous studies [6]. Figure 3.4 shows the Pareto frontier for multi-objective optimization of pump as a function of its main design variables: isentropic efficiency and inlet and outlet pressures. The range of values shown is limited by the problem constraints.

It is observed in Figure 3.4 that the total pump cost increases moderately as the pump exergy efficiency increases, and that the maximum pump exergy efficiency exists at design point *C* (0.94) while the pump cost rate is the greatest at this point (0.64 $/hr). Conversely, the minimum value for the pump cost rate occurs at design point *A* and is about 0.6 $/hr. Design point *C* is the optimal situation when exergy efficiency is the sole objective function, while design point *A* leads the optimum design when total cost rate of product is the sole objective function. Point *B* is the ideal solution of the multi-objective optimization because both objective functions are at their optimal values, that is, at higher exergy efficiency and lower total cost rate. Since this point is not located on the Pareto frontier, point *B* could be selected as one of the best solutions because it is close to the ideal solution.

The variations of pump inlet and outlet pressures and pump isentropic efficiency are illustrated in Figs. 3.5 through 3.7, where scatter distributions for the populations in the Pareto frontier are shown for each of the design parameters. The points in these figures are obtained from the developed Matlab code, and show how the design parameters change within their allowable ranges. In a genetic algorithm, a population that encodes candidate solutions to an optimization problem evolves toward better solutions. The scatter distribution of design parameters are within the range exhibiting good selections

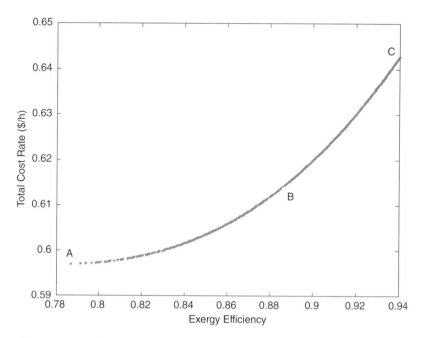

Figure 3.4 Pareto frontier for the optimization of a pump, highlighting the best trade-off among values for the objective functions.

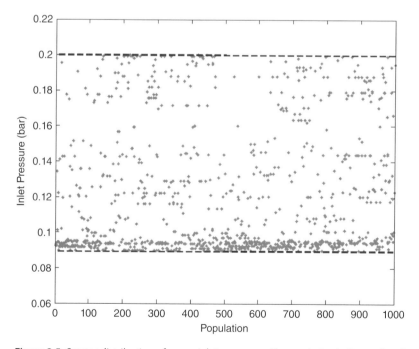

Figure 3.5 Scatter distribution of pump inlet pressure with population in Pareto frontier.

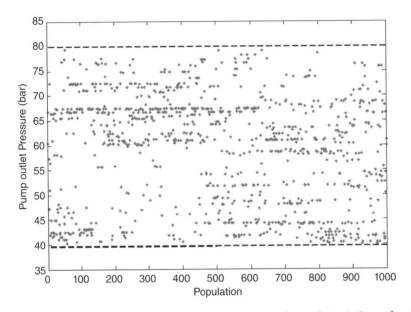

Figure 3.6 Scatter distribution of pump outlet pressure with population in Pareto frontier.

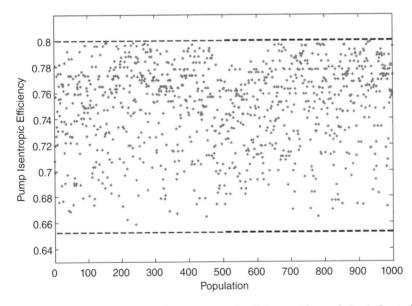

Figure 3.7 Scatter distribution of pump isentropic efficiency with population in Pareto frontier.

of these two parameters for optimization purposes. Note that the points in Figs. 3.5 to 3.7 are scattered almost randomly within their ranges, and not just near the boundaries. The selection of decision variables in real optimization activities is based on the scattered distribution of the decision variables, providing an efficient search for the best optimal solution of the objective functions.

3.5 Combustion Chamber

A combustion chamber (CC) is a common device in which thermal energy is produced by burning a fuel in air.

3.5.1 Modeling and Analysis of a Combustion Chamber

A schematic of a typical combustion chamber is shown in Figure 3.8. Fuel is injected into the combustion chamber at point 2 and air at point 1. Combustion chamber performance is dependent on many factors, including fuel type, flame temperature, insulation, and materials. The operation of a combustion chamber significantly affects not only system performance but also the environment and sustainability.

The optimization of a combustion chamber from an industrial gas turbine is described in this section. The modeling equations in Section 2.2.5 are used. The fuel for this CC is taken to be natural gas, for which the lower heating value (LHV) is 802,361 kJ/kmol and the enthalpy of formation is −74,872 kJ/kmol [5]. The operating pressure of the combustion chamber is 10 bar, which is the pressure of air exiting the compressors of most power plants, and the pressure loss across the chamber is negligible [6].

The exit temperature of the combustion chamber is one of its main design parameters, and significantly affects the performance of gas turbine power plants. Here, the temperature of the combustion gases exiting the combustion chamber is assumed to be 1400 K, due to the temperature limitation of gas turbine blades. Also, the inlet air temperature varies from 400 to 550 K and is dependent on both compressor pressure ratio and inlet air temperature. Therefore, the combustion chamber inlet air temperature is considered one of our design parameters. The fuel mass flow rate injected into the chamber is dependent on power generation and efficiency levels, for both full and part load conditions, and can vary between 1 and 2 kg/s. This parameter is also considered as a design parameter for the chamber.

Some simplifying assumptions are made to render the analysis in a more tractable way as follows:

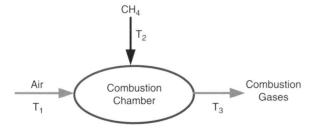

Figure 3.8 Schematic of a combustion chamber.

- The CC operates at steady state.
- The air and combustion gases both behave as ideal gas mixtures.
- Complete combustion occurs.
- Potential and kinetic effects are negligible [9].
- Heat loss from the CC is 2% of the lower heating value (LHV) of the fuel entering [10].

The exergy efficiency and cost rate of the system are considered the two objective functions for the optimization, with the exergy efficiency to be maximized and the cost rate minimized subject to applicable constraints. The exergy efficiency can be expressed as

$$\psi = \frac{\dot{Ex}_3}{\dot{Ex}_1 + \dot{Ex}_2} \tag{3.13}$$

where

$$\dot{Ex}_1 = \dot{m}_1(h_1 - h_0) - T_0(s_1 - s_0) \tag{3.14}$$

$$\dot{Ex}_2 = \dot{m}_{\text{fuel}} ex_f \tag{3.15}$$

Here, ex_f denotes the specific exergy of the fuel. To determine the chemical exergy of fuels, values are used of the exergy to energy ratio, $\Phi = ex_f/LHV$, provided by Szargut *et al.* [11]. For most hydrocarbons, this ratio is near unity. For instance, the ratio of chemical exergy to lower heat value for two common gaseous fuels, methane and hydrogen, are $\Phi_{CH_4} = 1.06$, and $\Phi_{H_2} = 0.985$. Also, the following correlation is applicable for a general liquid fuel $C_\alpha H_\beta N_\gamma O_\delta$, based on its atomic compositions [11]:

$$\Phi = 1.0401 + 0.1728\frac{\beta}{\alpha} + 0.0432\frac{\delta}{\alpha} + 0.2169\frac{\gamma}{\alpha}\left(1 - 2.062\frac{\beta}{\alpha}\right) \tag{3.16}$$

where $\frac{\beta}{\alpha}$, $\frac{\delta}{\alpha}$ and $\frac{\gamma}{\alpha}$ are the atomic ratios for the fuel.

The fuel being considered here is natural gas, which is composed mainly of CH_4 and modeled as pure methane for simplicity. The exergy at point 3 has both physical and chemical exergy. The physical exergy at this point can be calculated as:

$$\dot{Ex}_3 = \dot{m}_3(h_1 - h_0) - T_0(s_3 - s_0) \tag{3.17}$$

The chemical exergy at this point is the chemical exergy of the mixture of combustion gases. The chemical exergy of a mixture of N gases, all constituents of the reference environment, can be obtained as follows. We assume N different chambers. If each gas has a molar fraction of x_k and enters the chamber at T_0 and with a partial pressure $x_k P_o$, then each gas exits at the same temperature and a partial pressure x_k^e. Summing for all constituents, the chemical exergy per mole of the mixture can be calculated as follows:

$$\overline{ex}_{\text{ch}} = -RT_0 \sum x_k \ln\left(\frac{x_k^e}{x_k}\right) \tag{3.18}$$

By writing the logarithmic expression as $\ln(x_k^e) - \ln(x_k)$, this equation can be written as

$$\overline{ex}_{\text{ch}} = \sum x_k \overline{ex}_{\text{ch}}^k + RT_0 \sum x_k \ln(x_k) \tag{3.19}$$

Note that when a mixture including gaseous combustion products containing water vapor is cooled at a constant pressure, liquid water will form if the mixture temperature drops below the dew point, which is the saturation temperature corresponding to

Table 3.2 Physical constraints for optimization of a combustion chamber.

Constraint	Reason
$400\,^{\circ}C < T_{in} < 550\,^{\circ}C$	Temperature constraint
$1\,kg/s < \dot{m}_f < 2\,kg/s$	Mass flow rate constraint

the partial pressure of the water vapor. For example, if the combustion gas mixture is cooled to 25 °C at a fixed pressure of 1 atm, some condensation will occur. In this case we can model the situation at 25 °C as a gas phase containing saturated water vapor in equilibrium with a liquid water phase. To find the dew point temperature, we first calculate the partial pressure of water vapor. This can be done once all the flue gases from the combustion reaction are determined. Further details are provided in chapter 3 of reference [3].

3.5.1.1 Total Cost Rate

The total cost rate of the combustion chamber can be estimated as the cost rate of fuel $\dot{Z}_{Fuel}$ plus the cost rate of the combustion chamber $\dot{Z}_{CC}$ [12]:

$$\dot{Z}_{tot} = \dot{Z}_{CC} + \dot{Z}_{Fuel} \tag{3.20}$$

The combustion chamber cost rate can be expressed as

$$\dot{Z}_{CC} = \frac{Z_{CC} \times CRF \times \varphi}{N \times 3600} \tag{3.21}$$

Here, Z_{CC} is the combustion chamber purchase cost, which can be calculated as [12]:

$$Z_{cc}(\$) = 28.98\,\dot{m}_{air}(1 + e^{0.015(T_3 - 1540)}) \frac{1}{0.995 - p_3/p_1} \tag{3.22}$$

The fuel cost rate can be determined as

$$\dot{Z}_{Fuel} = c_f \dot{m}_f LHV \tag{3.23}$$

Here, c_f is the unit cost of natural gas (in $/MJ) and LHV is the lower heating value of the natural gas. Other economic parameters such as operating and maintenance costs, capital recovery factor (CRF), and years of operation are as described in previous sections. Here, the natural gas price is taken to be 0.003 $/MJ based on the average price of natural gas in Canada and the LHV of natural gas is taken to be 50 000 kJ/kg.

3.5.2 Decision Variables

The main design parameters or decision variables for optimization of the combustion chamber are the inlet air temperature and fuel mass flow rate.

3.5.3 Constraints

To formulate a meaningful optimization problem, two constraints are considered to ensure solutions are reasonable and realistic (see Table 3.2).

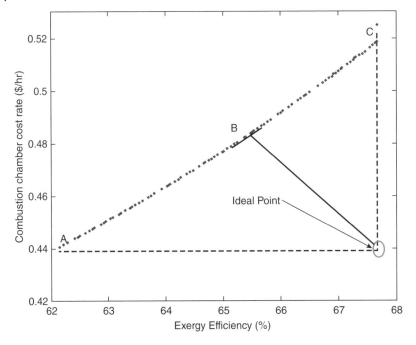

Figure 3.9 Pareto frontier for the optimization of a combustion chamber, highlighting the best trade-off among values for the objective functions.

3.5.4 Multi-objective Optimization

To determine the optimal design parameters for a combustion chamber, a modified version of a genetic algorithm developed with Matlab software is used. The Pareto frontier of this optimization problem is presented in Figure 3.9. A trade-off is evident between the total cost rate and exergy efficiency of the combustion chamber.

Point *A* has the lowest exergy efficiency and minimum cost, meaning that it is the optimum when cost is the sole objective function, while point *C* is the optimum when exergy efficiency is the sole objective function. As mentioned for other components previously, there is an ideal point in the Pareto frontier where both objective functions are at their desired optimal values. This point is seen not to be on the Pareto curve in Figure 3.16 and cannot be attained. However, the closest points to this point could be considered in the decision-making process as they exhibit reasonable exergy efficiencies at a moderate total cost rate. This point is shown as point *B* on the Pareto frontier.

Figures 3.10 and 3.11 show the distribution of the decision variables for optimum points on the Pareto frontier. The inlet air temperature is observed in Figure 3.10 to have a scattered distribution in its range, which means that the optimum air inlet temperature mainly depends on the objective function, and an increase in this parameter affects the objective functions. In Figure 3.11, on the other hand, all Pareto optimum points for minimum fuel mass flow rate are shown. In multi-objective optimization, if the design parameters are placed at their lower bound, we confirm that a decrease in fuel mass flow rate has a positive effect on both objective functions. In the case of the combustion chamber, a decrease in fuel injection into the combustion chamber results in an increase in exergy efficiency according to Equation 3.11. But when total cost rate is considered as the objective function, the lower the mass flow rate, the lower the total

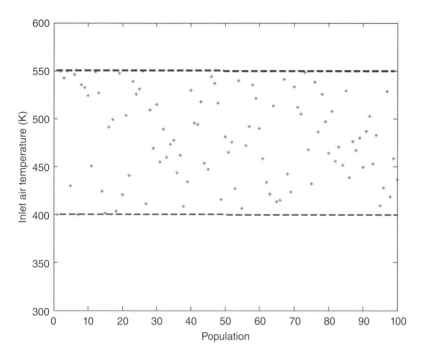

Figure 3.10 Distribution of inlet air temperature for points of the Pareto frontier.

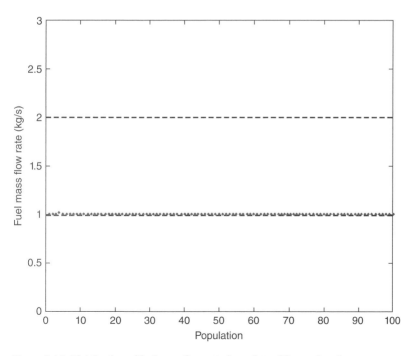

Figure 3.11 Distribution of fuel mass flow rate for points of Pareto frontier.

cost rate is expected to be. This explains why the scattered points of the Pareto frontier tend to be at their lower values.

3.6   Flat Plate Solar Collector

Flat plate solar collectors are used for various moderate to low temperature heating tasks. Solar designers have access to many models of flat plate collectors. Here we consider the model of a flat plate solar collector shown in Figure 3.12. Water enters the solar collector at T_{in}, is heated by solar energy exits at temperature T_{out}. The multi-objective optimization considered here seeks to maximize the exergy efficiency of the collector and minimize its total cost rate.

3.6.1   Modeling and Analysis of Collector

The rate of useful heat gain by the working fluid passing through the solar collector can be written as:

$$\dot{Q}_u = \dot{m}C_p(T_{out} - T_{in}) \tag{3.24}$$

where T_{out}, T_{in}, C_p and $\dot{m}$ are the water outlet temperature, inlet temperature, specific heat at constant pressure and mass flow rate. The Hottel-Whillier equation for the heat gained by the flat plate collector considering heat losses from the collector is calculated as [13]:

$$\dot{Q}_u = A_P F_R[(\tau\alpha)I - Q_L] \tag{3.25}$$

where T_0 is the ambient temperature and F_R is the heat removal factor, defined as

$$F_R = \frac{\dot{m}C_p}{U_l A_P}\left[1 - e^{\left\{-\frac{F'U_l A_P}{\dot{m}C_p}\right\}}\right] \tag{3.26}$$

Here, F' is collector efficiency factor which is approximately 0.914 for this case [13] and U_l is the overall collector loss coefficient, obtained from [13]. In Equation 3.21, $(\tau\alpha)$ is

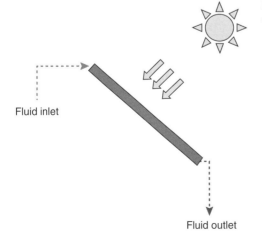

Figure 3.12 Schematic of a flat plate solar collector.

Fluid inlet

Fluid outlet

optical efficiency, I is solar radiation intensity and $\dot{Q}_L$ is the heat loss rate, defined as:

$$\dot{Q}_L = U_l(T_{in} - T_0) \tag{3.27}$$

The inlet exergy rate with solar radiation can be calculated as:

$$\dot{Ex}_{in,\,solar} = \eta_o I A_p \left(1 - \frac{T_o}{T_s}\right) \tag{3.28}$$

The exergy flow rate of the water input to the collector is

$$\dot{Ex}_{in,\,f} = \dot{m}_{water} C_p \left(T_{in} - T_o - T_o \ln \frac{T_{in}}{T_o}\right) \tag{3.29}$$

and the exergy flow rate of the hot water leaving the collector is:

$$\dot{Ex}_{out,\,f} = \dot{m}_{water} C_p \left(T_{out} - T_o - T_o \ln \frac{T_{out}}{T_o}\right) \tag{3.30}$$

In addition, the exergy loss rate due to heat leakage can be calculated as:

$$\dot{Ex}_l = U_l(T_{in} - T_o) \left(1 - \frac{T_o}{T_{in}}\right) \tag{3.31}$$

Using the exergy balance equation for a control volume around the flat plate collector, the exergy destruction rate of the collector can be expressed as:

$$\dot{Ex}_d = \dot{Ex}_{in,\,solar} + \dot{Ex}_{in,\,f} - \dot{Ex}_{out,\,f} - \dot{Ex}_l \tag{3.32}$$

The exergy efficiency of the solar flat plate collector can be expressed as:

$$\eta_{ex} = 1 - \frac{\dot{Ex}_d}{\dot{Ex}_{in,\,solar}} \tag{3.33}$$

For the economic aspects of the optimization, an economic model is used. The purchase cost of the collector is estimated as [14]:

$$Z_C(\$) = 235 A_p \tag{3.34}$$

where A_p is the collector area (in m^2).

The cost of the collector can be calculated by taking into account the maintenance factor (Φ) and capital recovery factor as follows:

$$Z_{FPC} = Z_C CRF \Phi \tag{3.35}$$

3.6.2 Decision Variables and Input Data

The main design parameters for optimization of the solar collector are its length, width, and mass flow rate. In the multi-objective optimization, these are considered the decision variables. The input data for modeling are listed in Table 3.3.

3.6.3 Constraints

Three constraints are considered here to ensure solutions are reasonable and realistic, as described in Table 3.4.

Table 3.3 Input data for the modeling and optimization.

Parameter	Value
U_l (J/kg.K)	4.67
I (W/m^2)	500
F'	0.91
$\tau\alpha$	0.84
T_{in} (K)	300
T_s (K)	4350
Φ	1.05

Table 3.4 Physical constraints for optimization of a solar collector.

Constraint	Reason
$0.05 \le \dot{m}_{water} \le 0.1$	Pumping limitation
$3 \le L \le 5$	Commercial availability
$1 \le W \le 3$	Commercial availability

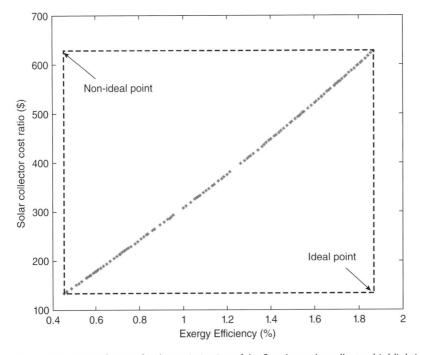

Figure 3.13 Pareto frontier for the optimization of the flat plate solar collector, highlighting the best trade-off among values for the objective functions.

3.6.4 Multi-objective Optimization

To identify the optimal design parameters for the flat plate solar collector, a modified version of a genetic algorithm developed with Matlab software is used. Figure 3.13 shows the Pareto frontier for the multi-objective optimization of the collector with mass flow rate, length, and width of the collector as the main design variables. The range of values shown is limited by the problem constraints. Both ideal and a non-ideal points are observed in this figure. The ideal point represents the non-feasible situation where the cost rate takes on its minimum value and the exergy efficiency its maximum value. The non-ideal point is where both objective functions are at undesired values. Note that each point on the Pareto frontier is an optimum solution and the final selection is up to the decision maker.

Figures 3.14 to 3.16 show the scattered distribution of the decision variables for various points of the Pareto front. It is clear that the optimum values for the collector length and width vary in their allowable range and mainly depend on the objective functions, while for the water mass flow rate the points tend to be at their higher values.

3.7 Ejector

An ejector is a non-moving component that can be used to change the pressure of a fluid flow. Operating on simple principles of gas dynamics, this component can replace the compressor in a vapor compression refrigeration cycle. An ejector can thereby increase the pressure of a working fluid without the usually required electrical or shaft power. Since using an ejector avoids a moving component in the refrigeration cycle, the vibration and noise produced by the compressor are avoided, and the operating and maintenance costs of the system are lower than for conventional vapor compression refrigeration cycles.

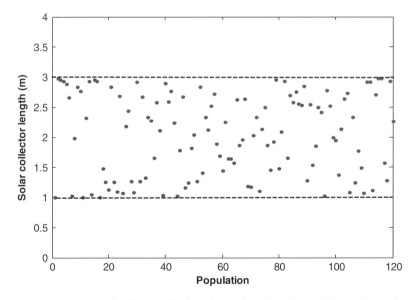

Figure 3.14 Scatter distribution of solar collector length with population in Pareto frontier.

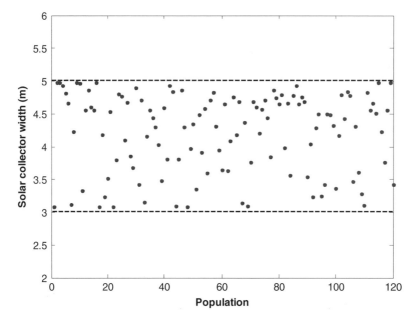

Figure 3.15 Scatter distribution of solar collector width with population in Pareto frontier.

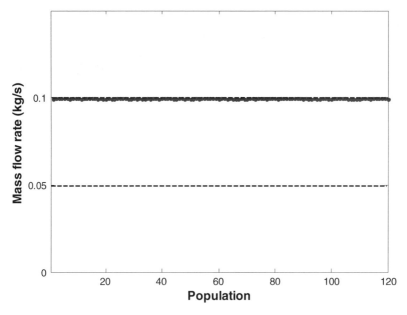

Figure 3.16 Scatter distribution of water mass flow rate in solar collector with population in Pareto frontier.

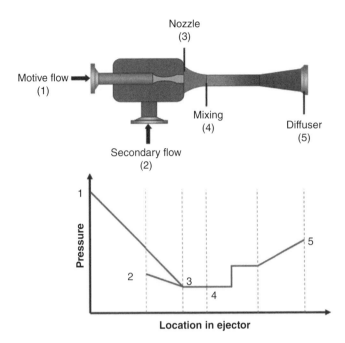

Figure 3.17 Schematic of an ejector (top) and a profile showing the variation of pressure with position in the ejector (bottom).

3.7.1 Modeling and Analysis of an Ejector

A typical ejector is illustrated in Figure 3.17. This ejector operates based on the interaction of two flows at different energy levels. The higher energy stream (i.e., the primary or motive flow) enters the ejector at point 1 while the lower energy stream (i.e., the secondary flow) is drawn in at point 2. The interaction of these two fluids in the ejector leads to a pressure increment without the use of other components.

The gas dynamics involved in the process help to explain the way the ejector operates. The motive flow expands through the converging-diverging nozzle and exits as a supersonic stream (point 3) with a significant pressure decrease. This creates a suction condition at the inlet of the mixing section, so the outlet flow of the evaporator (from the refrigeration cycle) is drawn into the ejector. The primary and secondary flows then mix in the mixing chamber (point 4). The mixed flow remains supersonic and enters a duct of constant cross-sectional area, in which the flow becomes subsonic and a normal shock occurs. The properties of the flow after the normal shock can be determined by finding the intersection of the Fanno and Rayleigh lines. Finally, the flow passes through a diffuser in which its pressure is raised to the pressure of the condenser of the refrigeration cycle (point 5).

The thermodynamic model and economic analysis of the ejector are now described. The following assumptions are made for these modeling and analysis studies:

- The velocities at the inlet and outlet of the ejector are assumed to be negligible.
- Friction and mixing losses in the nozzle, the mixing chamber, and the diffuser are taken into account in the efficiency for each component.

- The mixing chamber operates at constant pressure.
- The ejector does not exchange heat with the surroundings.

Considering the fluid velocity u at the inlet of ejector to be zero, the outlet velocity of the nozzle can be calculated as

$$u_3 = \sqrt{(2\eta_n(h_1 - h_3))} \tag{3.36}$$

The entrainment ratio is defined as

$$\mu = \frac{\dot{m}_{sf}}{\dot{m}_{mf}} \tag{3.37}$$

Applying the principles of conservation of mass and momentum to the mixing chamber yields:

$$\dot{m}_{mf}h_3 + \dot{m}_{sf}h_2 = (\dot{m}_{mf} + \dot{m}_{sf})h_{4s} \tag{3.38}$$

Neglecting the secondary flow velocity u_2, the isentropic velocity of the mixing outlet flow can be calculated as follows:

$$u_{12s} = \frac{u_3}{1 + \mu} \tag{3.39}$$

The mixing efficiency can be defined as

$$\eta_m = \frac{u_4^2}{u_{4s}^2} \tag{3.40}$$

The velocity at the mixing section outlet flow can be expressed as

$$u_4 = u_3 \frac{\sqrt{\eta_m}}{1 + \mu} \tag{3.41}$$

and the outlet specific enthalpy of the mixing section as

$$h_4 = \frac{h_1 + \mu h_2}{1 + \mu} - \frac{u_4^2}{2} \tag{3.42}$$

The kinetic energy in the flow is converted to pressure in the diffuser. Neglecting the outlet velocity of the diffuser one can calculate the outlet specific enthalpy as follows:

$$h_5 = h_4 + \frac{h_{5s} - h_4}{\eta_d} \tag{3.43}$$

where h_{5s} represents the isentropic specific enthalpy of the diffuser outlet flow, that is, the specific enthalpy of the diffuser outlet flow after an isentropic process.

The entrainment ratio, μ, can be written as follows:

$$\mu = \sqrt{\eta_n \eta_m \eta_d \frac{(h_1 - h_{3s})}{(h_{5s} - h_4)} - 1} \tag{3.44}$$

As shown in Figure 3.18, evaluation of the entrainment ratio for the ejector requires a trial-and-error iterative process. With this approach, the entrainment ratio is guessed and after estimation of several parameters it is calculated using Equation 3.64. If the relative error exceeds the stopping criterion, the previous value of the entrainment ratio is automatically replaced with the new one and the procedure is repeated until a reasonable stopping criterion is met.

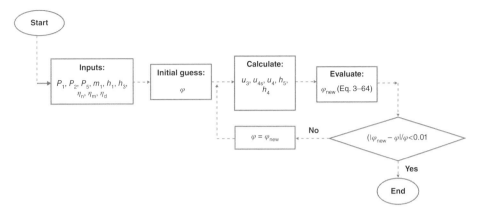

Figure 3.18 Flowchart for an evaluation procedure for the entrainment ratio of the ejector.

Table 3.5 Decision variables for the ejector and their ranges.

Parameter	Range	Reason
P_{motive} (kPa)	[2800, 3975]	Commercially available
$T_{motive, rise}$ (K)	[0, 10]	Commercially available
$P_{suction}$ (kPa)	[600, 770]	Pinch point limitation
$P_{discharge}$ (kPa)	[180, 290]	Pinch point limitation

The ejector equipment purchase cost of an ejector can be expressed for economic modeling of the component [15] as follows:

$$Z_{ejector} = 1000 \times 16.14 \times 0.989 \times \left(m_1 \times \left(\frac{T_1}{P_1 \times 0.001} \right)^{0.05} \right) \times (P_5^{-0.75})$$

The cost of the ejector, taking into account the maintenance factor Φ and the capital recovery factor, can be determined as

$$Z_{ejector} = Z_{ej} \, CRF \, \Phi$$

3.7.2 Decision Variables and Constraints

The main design parameters or decision variables for optimization of the ejector are listed in Table 3.5 with their allowable ranges.

3.7.3 Objective Functions and Optimization

Various objective functions can be considered depending on the problem considered. Here, we choose to maximize the exergy efficiency of the system while minimizing the cost of the system. These objective functions are dependent on design parameters of the system and can be expressed as:

$$OF\,I : \eta_{ex} = 1 - \frac{\dot{Ex}_D}{\dot{Ex}_1}$$

$$OF\,II : Z_{ejector} = Z_{ejector} \, CRF \, \Phi$$

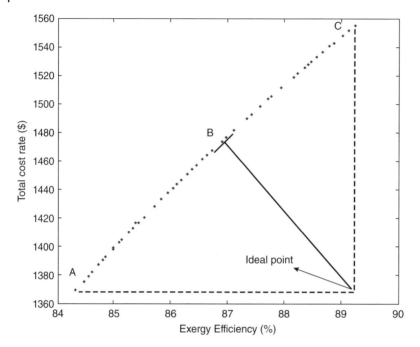

Figure 3.19 Pareto frontier of the ejector.

The Pareto frontier for the ejector is shown in Figure 3.19. All points of the Pareto frontier are optimal solutions and can be considered for selection by a decision maker. However, there is an optimum solution that may be the best compared with others. This point is named the ideal point in Figure 3.19. Therefore, it is beneficial to find the nearest point to the ideal point. That point is labeled point *B* on the Pareto curve. In this figure, there are two other important points, each of which can be obtained when single objective optimization is conducted. Point A is the point when cost is the sole objective function while point *C* is the optimal point when exergy efficiency is the single objective. When we perform multi-objective optimization, any combination of these two objective functions can be selected and there is no special preference unless we find some points closest to the ideal point.

To enhance our understanding of the multi-objective optimization of the ejector, the variations of each of the design parameters over their allowable ranges are determined. Figure 3.20 shows the variation of the motive flow pressure for points of the Pareto front. As is evident for all optimum points, whether considering the minimum cost rate or maximum exergy efficiency or even any combination of these two objective functions, this variable should have its maximum available value. Figure 3.21 also shows the variation of the motive flow temperature rise, which is intended to superheat the motive flow before it enters the ejector. This variable is seen to take on its minimum value, meaning that superheating of the motive flow is not recommended as an enhancement of the ejector cost rate and exergy efficiency. Figure 3.22 shows the variation of the suction flow for point of the Pareto front. From the distribution of this variable, it is concluded that the optimum value of this variable depends significantly on the objective functions investigated. Figure 3.23 shows the variation of the discharge pressure for points of the

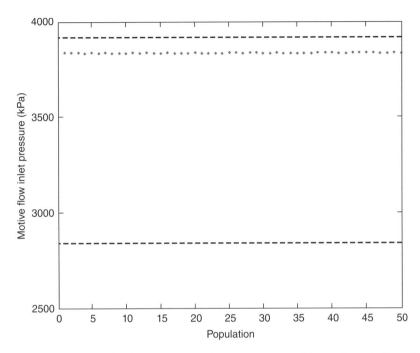

Figure 3.20 Scatter distribution of motive flow inlet pressure with population in Pareto frontier.

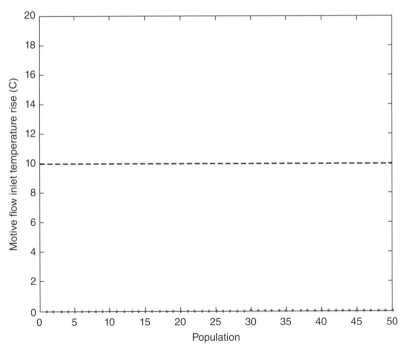

Figure 3.21 Scatter distribution of motive flow inlet temperature with population in Pareto frontier.

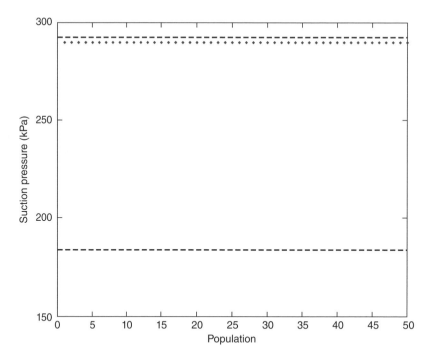

Figure 3.22 Scatter distribution of suction pressure with population in Pareto frontier.

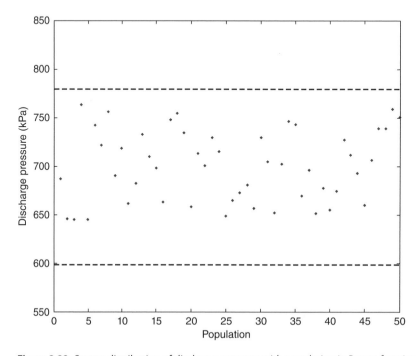

Figure 3.23 Scatter distribution of discharge pressure with population in Pareto frontier.

Pareto front. As shown for all the optimum points, this variable exhibits its maximum available value.

3.8 Concluding Remarks

In this chapter, the modeling and optimization of some of the main thermal components used in energy systems are described. We first model each thermal component, invoking reasonable simplifying assumptions as required, and then define different objective functions for optimization purposes. Appropriate constraints are considered in order to ensure realistic optimal design parameters are obtained. An evolutionary algorithm-based multi-objective optimization is then applied to each thermal component and a Pareto curve is obtained. The variation of each of the design parameters over their allowable ranges is also shown to understand better how the parameters vary within their domains. These thermal components are used in subsequent chapters as part of various applications, ranging from advanced power plants to fuel cells and hydrogen production units. The particular importance of this chapter is that it helps determine how each device operates optimally, an outcome that will help in selecting the proper design parameters for more sophisticated and complex energy systems.

References

1 Ahmadi, P., Dincer, I. and Rosen, M. A. (2013) Development and assessment of an integrated biomass-based multi-generation energy system. *Energy* **56**:155–166.

2 Ahmadi, P., Dincer, I. and Rosen, M. A. (2013) Performance assessment and optimization of a novel integrated multigeneration system for residential buildings. *Energy and Buildings* **67**:568–578.

3 Dincer, I.and Rosen, M. A. (2012) *Exergy: Energy, Environment and Sustainable Development*, Second Edition, Elsevier.

4 Kotas, T. J. (1995) *The Exergy Method of Thermal Plant Analysis*, Krieger Melbourne, FL.

5 Bejan, A. and Moran, M. J. (1996) *Thermal Design and Optimization*, John Wiley & Sons, Inc., New York.

6 Ahmadi, P., Dincer, I. and Rosen, M. A. (2011) Exergy, exergoeconomic and environmental analyses and evolutionary algorithm based multi-objective optimization of combined cycle power plants. *Energy* **36**:5886–5898.

7 Ghaebi, H., Saidi, M. and Ahmadi, P. (2012) Exergoeconomic optimization of a trigeneration system for heating, cooling and power production purpose based on TRR method and using evolutionary algorithm. *Applied Thermal Engineering* **36**:113–125.

8 Ahmadi, P., Dincer, I. and Rosen, M. A. (2014) Thermoeconomic multi-objective optimization of a novel biomass-based integrated energy system. *Energy* **68**:958–970.

9 Ahmadi, P., Dincer, I. and Rosen, M. A. (2013) Thermodynamic modeling and multi-objective evolutionary-based optimization of a new multigeneration energy system. *Energy Conversion and Management* **76**:282–300.

10 Ahmadi, P. and Dincer, I. (2011) Thermodynamic analysis and thermoeconomic optimization of a dual pressure combined cycle power plant with a supplementary firing unit. *Energy Conversion and Management* **52**:2296–2308.

11 Karman, D. (2006) Life-cycle analysis of GHG emissions for CNG and diesel buses in Beijing. In *EIC Climate Change Technology, 2006 IEEE*. IEEE; 2006: 1–6.

12 Roosen, P., Uhlenbruck, S. and Lucas, K. (2003) Pareto optimization of a combined cycle power system as a decision support tool for trading off investment vs. operating costs. *International Journal of Thermal Sciences* **42**:553–560.

13 Farahat, S., Sarhaddi, F. and Ajam, H. (2009) Exergetic optimization of flat plate solar collectors. *Renewable Energy* **34**:1169–1174.

14 Ahmadi, P., Dincer, I. and Rosen, M. A. (2015) Multi-objective optimization of an ocean thermal energy conversion system for hydrogen production. *International Journal of Hydrogen Energy* **40**:7601–7608.

15 Mabrouk, A. A., Nafey, A. and Fath, H. (2007) Thermoeconomic analysis of some existing desalination processes. *Desalination* **205**:354–373.

Study Questions/Problems

1 If you are given a device, how would you determine the appropriate decision variable(s) for optimization?

2 Repeat the multi-objective optimization of the compressor considered in section 3.2, but with the compressor cost and compressor required power as the two objective functions. Obtain the Pareto frontier.

3 Carbon dioxide enters an adiabatic compressor at 100 kPa and 300 K at a rate of 0.5 kg/s. If the compressor pressure ratio and the compressor isentropic efficiency are considered as two design parameters, what is the minimum power required for compression?

4 Consider a combustion chamber and a gas turbine as two components. The gas turbine can generate 5 MW of electricity and the gas turbine inlet temperature is fixed at 1400 K. Flue gases exit the gas turbine at atmospheric pressure. The fuel injected into the combustion chamber is natural gas with LHV = 50 000 kJ/kg. There is no pressure drop across the combustion chamber and both air and the flue gas can be treated as ideal gas mixtures.
   a) Apply a multi-objective optimization considering the exergy efficiency, the cost of electricity (in terms of kWh), and the CO_2 emissions (in terms of kg/MWh) as the objective functions. Use reasonable constraints and proper design parameters.
   b) Perform a sensitivity analysis to determine the effect of changing the interest rate and provide the Pareto frontier when the interest rate is $i = 5\%$, $i = 10\%$ and $i = 15\%$.
   c) Make recommendations for enhancing performance, environmental stewardship and sustainability.

5 How does varying the fuel cost affect the optimization for the combustion chamber considered in section 3.5?

6 In modeling a combustion chamber, how does the combustion equation change if we consider the complete combustion instead of actual combustion? How do the greenhouse gas emissions vary as the combustion chamber inlet temperature increases?

7 Consider the simple gas turbine power plant shown in Figure 3.24. Air at ambient conditions enters the air compressor at point 1 and exits after compression at point 2. The compressed air enters the combustion chamber (CC) into which fuel is injected, and hot combustion gases exit at point 3 and expand in the gas turbine to produce shaft power. The expanded hot gas exits the gas turbine at point 4. Natural gas (with a volumetric composition of $CH_4 = 96.57\%$, $C_2H_6 = 2.63\%$, $C_3H_8 = 0.1\%$, $C_4H_{10} = 0.7\%$) is injected into the combustion chamber. Consider the design parameters to be: the compressor pressure ratio, the gas turbine inlet temperature, the compressor isentropic efficiency, and the gas turbine isentropic efficiency. If the gas turbine power plant generates 10 MW of electricity,
a) What are the optimal design parameters when the exergy efficiency is maximized?
b) Calculate the exergy destruction rate and exergy efficiency for each component at their optimal values and discuss the meaning and uses of the results.

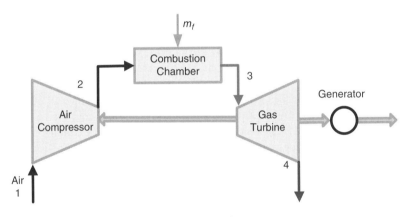

Figure 3.24 Schematic of a gas turbine power plant.

4

Modeling and Optimization of Heat Exchangers

4.1 Introduction

In industrial facilities, many operations depend on heating and cooling processes. For example, the liquid entering a distillation column is heated to enable the distillation process to occur, or a process liquid is cooled so it can be properly stored. Such heating and cooling is often done by transferring heat from one fluid to another in devices called heat exchangers.

To understand how heat is transferred, it is important to understand what heat is. Heat is a form of energy that is associated with the movement of molecules in a material. This energy is usually manifested and measured as temperature. For heat transfer to occur, there must be a difference in temperature. In the presence of a temperature difference, heat is transferred from materials with higher temperatures to materials with lower temperatures.

The heat transfer commonly occurs in heat exchangers. A heat exchanger is device that promotes heat transfer from one medium to another. The hot and cold media may be separated by a solid wall to prevent mixing or they may be in direct contact. Heat exchangers are widely used in space heating, refrigeration, air conditioning, power generation, chemical and petrochemical processing, petroleum refining, natural gas processing, sewage treatment, and energy storage. A long-time example of a heat exchanger is found in an internal combustion engine, where a circulating fluid known as the engine coolant flows through radiator coils while air flows over the coils, causing the coolant to cool and heating the incoming air.

The heat transfer process in heat exchangers usually involves convective heat transfer in each fluid and conduction through the wall separating the two fluids. In the analysis of heat exchangers, it is convenient to utilize the overall heat transfer coefficient U, which integrates the contributions of all heat transfer effects. The rate of heat transfer between the two fluids at a location in a heat exchanger depends on the magnitude of the temperature difference at that point, and varies along the length of a heat exchanger. In the design and analysis of heat exchangers, it is usually convenient and conventional to utilize the logarithmic mean temperature difference (LMTD), which is an equivalent mean temperature difference between the two fluids for the entire heat exchanger.

In this chapter, we begin by describing various types of heat exchangers and discuss their applications in various industries. We then consider some of the widely used heat exchangers in power generation, combined heat and power, and air conditioning systems, in order to comprehensively model and optimize them. An evolutionary algorithm based optimization is applied to determine the optimal design parameters of each heat

Optimization of Energy Systems, First Edition. Ibrahim Dincer, Marc A. Rosen, and Pouria Ahmadi.
© 2017 John Wiley & Sons Ltd. Published 2017 by John Wiley & Sons Ltd.

exchanger, depending on the objective functions considered, which range from heat exchanger efficiency to its size and cost. To enhance understanding of the design criteria, a sensitivity analysis is conducted to see how each objective function varies when small changes are applied to selected design parameters. We complete the chapter with closing remarks on the efficient designs of heat exchangers.

4.2 Types of Heat Exchangers

Although widespread in application, heat exchangers are usually categorized based on transfer phenomena, configuration, flow directions, and heat transfer mechanisms. Other categorization bases are also used. A listing and explanation of the main heat exchanger categorizations follows:

- **Transfer phenomena**
  - Direct contact: hot and cold flows are in direct contact (e.g., cooling towers)
  - Indirect contact: hot and cold flows remain separate(e.g., heat recovery steam generators)

- **Surface compatibility**
  Compact: the surface area to volume ratio exceeds $700 \text{ m}^2/\text{m}^3$
  Non-compact: the surface area to volume ratio is less than $700 \text{ m}^2/\text{m}^3$

- **Arrangement**
  - Parallel flow: hot and cold flows move in the same direction
  - Counter flow: hot and cold flows move in opposite directions
  - Cross flow: one flow moves across the other at a 90 degree or other angle

- **Configuration**
  - Shell and tube heat exchanger
  - Plate heat exchanger
  - Tubular heat exchanger
  - Spiral heat exchanger

- **Heat transfer mechanism**
  - Single-phase convection on both sides
  - Single-phase convection on one side and two-phase on the other
  - Two-phase convection on both sides

- **Heat transfer surface**
  - Primary surfaces: the main surfaces separating the hot and cold flows
  - Secondary surfaces: surfaces mounted on the primary surfaces to increase heat transfer area (such as fins)

Different applications typically require different types of heat exchangers and associated equipment. The three classifications of heat exchangers according to flow arrangement are quite fundamental. Parallel flow and counter flow concentric heat exchangers are shown in Figure 4.1. In parallel flow heat exchangers, the two fluids enter the exchanger at the same end, and travel in parallel to the other end. In counter flow heat exchangers, the fluids enter the exchanger from opposite ends. The counter flow design is the most efficient heat exchanger, in that it can transfer the most heat from the hot medium to the

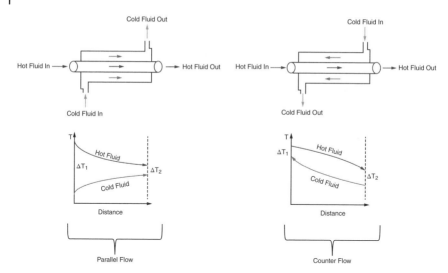

Figure 4.1 Two basic heat exchanger types and their temperature profiles for a double-pipe heat exchanger.

cold medium because the average temperature difference along any unit length is greater than with other configurations. The temperature variations along the heat exchanger for these two arrangements are also shown in Figure 4.1.

The rate of heat transfer between the two fluids is one of the most important criteria in heat exchanger design. Compact heat exchangers are typically designed with a large heat transfer surface area per unit volume, so as to realize high heat rates. By permitting a high heat transfer rate in a small volume, compact heat exchangers find uses in applications with space (especially volume) limitations. In a cross flow heat exchanger, the fluids travel roughly perpendicular to one another through the exchanger. To achieve good efficiencies, heat exchangers are designed to maximize the surface area of the wall between the two fluids, while minimizing the resistance to fluid flow through the exchanger. The performance of heat exchangers can be augmented by the addition of fins or corrugations in one or both directions to increase surface area. The large surface area in compact heat exchangers is obtained by attaching tightly packed thin plates to the walls separating the two fluids. Compact heat exchangers are commonly used in applications involving gas-to-gas and gas-to-liquid heat transfer to counteract the low heat transfer coefficients associated with gas flows, effectively by increasing the surface area. The car radiator is an example of a compact heat exchanger (see Figure 4.2). In a car radiator, which is a water-to-air compact heat exchanger, fins are attached to the air side of the tube surface.

A shell and tube heat exchanger (Figure 4.3) is another important class of heat exchanger. It is commonly utilized in power generation, petrochemical, and chemical plants, and is particularly suited to high pressure applications. This type of heat exchanger consists of a shell with tubes inside. One fluid runs through the tubes, while the other flows over the tubes through the shell. Heat transfer between the two fluids occurs through the walls of the tubes. The set of tubes, that is, the tube bundle, may consist of plain or finned tubes. Baffles are commonly placed in the shell side to force the shell side fluid to flow across the shell in order to increase heat transfer and to maintain uniform spacing between the tubes. Despite their widespread use, shell and

Figure 4.2 A car radiator, which is essentially a water-to-gas compact heat exchanger. *Source:* Courtesy of PRC, USA.

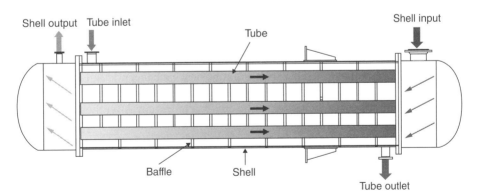

Figure 4.3 A shell and tube heat exchanger.

tube heat exchangers are not suitable for use in applications with tight volume or weight restrictions, such as in automobiles or aircraft, because of their large size and weight.

The plate fin heat exchanger (Figure 4.4) is composed of multiple, thin, slightly separated plates that have large surface areas and fluid flow passages for heat transfer. This stacked-plate arrangement is more effective, in a given space, than the shell and tube heat exchanger. In HVAC applications, large heat exchangers of this type are

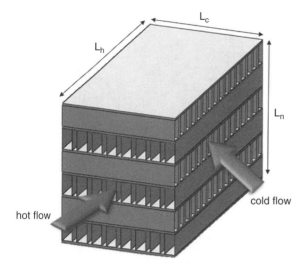

Figure 4.4 Plate fin heat exchanger. *Source*: Adapted from Ahmadi 2011.

called plate-and-frame; when used in open loops, these heat exchangers are normally of the gasket type to allow periodic disassembly, cleaning, and inspection. There are many types of permanently bonded plate heat exchangers, such as dip-brazed, vacuum-brazed, and welded plate varieties, and they are often specified for closed-loop applications such as refrigeration. Plate heat exchangers also differ in the types of plates used and their configurations.

4.3 Modeling and Optimization of Shell and Tube Heat Exchangers

Shell and tube heat exchangers are used to transfer heat between two or more fluids, between a solid surface and a fluid, or between solid particulates and a fluid, at various temperatures. Heat exchangers usually experience no external heat and work interactions. Some important parameters in shell and tube heat exchanger design include tube number, length and arrangement, and baffle spacing. Here, we model shell and tube heat exchangers and optimize them with multi-objective optimization. After thermodynamic modeling of an industrial shell and tube heat recovery heat exchanger using the ε-NTU method and the Bell–Delaware approach, the heat exchanger is optimized. Exergy efficiency, total cost rate and total exergy destruction rate are considered as our objective functions and relevant constraints are applied to ensure reasonable design parameters. A genetic algorithm technique is employed to provide a set of Pareto optimal solutions. A sensitivity analysis is performed to assess the impact of a change in optimum values of exergy efficiency, total exergy destruction rate, and total cost due to changes in design variables.

4.3.1 Modeling and Simulation

An *E* type TEMA (Tubular Exchanger Manufacturers Association) shell is considered to demonstrate the modeling and simulation of shell and tube heat exchangers. Developed by manufacturers of heat exchangers, TEMA is a set of standards defining design/manufacturing parameters of shell and tube heat exchangers. TEMA standards provide a recognized approach for users and permit design comparisons for varied applications [1].

The heat exchanger effectiveness for this type of shell can be expressed as follows [2]:

$$\varepsilon = \frac{2}{(1+C^*) + (1+C^{*2})^{0.5} \coth\left(\frac{NTU}{2}(1+C^{*2})^{0.5}\right)} \tag{4.1}$$

where C^* denotes the heat capacity ratio and NTU the number of transfer units. These terms are defined below:

$$C^* = \frac{C_{min}}{C_{max}} = \frac{min[C_s, C_t]}{max[C_s, C_t]} = \frac{min\{[\dot{m}C_p]_s, [\dot{m}C_p]_t\}}{max\{[\dot{m}C_p]_s, [\dot{m}C_p]_t\}} \tag{4.2}$$

$$NTU = \frac{UA_t}{C_{min}} \tag{4.3}$$

where s and t denote the shell and tube side, C_p is the specific heat at constant pressure, A_t is the total tube outside heat transfer surface area, and U is the overall heat transfer coefficient. Furthermore,

$$A_t = \pi L d_o N_t \tag{4.4}$$

$$U = \frac{1}{\frac{1}{h_o} + R_{o,f} + \frac{d_o \ln\left(\frac{d_o}{d_i}\right)}{2k_w} + R_{i,f}\frac{d_o}{d_i} + \frac{1}{h_i}\frac{d_o}{d_i}} \tag{4.5}$$

Here, L, N_t, d_i, d_o, $R_{i,f}$, $R_{o,f}$, k_w respectively denote tube length, tube number, inside and outside tube diameter, fouling resistance on tube and shell sides, and thermal conductivity of tube wall. The tube side heat transfer coefficient h_i is expressed as follows [2]:

$$h_i = h_t = \left(\frac{k_t}{d_i}\right) \times 0.024 Re_t^{0.8} Pr_t^{0.4} \qquad for \qquad 2500 < Re_t < 1.24 \times 10^5 \tag{4.6}$$

where k_t and Pr_t denote the tube side fluid thermal conductivity and Prandtl number, respectively, while Re_t denotes the Reynolds number, defined as:

$$Re_t = \frac{\dot{m}_t d_i}{\mu_t A_{o,t}} \tag{4.7}$$

Here, $\dot{m}_t$ denotes the tube side mass flow rate, μ_t the tube side dynamic viscosity, and $A_{o,t}$ the tube side flow cross section area per pass, which is expressible as:

$$A_{o,t} = 0.25\pi d_i^2 \frac{N_t}{n_p} \tag{4.8}$$

Here, n_p is the number of tube passes. The shell diameter can be determined as follows [3]:

$$D_s = 0.637 p_t \sqrt{(\pi N_t)\frac{CL}{CTP}} \tag{4.9}$$

where P_t is tube pitch (in meters) and CL is the layout constant, which is equal to unity for 45° and 90° tube arrangements and 0.87 for 30° and 60° tube arrangements. CTP is the tube count calculation constant which takes on values of 0.93, 0.9, and 0.85 for single pass, two pass, and three pass tubes, respectively [4]. Furthermore, the Bell–Delaware method is used to calculate the shell side heat transfer and friction factor coefficients. These can be expressed as:

$$h_o = h_s = h_{id} J_c J_l J_b J_s J_r \tag{4.10}$$

where h_{id} denotes the heat transfer coefficient for pure cross flow over the tube bundle calculated at the Reynolds number at the centreline of the shell. This is expressible as follows:

$$h_{id} = J_s C_{p,s} \left(\frac{\dot{m}_s}{A_s} \right) \left(\frac{k_s}{C_{p,s} \mu_s} \right)^{\frac{2}{3}} \left(\frac{\mu_s}{\mu_{s,w}} \right)^{0.14} \tag{4.11}$$

Here, J_s denotes the ideal tube bank Colburn factor, A_s the cross flow area at or near the shell centerline, and $\mu\_s/\mu\_(s, w)$ the ratio of viscosities at the average temperature and wall temperature on the shell side.

The shell side pressure drop contains three terms, including the cross flow section pressure drop, the inlet and outlet pressure drops, and the window section pressure drop. Details on the Colburn factor, friction factor, and cross flow area at or near the shell centreline can be found elsewhere [4]. These factors depend on the tube arrangement and Reynolds number. In Equation 4.10, J_c is the correction factor for a baffle configuration (e.g., baffle cut and spacing), J_l is the correction factor for baffle leakage effects, which considers both shell-to-baffle and tube-to-baffle hole leakages, J_b is the correction factor for bundle and pass partition bypass streams, which depend on the flow bypass area and the number of sealing strips, J_s is the correction factor for bigger baffle spacing at the shell inlet and outlet sections, and J_r is the correction factor for the adverse temperature gradient in laminar flows. The pressure drop on the tube side is expressed as:

$$\Delta P_t = \frac{\dot{m}_t^2}{2 \rho_t A_{o,t}^2} \left[\frac{4 f_t L}{d_i} + (1 - \sigma^2 + K_c) - (1 - \sigma^2 - K_e) \right] n_p \tag{4.12}$$

where K_c and K_e respectively are the inlet and outlet pressure loss coefficients for a multiple circular tube core. Also, σ is the ratio of minimum free flow area to frontal area (given in Table. 4.1), and f_t is the friction factor, which is calculated as follows [2]:

$$f_t = 0.00128 + 0.1143 (Re_t)^{-0.311} \tag{4.13}$$

This equation is valid when $4000 < Re_t < 10^7$.

Table 4.1 Ratio of minimum free flow area to frontal area of shell and tube heat exchanger.

Ratio	Arrangement		
	30° triangular	45° rotated square	90° rotated square
σ	$\dfrac{(p_t - d_0)}{p_t}$	$\dfrac{\sqrt{2} p_t - d_0}{\sqrt{2} p_t}$ if $\dfrac{p_t}{d_o} \geq 1.7$ $\dfrac{2(p_t - d_0)}{\sqrt{2} p_t}$ if $\dfrac{p_t}{d_o} \leq 1.7$	$\dfrac{(p_t - d_0)}{p_t}$

4.3.2 Optimization

A multi-objective optimization method based on an evolutionary algorithm is applied to the shell and tube heat exchanger. Objective functions, design parameters and constraints, and the overall optimization process are described in this section.

4.3.2.1 Definition of Objective Functions

Two objective functions are considered here for multi-objective optimization: exergy efficiency (to be maximized) and total cost rate of product (to be minimized). The objective functions in this analysis can be expressed as follows:

Total cost rate: The total cost rate includes the cost of the heat exchanger, correlated here with the device's heat transfer surface area, and the operating cost for pumping power. Thus we can write:

$$C_{\text{total}} = C_{\text{in}} + C_{\text{op}} \tag{4.14}$$

The monetary units used here are 2015 US dollars. The investment cost (in US\$) for a stainless steel shell and tube heat exchanger can be expressed as a function of surface area as [5, 6]:

$$C_{\text{in}} (\$) = 8500 + 409A_t^{0.85} \tag{4.15}$$

where A_t is the total tube outside heat transfer area. The total discounted operating cost related to the pumping power to overcome friction losses can be calculated as follows:

$$C_{\text{op}} = \sum_{k=1}^{n_y} \frac{C_o}{(1 + i)^k} \tag{4.16}$$

where

$$C_o = Pk_{\text{el}}\tau \tag{4.17}$$

$$P = \frac{1}{\eta} \left(\frac{\dot{m}_t}{\rho_t} \Delta P_t + \frac{\dot{m}_s}{\rho_s} \Delta P_s \right) \tag{4.18}$$

Here, n_y denotes the equipment lifetime in years, i the annual interest rate, k_{el} the electricity price, τ the hours of operation per year and η the isentropic efficiency.

Exergy efficiency: The Exergy Efficiency for the Shell and Tube Heat Exchanger Is Defined As

$$\psi = 1 - \frac{\dot{Ex}_D}{\sum \dot{Ex}_{\text{in}}} \tag{4.19}$$

where $\dot{Ex}_D$ is the exergy destruction rate.

4.3.2.2 Decision Variables

The following decision variables (design parameters) are selected for this study: tube arrangement, tube diameters, tube pitch ratio (p_t/d_o), tube length, number of tubes, baffle spacing ratio ($L\_bc/D\_(s,i)$), and baffle cut ratio ($BC/D\_(s,i)$). Although the decision variables may be varied in the optimization procedure, each is normally required to be within a reasonable range. Such constraints, based on earlier reports, are listed in Table 4.2.

Table 4.2 Design parameters for shell and tube heat exchanger, their ranges of variation and their change step size.

Decision variable	Lower bound	Upper bound	Step size
Tube arrangement	30°, 45°, 90°	–	1
Tube inside diameter (m)	0.0112	0.0153	–
Tube outside diameter (m)	0.0126	0.022	–
p_t/d_o	1.25	2	0.001
Tube length (m)	3	8	0.001
Tube number	100	600	1
Baffle cut ratio	0.19	0.32	1
Baffle spacing ratio	0.2	1.4	0.001

Table 4.3 Fluid and device properties for the shell and tube heat exchanger (input data for the model).

Property	Shell side (hot water)	Tube side (cold water)
Density of fluid (kg/m³)	980	995
Specific heat of fluid (J/kg K)	4180	4120
Viscosity of fluid (Pa s)	0.000672	0.000695
Fouling factor of heat exchanger surface (m²W/K)	0.000065	0.000074
Thermal conductivity of fluid (W/m K)	0.56	0.634

4.3.3 Case Study

The optimum heat exchanger configuration is obtained for an oil cooler shell and tube heat recovery heat exchanger in the Sarcheshmeh copper production power plant located in the south of Kerman city, Kerman, Iran. The objectives in this case study are to maximize the exergy efficiency while minimizing the total cost rate. Hot water (the hot stream) enters the shell side of the heat exchanger at a mass flow rate of 8.1 kg/s and temperature of 78.3°C. Water (the cold stream) enters the tube side at a mass flow rate of 12.5 kg/s and temperature of 30°C. Values for several fluid and device properties are listed in Table 4.3.

For this shell and tube heat exchanger case study, the equipment lifetime is assumed to be $n_y = 10$ years, the interest rate is assumed to be $i = 10\%$, the electricity price is 0.15 $/kWh, the duration of operation is assumed to be $\tau = 7500$ h/year, and the heat exchanger efficiency is taken to be $\eta = 0.6$. The design variables include three possible tube arrangements (30°, 45°, 90°) and 20 standard tube types. The inner and outer diameters of the 20 tube types are listed in Table 4.4 [6].

4.3.4 Model Verification

To ensure the accuracy and correctness of the model, which is incorporated in developed simulation code, the modeling results are compared to the corresponding results reported in reference [2]. This comparison (see Table 4.5) indicates good agreement

Table 4.4 Inner and outer diameters of 20 standard tubes for shell and tube heat exchanger.

Diameter	Value (in)
Outer	1/2 5/8 5/8 5/8 5/8 5/8 5/8 5/8 5/8 3/4 3/4 3/4 3/4 3/4 3/4 3/4 3/4 3/4 3/4 7/8
Inner	0.44 0.40 0.43 0.48 0.49 0.50 0.52 0.54 0.55 0.48 0.51 0.53 0.56 0.58 0.60 0.62 0.63 0.35 0.68 0.67

Table 4.5 Comparison of the modeling output and the results from reference.

Variable	Results from [2]	Present study	Relative difference magnitude (%)
Thermal effectiveness (−)	0.155	0.1599	2.83
Total cost ($)	74,598	74,112	0.65
Tube side pressure drop (kPa)	17.58	17.66	0.45
Shell side pressure drop (kPa)	112	111	0.87
Heat transfer rate (kW)	393.6	404.6	2.78
Tube side heat transfer coefficient (W/m^2 K)	7837	7838	0.01
Shell side heat transfer coefficient (W/m^2 K)	698.8	730.2	4.45

Source: Shah 2003. Reproduced with permission of Wiley.

between the simulation and the reported results; observed differences are of acceptable magnitudes.

4.3.5 Optimization Results

The results of the optimization are presented. As mentioned before, exergy efficiency and total cost rate of the shell and tube heat exchanger are the two objective functions considered, and seven design parameters are selected: tube arrangement, tube diameters, tube pitch ratio, tube length, tube number, baffle spacing ratio, and baffle cut ratio. The design parameters and the ranges of their variations are listed in Table 4.2. The number of iterations for finding the global optimum over the entire search domain is 8.2×10^{15}. The genetic algorithm optimization is performed for 200 generations, using a search population size of $M = 100$ individuals, a crossover probability of $p_c = 0.9$, a gene mutation probability of $p_m = 0.035$ and a controlled elitism value $c = 0.55$.

Figure 4.5 shows the Pareto frontier solution for the shell and tube heat exchanger obtained via multi-objective optimization using the objective functions in Equations 4.14 and 4.19. Geometrical changes that increase exergy efficiency usually lead to an increase in the total cost rate, and vice versa. This suggests the need for multi-objective optimization techniques in the case of a shell and tube heat exchanger.

As shown in Figure 4.5, the maximum exergy efficiency exists at design point A (0.92), while the total cost of products is the greatest at this point ($57 359). Also, the minimum value for the total cost, which is about $14 337, occurs at design point C. Design point C is the optimal situation when the total cost is the sole objective function, while design point A is the optimum point when exergy efficiency is the sole objective function. In multi-objective optimization, a process of decision-making for selection of the final optimal solution from the available solutions is required. It can be clearly seen that

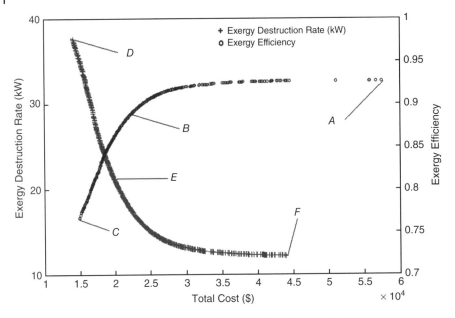

Figure 4.5 Pareto frontiers, showing best trade-offs for the objective functions.

it is not feasible to have both objectives at their optimum point simultaneously and, as shown in Figure. 4.5, the ideal point is not a solution located on the Pareto frontier. Since each point in the Pareto solutions obtained in multi-objective optimization can be utilized as the optimized point, the selection of the optimum solution depends on the preferences and criteria of the decision maker, suggesting that each may select a different point as the optimum solution depending on his/her needs and preferences. Table 4.6 shows all the design parameters for points *A-C*. As shown in Figure 4.5, the optimized values for exergy efficiency on the Pareto frontier range between 75% and 92%. To provide a good balance between exergy efficiency and total cost rate, a curve is fitted to the optimized points obtained from the evolutionary algorithm, and the expression for this fitted curve is:

$$C_{total} (\$) = \frac{5.848\psi^2 + 384\psi - 5215}{\psi^2 - 3.48\psi - 97.95} \times 10^3 \tag{4.20}$$

Table 4.6 Optimum values of exergy efficiency, total cost and exergy destruction rate for design points *A* to *F* in Pareto optimal fronts.

Quantity	Design point					
	A	*B*	*C*	*D*	*E*	*F*
Exergy efficiency	0.92	0.89	0.76	0.7686	0.86	0.92
Total cost ($)	57359	22090	14337	14397	19995	43959
Exergy destruction rate (kW)	12.15	17.97	37.95	37.44	21.43	12.19

This equation is valid when the exergy efficiency varies between 75% and 92%. To study the variation of thermodynamic characteristics, several points (*A* to *F*) on the Pareto frontier are considered. Table 4.6 gives the total cost of the heat exchanger, the total exergy destruction rate, and the exergy efficiency. From point *A* to point *C* in this table, both the total cost rates of the system and the exergy efficiencies decrease. As previously stated, point *A* is preferred when exergy efficiency is the sole objective function, and design point *C* when the total cost is the sole objective function. Design point *F* is also preferred when the total exergy destruction rate is the sole objective function.

4.3.6 Sensitivity Analysis Results

Sensitivity analyses are performed to better understand the results of the multi-objective optimization. The effects of varying each of the design parameters for point *A-F* on both objective functions are investigated. An increase in P_t/d_o results in a decrease in both exergy efficiency and total cost for all design points *A-C* (see Figure 4.6). However, an increase in this parameter results in an increase in the exergy destruction rate for all points on the Pareto frontier from *D* to *F* in Figure 4.7a. Therefore, variations of tube pitch ratio cause a conflict between the two objectives.

The results also show that an increase in the tube length leads to an increase in both exergy efficiency and total cost for all design points *A-C* (Figure 4.6b), and a decrease in exergy destruction ratefor all points from *D* to *F* (Figure 4.7b). Therefore, variations in tube length cause a conflict between two objective functions.

Like tube length, both exergy efficiency and the total cost increase with an increase in the tube number (see Figure 4.6c). Moreover, an increase in this parameter leads to a decrease in exergy destruction rate for all points from D to F in Figure 4.7c. Therefore, the tube number causes a conflict between two objective functions. As shown in Figure 4.6d and 4.7d, an increase in the baffle spacing ratio (Lbc/Ds,i) creates a conflict between two sets of objective functions (exergy efficiency and total cost as the first set and exergy destruction rate and total cost as the second set). In addition, an increase in baffle cut ratio (BC/Ds,i) leads to a decrease in both exergy efficiency and total cost as shown in Figure 4.6e, although this change is minor. Furthermore, an increase in this parameter results in an insignificant change in exergy destruction rate, as observed in Figure 4.7e.

4.4 Modeling and Optimization of Cross Flow Plate Fin Heat Exchangers

Cross flow plate-fin heat exchangers are widely used in gas-to-gas applications where a large surface area is required. Such applications occur in cryogenics, micro-turbines, automobiles, chemical process plants, and aeronautical devices. A plate-fin heat exchanger (Figure 4.8) has a high thermal effectiveness for three main reasons:

- fins are employed on both the sides to interrupt boundary layer growth,
- it has a large heat transfer surface area per unit volume, and
- it has a good thermal conduction due to the small thickness of the plate.

Together, these factors lead to a low space requirement, weight, energy requirement, and cost, and result in a compact heat exchanger. However these advantageous characteristics come at the expense of high frictional losses (i.e., pressure drops).

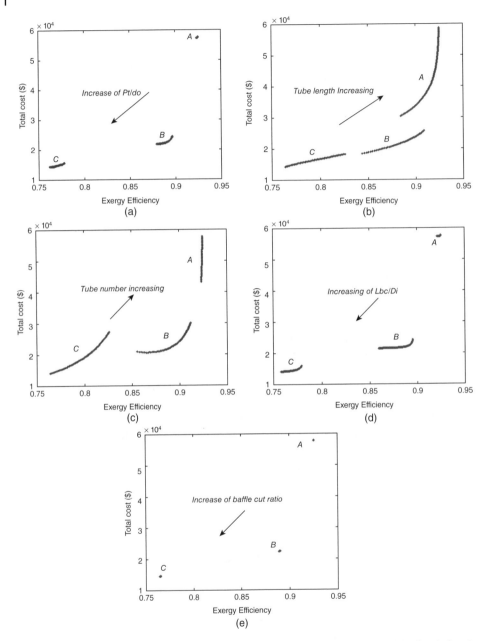

Figure 4.6 Effect of varying design parameters on total cost and exergy efficiency: (a) tube pitch ratio, (b) tube length, (c) tube number, (d) baffle spacing ratio, (e) baffle cut ratio.

Rectangular offset strip fins are shown in Figure 4.9. Such offset strip fins have high compactness, heat transfer efficiency, and reliability, and are widely employed in heat exchangers for cooling systems of aircraft, automobiles, and HVAC devices. Offset strip fins typically have superior heat transfer performance to plain fins. Furthermore, offset strip fins usually exhibit higher strength and reliability than louver fins.

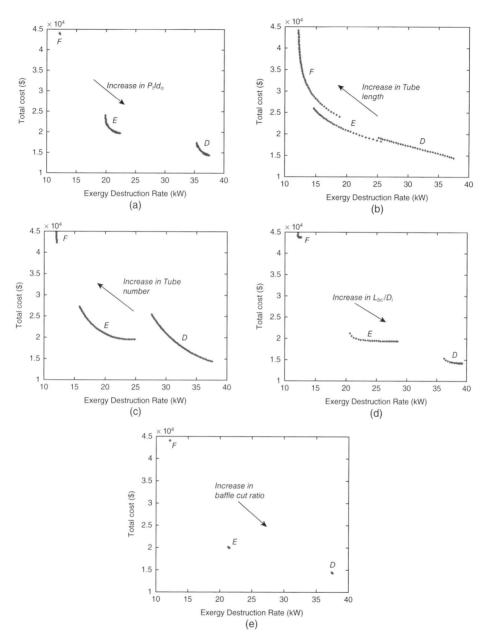

Figure 4.7 Effect of varying design parameters on total cost and total exergy destruction rate: (a) tube pitch ratio, (b) tube length, (c) tube number, (d) baffle spacing ratio, (e) baffle cut ratio.

4.4.1 Modeling and Simulation

To model and simulate the plate in order to determine the optimal heat exchanger, the ε-NTU method is emloyed to predict the heat exchanger performance. Thus, the effectiveness of the cross flow heat exchanger with both fluids unmixed can be expressed

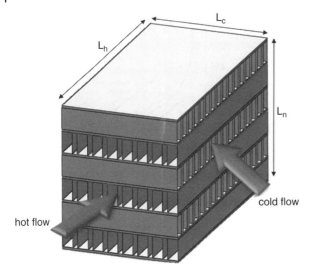

Figure 4.8 Plate fin heat exchanger.

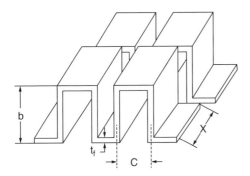

Figure 4.9 Typical rectangular offset strip fin core. *Source*: Adapted from Ahmadi 2011.

as [1]:

$$\varepsilon = 1 - \exp[-(1 + C^*)\text{NTU}] \times \left\{ I_0(2\text{NTU}\sqrt{C^*}) + \sqrt{C^*}I_1(2\text{NTU}\sqrt{C^*}) \right.$$

$$\left. - \frac{1 - C^*}{C^*} \sum_{n=2}^{\infty} C^{*\frac{n}{2}} I_n(2\text{NTU}\sqrt{C^*}) \right\} \quad (4.21)$$

where I is the modified Bessel function. The number of transfer units (NTU) and heat capacity ratio (C^*) are defined as follows [1]:

$$NTU_{max} = UA/C_{min} \quad (4.22)$$

$$C^* = \frac{C_{min}}{C_{max}} \quad (4.23)$$

Here, U denotes the overall heat transfer coefficient and A the total heat transfer surface area. These terms are expressible as follows:

$$U = \frac{1}{\frac{1}{h_0 \eta_{s,c}} + \frac{1}{\frac{A_h}{A_c}(h_h \eta_{s,h})}} \quad (4.24)$$

and

$$A = (\beta V_p)_c + (\beta V_p)_h \tag{4.25}$$

Here, h denotes the heat transfer coefficient and β the heat transfer surface area per unit volume. Also, η_s is the overall surface efficiency, defined as

$$\eta_s = 1 - \frac{A_f}{A_{\text{cell}}}(1 - \eta_f) \tag{4.26}$$

where A_f is the fin heat transfer area and η_f the efficiency of a single fin, expressible as

$$\eta_f = \frac{\tanh(ml)}{ml} \tag{4.27}$$

where $m = \sqrt{2h/k_f t_f}$, $l = \frac{b}{2}$ and k_f is the fin thermal conductivity. To evaluate the actual exchanger effectiveness, a reduction in cross flow exchanger effectiveness ($\Delta\varepsilon/\varepsilon$) due to longitudinal wall heat conduction is considered by interpolation of the tabular results given elsewhere [1]. In addition, the pressure drop is expressed as

$$\Delta P = \frac{G^2}{2} v_{in} \left[(1 + \sigma^2) \left(\frac{v_{out}}{v_{in}} - 1 \right) + f \frac{4L}{D_h} \frac{v_{avg}}{v_{in}} \right] \tag{4.28}$$

Here, σ the ratio of minimum free flow area to frontal area and f is the friction factor. The number of entropy generation units is defined as follows [7]:

$$NS = \frac{\dot{S}}{C_{\max}} \tag{4.29}$$

where $\dot{S}$ is the entropy generation rate, expressed as

$$\dot{S} = \Delta S_c + \Delta S_h \tag{4.30}$$

where

$$\Delta S = \dot{m} \left[C_p \ln \left(\frac{T_{out}}{T_{in}} \right) - R \ln \left(\frac{P_{out}}{P_{in}} \right) \right] \tag{4.31}$$

4.4.2 Optimization

For optimization purposes, entropy generation, total annual cost, heat exchanger effectiveness, heat exchanger exergy efficiency, and heat exchanger exergy destruction rate are considered as objective functions. The total annual cost includes the levelized investment cost (the annualized cost of the heat transfer surface area). These terms are expressible as

$$C_{\text{total}} = aC_{\text{in}} + C_{\text{op}} \tag{4.32}$$

$$C_{\text{in}} = C_A A^n \tag{4.33}$$

$$C_{\text{op}} = \left(k_{\text{el}} \tau \frac{\Delta P V_t}{\eta} \right)_c + \left(k_{\text{el}} \tau \frac{\Delta P V_t}{\eta} \right)_h \tag{4.34}$$

where C_A and k_{el} are the heat exchanger investment cost per unit surface area and the electricity unit cost respectively, n is a constant, and τ is the annual number of operational hours of the heat exchanger. The latter quantity is 6000 hr for this heat exchanger in this chapter. ΔP, V_t and η denote pressure drop, volume flow rate and compressor

Table 4.7 Design parameters, their ranges of variation and their change steps.

Variable	Lower value	Upper value	Change step
Fin pitch (mm)	1	2.5	0.010
Fin height (mm)	2.5	8	0.010
Fin length (mm)	2	3.5	0.010
Hot stream flow length (m)	0.2	0.4	0.001
Cold stream flow length (m)	0.7	1.2	0.001
No-flow length (m)	0.2	0.4	0.001

isentropic efficiency, respectively. Also, a is the annual cost coefficient, which is calculated as follows:

$$a = \frac{i}{1 - (1+i)^{-y}} \qquad (4.35)$$

where i and y are the interest rate and the depreciation time respectively.

4.4.2.1 Decision Variables

The following decision variables (design parameters) are selected for this study: fin pitch (c), fin height (b), fin offset length (x), cold stream flow length (L_c), no-flow length (L_n), and hot stream flow length (L_h). Moreover, constraints are introduced to ensure that α, δ, γ are in the range of $0.134 < \alpha < 0.997$, $0.012 < \delta < 0.048$ and $0.041 < \gamma < 0.121$, where $\alpha = c/b$, $\delta = t_f/x$ and $\gamma = t_f/c$. Note that the relations for friction coefficient and Colburn factor used here are valid over these ranges, which also can be considered as constraints. Although the decision variables may be varied in the optimization procedure, each is normally required to be within a reasonable range. Such constraints, based on earlier reports, are listed in Table 4.7.

4.4.3 Case Study

The plate fin heat exchanger (PFHE) optimum design parameters are obtained for a gas furnace in Almas Kavir tile factory located in northwest of Kerman city, Kerman, Iran. A schematic of the furnace, including the heat exchanger, is shown in Figure 4.10. The furnace temperature is 380 K at initial process stages and about 1200 K during the final stages. The hot stream exits from the middle stages of the furnace with a mass flow rate of 1.8 kg/s and enters the heat exchanger at 658.15 K. The fresh air, which is the cold stream, enters the heat exchanger with a mass flow rate of 2 kg/s at 306.15 K. The PFHE is made of stainless steel, with thermal conductivity $k_w = 18$ W/mK. The operating conditions and the cost function constant values are listed in Table 4.8. Moreover, the thermophysical properties of air (e.g., Prandtl number, viscosity, and specific heat) are considered as temperature dependent.

4.4.4 Model Verification

In order to verify the modeling results, the simulation outputs are compared with the corresponding results reported in the literature. A comparison is made between our modeling results and the corresponding values from reference [2], for the input values

Figure 4.10 A tile furnace with a plate fin heat exchanger used as a preheater.

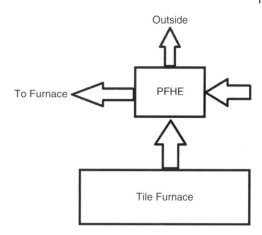

Table 4.8 Operating conditions of PFHE in the tile furnace.

Mass flow rate of hot flow (kg/s)	1.8
Mass flow rate of cold flow (kg/s)	2
Hot flow inlet temperature (K)	658.15
Cold flow inlet temperature (K)	306.15
Inlet pressure (hot side) (kPa)	180
Inlet pressure (cold side) (kPa)	120
Heat exchanger price per unit area ($/m²)	100
Exponent of nonlinear increase with area increase	0.6
Hours of operation per year (h/yr)	6000
Unit price of electrical energy ($/MWh)	25
Compressor isentropic efficiency	0.67

in Table 4.9. The results show that the differences between these two modeling output results are acceptable for the two objective functions. Furthermore, the optimization results obtained using the procedure presented in this paper, for the same case study and the same input values as in reference [8], are listed in Table 4.10. The optimization results show a 14.3% decrease in the total annual cost at a fixed number of entropy generation units, as well as a 3.5% decrease in the number of entropy generation units for a fixed total annual cost, which constitutes a significant improvement.

4.4.5 Optimization Results

The optimization results are now presented. To minimize the number of entropy generation units and the total annual cost, and to maximize the heat exchanger exergy efficiency and the heat exchanger effectiveness, six design parameters are chosen for consideration. They are: fin pitch, fin height, fin offset length, cold stream flow length, no-flow length, and hot stream flow length. Design parameters and the ranges of their variation are listed in Table 4.7, as given previously. The number of iterations for determining the global extremum in the entire search domain is about 2.9×10^{14}. The system is optimized for a depreciation time $y = 10$ years and an interest rate $i = 0.1$. The genetic

Table 4.9 Comparison of modeling output and corresponding results from reference.

Output parameter	Value from [2]	Present study	Relative difference between present study and [2] (%)
J_h (−)	0.017	0.0175	2.88
f_h (−)	0.067	0.0684	2.242
J_c (−)	0.0134	0.0149	11.19
f_c (−)	0.0534	0.055	2.996
h_h (W/m² K)	0.0534	370.50	2.68
h_c (W/m² K)	360.83	371.49	11.19
ε (−)	0.838	0.8444	0.75
C_{total} ($)	1518.78	1530.82	0.79
NS	0.1304	0.1345	3.14
ΔP_{total} (kPa)	17.425	16.84	−3.357
A_{total} (m²)	169.208	174.3	3.01

Source: Shah 2003. Reproduced with permission of Wiley.

Table 4.10 Input values from reference.

Mass flow rate of hot flow (m³/s)	1.2
Mass flow rate of cold flow (m³/s)	0.6
Hot flow inlet temperature (K)	513.15
Cold flow inlet temperature (K)	277.15
Inlet pressure for hot side (kPa)	110
Inlet pressure for cold side (kPa)	110
Fin material	Aluminum
Heat exchanger price per unit area ($/m²)	100
Exponent of nonlinear increase with area increase	0.6
Hours of operation per year (h/year)	6500
Unit price of electrical energy ($/MWh)	30
Compressor isentropic efficiency	0.5

Source: Xie 2008. Reproduced with permission of Elsevier.

algorithm optimization is performed for 250 generations, using a search population size of $M = 100$ individuals, a crossover probability of $p_c = 0.9$, a gene mutation probability of $p_m = 0.035$ and a controlled elitism value $c = 0.55$. The resulting Pareto optimal frontier is shown in Figure. 4.11, which identifies the conflict between the two objective functions: number of entropy generation units and total annual cost. Any geometrical changes that decrease the number of entropy generation units lead to an increase in the total annual cost and vice versa.

The results demonstrate the importance of multi-objective optimization techniques in the optimal design of a PFHE. As shown in Figure 4.11, the minimum number of entropy generation units exists at design point A (0.0939), while the total annual cost

Table 4.11 Optimized values for selected design parameters on the Pareto frontier in Figure 4.11 based on multi-objective optimization.

Objective function	A	B	C	D	E
NS	0.093	0.097	0.104	0.114	0.130
Annual cost of heat exchanger ($/year)	4031	2754	1925	1327	1031

has the highest value at this point. Conversely, the minimum total annual cost occurs at design point E ($1031.0), with has the largest number of entropy generation units (0.13). Design point A is the optimal solution when the number of entropy generation units is a single objective function, while design point E is the optimum condition at which the total annual cost is a single objective function.

The optimum values of the two objectives for five typical points from A to E (Pareto optimal frontiers) are listed in Table 4.11 for the input values in Table 4.7.

To provide a good understanding of the relation between the numbers of entropy generation and total cost, a curve is fitted to the optimized points obtained from the evolutionary algorithm. The expression for this fitted curve follows:

$$C_{total} (\$) = \frac{-2.81NS^3 - 4.31NS^2 + 1.72NS - 0.48}{NS^2 + 21.84NS - 1.86} \times 10^3 \qquad (4.36)$$

This equation is valid when the entropy generation varies between 0.094 and 0.13. An interesting point is that, when a numerical value is provided for the number of entropy generation units in the validity range, Equation 4.36 provides the minimum total annual cost for that optimal point along with other optimal design parameters. In addition, the change in effectiveness versus annual cost for the set of optimal points (Figure 4.11) is shown in Figure 4.12. It can be seen there that considering entropy generation units

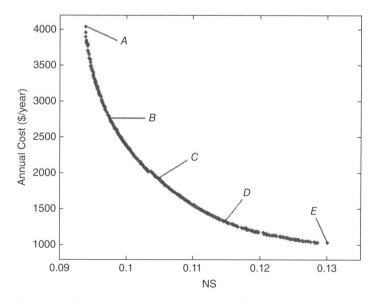

Figure 4.11 Pareto frontier, showing best trade-off among objective functions.

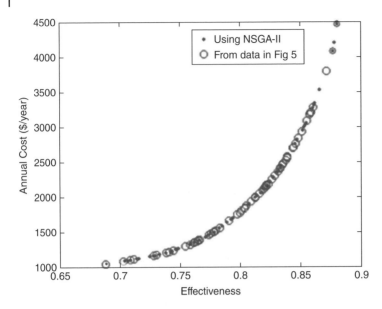

Figure 4.12 Distribution of annual cost versus effectiveness for points of Pareto frontier in Figure 4.11 and using NSGA-II.

allows a unique objective function to be used for thermodynamic optimization. Moreover, in Figure 4.12, the Pareto optimal front for another two objective functions is shown (i.e., annual cost and effectiveness). This figure shows that any geometric changes which increase the effectiveness lead to an increase in the total annual cost and vice versa. The results of Figure 4.12 show that by considering the entropy generation and annual cost we can predict the optimization results when effectiveness and annual cost are taken to be the two objective functions.

Furthermore, the Pareto optimal frontiers for the heat exchanger exergy efficiency and exergy destruction are shown in Figure 4.13 as a function of annual cost rate. This figure shows that when the exergy efficiency of the heat exchanger rises, the total cost of the heat exchanger increases. This suggests that a higher heat exchanger exergy efficiency leads to more efficient heat exchangers from both thermodynamic and thermoeconomic points of view. The same conclusion can be obtained by considering another two objective functions (from potential choices such as exergy destruction rate, exergy efficiency, and annual cost), as shown in Figure 4.13. It is clear from this figure that, when the exergy destruction rate of the heat exchanger increases, the total cost decreases. This is due to the fact that the design parameters are not selected for the best case, in which the exergy destruction is low. But, by decreasing the exergy destruction rate, the total annual cost increases. This observation reveals that irreversibility, like pressure drop and high temperature difference between cold and hot streams, plays a key role in exergy destruction. Thus, a more efficient heat exchanger leads to a heat exchanger with a higher total cost rate.

4.4.6 Sensitivity Analysis Results

For better insight into this analysis, the distribution of decision variables for the optimal points on the Pareto frontier in Figure 4.11 are shown in Figures 4.14a to Figure 4.14f.

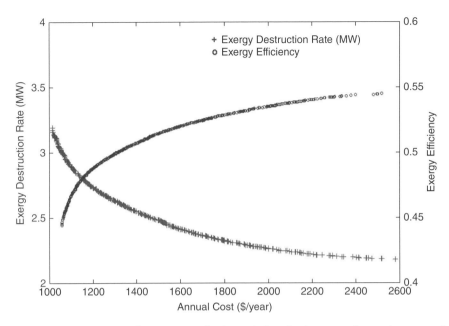

Figure 4.13 Distribution of Pareto optimal points solutions for the exergy destruction rate and exergy efficiency versus annual cost using NSGA-II.

The lower and upper bounds of the variables are illustrated by the dotted lines. The following points are noted for the optimal variables in Figure 4.14.

1) Values for the fin pitch and plate spacing are distributed equally across the entire allowable domain.
2) The numerical values of the fin offset length, cold stream flow length, hot stream flow length, and no-flow length are at their maximum levels.

The optimum values of the fin pitch and the plate spacing, which have scattered distributions throughout their allowable domains, show that these parameters have significant effects on the tension between the lower values of the number of entropy generation units and the total annual cost.

   The fin offset length, the cold stream flow length, the no-flow length, and the hot stream flow length, which are at their maximum values, show that these parameters have no effect on the conflict between the two objective functions. The variation of optimum value of number of entropy generation units with the total annual cost for various values of optimum design parameters in cases *A-E* (Pareto frontier) are shown in Figures 4.15a to 4.15f. It is observed that the variation of the two objective functions at other points on the Pareto optimal frontier displays the same trend as the five points *A-E*. An increase in the fin pitch results in an increase in the number of entropy generation units while the total annual cost decreases for all of the design points *A-E* (see Figure 4.15a). Therefore, variations of fin pitch cause a conflict between the two objective functions, and values of fin pitch have a distribution over the full allowable domain (see Figure 4.14a). As for fin pitch, by increasing the fin height, the number of entropy generation units increases while the total annual cost decreases for all design points *A-E* (see Figure 4.15b). Consequently, the fin height causes a conflict between two objective functions and the fin height values obtained on the Pareto optimal front exhibit a scattered distribution, as

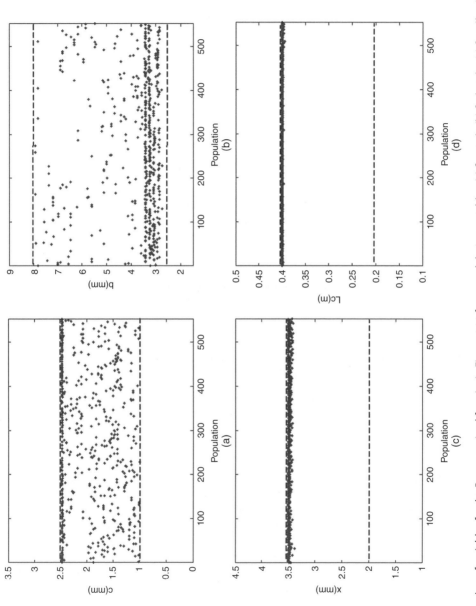

Figure 4.14 Scattering of variables for the Pareto optimal frontier in Figure 4.11 for several design variables: (a) fin pitch, (b) fin height, (c) fin offset length, (d) cold stream flow length, (e) no-flow length, and (f) hot stream flow length.

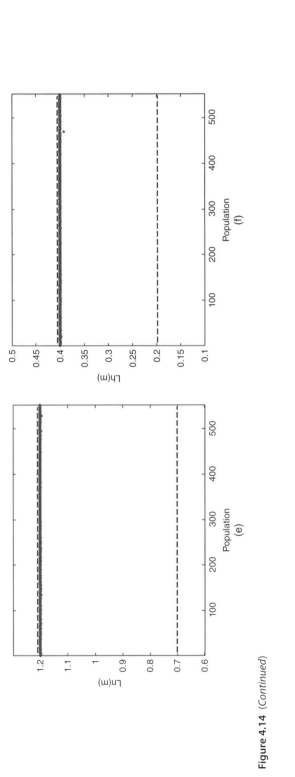

Figure 4.14 (*Continued*)

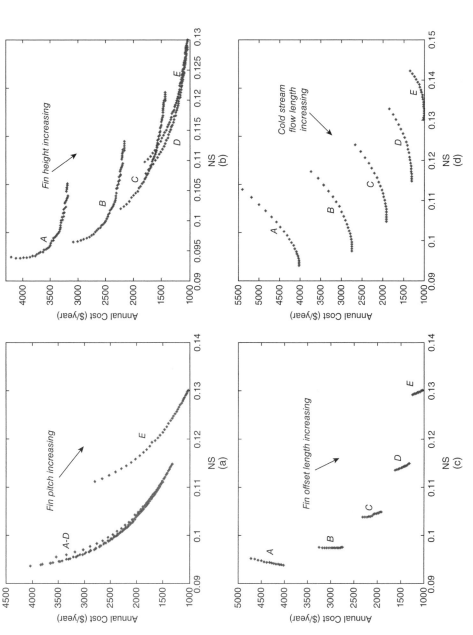

Figure 4.15 Variation of number of entropy generation units with annual cost for six design parameters in five cases of points A-E in Figure 4.11 (a) fin pitch, (b) fin height, (c) fin offset length, (d) cold stream flow length, (e) no-flow length, and (f) hot stream flow length.

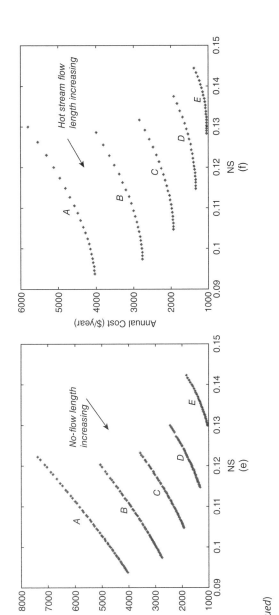

Figure 4.15 (*Continued*)

is evident in Figure 4.14b. In addition, an increase in fin length leads to an increase in the number of entropy generation units and a decrease in the total annual cost (see Figure 4.15c). However Figure 4.15c shows that there is a region in design points *A* to *B* that has no effect on the conflict between the objective functions.

An increase in the cold stream flow length and/or the hot stream flow length results in a decrease in both the number of entropy generation units and the total annual cost (see Figures 4.15d and 4.15f). In fact, the investment cost of the heat exchangers increases with increases in these two parameters, due to the increment in the thermal surface area. However, the operational cost decreases due to the fact that in this case the pressure drop declines when these two parameters increase. In this case, therefore, increasing these two parameters causes the total annual cost first to decrease and then to reach to a fixed value. As a consequence, the maximum value of these parameters is desired to improve both objective functions simultaneously (see Figures 4.14d and 4.14f). An increase in the no-flow length has the same effect as increasing the cold and hot stream flow lengths, which reduces both the number of entropy generation units and the total annual cost (Figure 4.15e). As a result, the no-flow length has no effect on conflict between the two objective functions. The distribution of the no-flow length around its maximum value verifies this point (Figure 4.14e).

4.5 Modeling and Optimization of Heat Recovery Steam Generators

The higher efficiency of combined cycle power plants (CCPPs) in comparison with Brayton or Rankine cycles has made this form of power generation quite attractive. The heat recovery steam generator (HRSG) is an important part of a combined cycle power plant as it allows recovery of energy from the turbine exhaust gases. An HRSG may contain up to three pressure levels: low pressure (LP), intermediate pressure (IP), and high pressure (HP). Each pressure level includes three main groups of heat exchangers or heating elements: economizer, evaporator, and superheater. When the turbine exhaust gases pass over the HRSG heating elements, water inside the tubes is heated by the hot gases and vaporized. The steam produced expands in a steam turbine, generating shaft power. Water preheating and evaporation occur in the economizer(s) and the evaporator, respectively. After separating the liquid water and steam in a drum, the water returns to the evaporator down comers while the steam enters the superheater. The optimal design of such a thermal system is of great interest, especially due to extensive use of combined cycles as a preferred method of generating electricity. These systems are designed mainly based on the gas turbine exhaust mass flow rate and temperature. The design of the HRSG significantly affects the overall cycle efficiency and power output, and thus merits optimization. An optimal design provides a cost effective system with minimum total cost. Here, the simulation and optimal design of HRSGs with and without duct burners are performed. The allowable arrangements of heat transfer elements are selected.

4.5.1 Modeling and Simulation

Here, the optimum physical and thermal design parameters of the system are determined, and a simulation program is developed. The temperature profile along the HRSG,

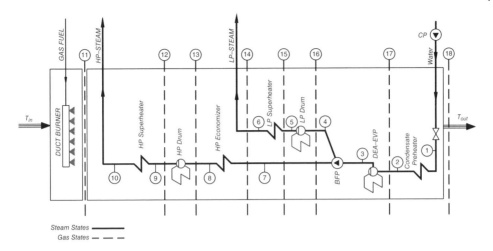

Figure 4.16 Schematic of an HRSG with dual pressure level and supplementary firing.

the input and output exergies, and the heat transfer surface area of each heating element are determined to study HRSG performance. The energy balance equations for various parts of the system (see Figure 4.16) are as follows:

High pressure superheater

$$\dot{m}_g C_p (T_{11} - T_{12}) = \dot{m}_{s,HP}(h_{10} - h_9) \tag{4.37}$$

High pressure evaporator

$$\dot{m}_g C_p (T_{12} - T_{13}) = \dot{m}_{s,HP}(h_9 - h_8) \tag{4.38}$$

High pressure economizer

$$\dot{m}_g C_p (T_{13} - T_{14}) = \dot{m}_{s,HP}(h_8 - h_7) \tag{4.39}$$

Low pressure superheater

$$\dot{m}_g C_p (T_{14} - T_{15}) = \dot{m}_{s,LP}(h_6 - 5) \tag{4.40}$$

Low pressure evaporator

$$\dot{m}_g C_p (T_{15} - T_{16}) = \dot{m}_{s,LP}(h_5 - h_4) \tag{4.41}$$

Deaerator evaporator

$$\dot{m}_g C_p (T_{16} - T_{17}) = \dot{m}_{s,LP}(h_3 - h_2) \tag{4.42}$$

Condensate preheater

$$\dot{m}_g C_p (T_{17} - T_{18}) = \dot{m}_{s,LP}(h_2 - h_1) \tag{4.43}$$

Energy and mass balance equations are numerically solved and the temperature profiles for the gas and water/steam sides of the HRSG are calculated.

From the system exergy balance equation, the exergy destruction rate in the HRSG is computed as

$$\dot{I} = \dot{E}x_i - \dot{E}x_e \tag{4.44}$$

The inlet exergy to the HRSG is the corresponding exergy of the gas turbine exhaust. The exergies of the inlet water streams, which are assumed to be at ambient temperature (the dead state), are zero.

The output exergy from the HRSG is in two forms: the steam generated in the evaporator, which is sent to the superheater, and the flue gas exiting through the stack. Thus,

$$\dot{I} = \dot{Ex}_g - \dot{Ex}_s - \dot{Ex}_f \tag{4.45}$$

Expressions for the key exergy flows related to the HRSG are described below.

Hot inlet gas exergy

$$\dot{Ex}_g = \dot{m}_g \left\{ C_p(T_g - T_0) - T_0 \left[C_p \text{Ln}\left(\frac{T_g}{T_0}\right) + R\text{Ln}\left(\frac{P_g}{P_0}\right) \right] \right\} \tag{4.46}$$

Here, the last term is negligible due to the similar magnitudes of the gas flow pressure P_g and atmospheric pressure P_0.

Steam exergy

$$\dot{Ex}_g = \sum \dot{m}[(h - h_0) - T_0(s - s_0)] \tag{4.47}$$

Flue gas exergy

$$\dot{Ex}_f = \dot{m}_g[C_p(T_{18} - T_0) - T_0(s_{18} - s_0)] \tag{4.48}$$

Condensate preheater inlet water exergy

$$\dot{Ex}_{s,\text{cond}} = \dot{m}_w[(h_1 - h_0) - T_0(s_1 - s_0)] \tag{4.49}$$

In the above expressions, the specific heat capacity is computed as follows [9]:

$$C_p(T) = 0.991 + \frac{6.99}{10^5}T + \frac{2.71}{10^7}T^2 - \frac{1.22}{10^{10}}T^3 \tag{4.50}$$

The required heat transfer surface area is determined from

$$Q = UA\Delta T_m \tag{4.51}$$

where ΔT_m is an appropriate mean temperature difference. The overall heat transfer coefficient U is sometimes treated as constant. Here, however, the variations of heat transfer coefficients with changes in decision variables (such as steam mass flow rate) are taken into account using an in-house developed software program [1]. A key design parameter in each pressure level of the HRSG is the pinch point. A large pinch point temperature corresponds to a small exchanger heat transfer surface area and a relatively low capital cost for the recovery system, for a fixed rate of energy exchange. A small pinch point temperature corresponds to a larger heat transfer surface area and more costly system. Using international prices, the average investment cost of the equipment for each heating element is computed based on the data given in Table 4.12.

The capital cost of the HRSG can be determined as follows:

$$C_c = (K_{EVP}A_{EVP} + K_{ECO}A_{ECO} + K_{SH}A_{SH} + K_{DEA}A_{DEA})_{LP}$$
$$+ (K_{EVP}A_{EVP} + K_{ECO}A_{ECO} + K_{SH}A_{SH})_{HP} \tag{4.52}$$

The duct burner investment cost can be expressed based on its heating capacity as follows [11]:

$$C_{\text{duct burner}} = 700\dot{m}_f LHV\eta_{\text{comb}} \tag{4.53}$$

where η_{comb} is the combustion efficiency.

Table 4.12 Average investment costs of HRSG equipment.

Element	Economizer K_{ECO} ($/m²)	Evaporator K_{EVP} ($/m²)	Superheater K_{SUP} ($/m²)	Deaerator evaporator $K_{DEA-EVP}$ ($/m²)
Price	34.9	45.7	96.2	41.1

Source: Casarosa 2004. Reproduced with permission of Elsevier.

4.5.2 Optimization

In this heat exchanger, the total annualized cost per unit of steam produced exergy $(C_T/\dot{Ex}_s)$ is selected as the objective function. The total annualized cost of the plant is obtained as follows:

$$C_T = aC_c + C_{op} + C_E\dot{I}t_{op} \tag{4.54}$$

The first term (aC_c) represents the capital or investment cost, where a is the annual recovery factor. This economic parameter depends on the interest rate i and the equipment lifetime k, which is expressible in the form $(i(1 + i)^\wedge k)/((1 + i)^\wedge k - 1)$. The second term C_{op} is the operating cost, which may be used in the case of a running a duct burner where there is fuel consumption in the HRSG. The third term $(C_E\dot{I}t_{op})$ is the corresponding cost of exergy destruction. Here, C_E is the exergy unit price ($/kJ),$\dot{I}$, is the rate of exergy destruction (kJ/s), and t_{op} is the annual working hours of the system. Therefore, the objective function contains decision variables for which optimum values are the optimum system design parameters. These values minimize the total cost per unit of produced steam exergy, that is, the objective function.

4.5.2.1 Decision Variables

The decision variables in this problem are the HP and LP drum pressure levels, the HP and LP pinch point temperature differences, the high pressure and low pressure side mass flow rate ratios of steam at each pressure level, and the duct burner fuel consumption mass flow rate. The ranges of variations for these design parameters are listed in Table 4.13. To avoid water vapor condensation and stack corrosion, the exit stack gas temperature is assumed to be higher than 120°C.

4.5.3 Case Study

The input parameters for the HRSG we are examining are specified in this section. The fuel supplied to the duct burner is natural gas with LHV = 50 000 kJ/kg. The fuel price is

Table 4.13 Design parameters and their ranges of variation.

Symbol	Parameter	Unit	Lower value	Upper value
P_{HP}	Pressure	MPa	5	18
P_{LP}	Pressure	MPa	0.4	7
PP_{HP}	Temperature	K	4	30
PP_{HP}	Temperature	K	4	30
$m_{s,HP}$	Mass flow ratio	%	0.1	0.9
$\dot{m}_{DB}$	Duct burner mass flow rate	kg/s	0.5	1.2

Table 4.14 Effects of exergy unit cost on optimum design parameter values for the HRSG with a duct burner, for an inlet gas temperature of 773 K.

C_E [$/kWh]	High pressure level design parameters				Low pressure level design parameters			η_{ex} [%]	C_T/Ex_s
	Pressure P [bar]	Pinch temp., PP [K]	Steam mass flow rate ratio, $\dot{m}_r$	Burner mass flow rate [kg/s]	Pressure P [bar]	Pinch temp., PP [K]	Steam mass flow rate ratio, $\dot{m}_r$		
0.02	128.4	15.0	0.81	0.73	40.23	25.00	0.186	0.83	62.92
0.03	136.8	14.0	0.81	0.717	41.21	23.21	0.190	0.83	77.14
0.04	144.3	13.3	0.80	0.702	42.48	21.38	0.195	0.83	89.43
0.05	156.4	12.3	0.80	0.6976	43.56	20.727	0.195	0.83	102.19
0.06	158.9	12.0	0.80	0.6966	44.01	20.21	0.195	0.83	115.23

taken to be $C_f = 0.003$ $/MJ. The annual number of working hours for the HRSG is 7000 hr. Considering the value of interest rate i and k to be 12% and 20 years, respectively, the value for annual recovery factor a is 13.39%. The exergy unit price C_E is 0.03 $/kWh, which represents the average selling price of electricity in Iran.

4.5.4 Modeling Verification

To verify the modeling results, they are compared to the corresponding measured values obtained from an actual operating HRSG in the north of Iran, that is, the HRSG in the Neka combined cycle power plant, which has a net power output of 420 MW [12]. The input values for thermal modeling of the HRSG at the Neka CCPP are listed in Table 4.14. The gas temperature variation for the Neka dual pressure HRSG with a duct burner is obtained using a simulation program and the corresponding measured values are shown and compared in Figure 4.17. The average of difference between the numerical and the measured values of parameters at various sections of the HRSG is about 1.14%, with the maximum of 1.36% occurring in the LP superheater. This helps verify that the simulation code developed to model the thermal performance of HRSG performs reasonably accurately. Furthermore, the results obtained with the simulation program demonstrate an acceptable level of agreement with the outputs of a commercial software program.

4.5.5 Optimization Results

The optimization results are now presented and explained. The genetic algorithm optimization is applied to obtain the HRSG optimum design parameters. Two HRSG cases are considered: with and without a duct burner. Figure 4.18 shows the convergence of the objective function with number of generations (50 in our case, beyond which there is no noticeable change in the value of the objective function). The numerical value of the objective function is greater for the HRSG with the duct burner due to that device's higher capital and operational costs.

The unit exergy price is an effective parameter that plays a key role in thermoeconomic optimization, in that it relates the effect of exergy efficiency to the total cost of the system. The HRSG second law efficiency (e.g., exergy efficiency) is defined as the ratio of the

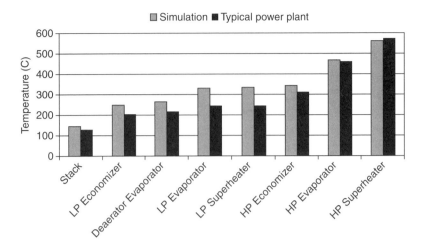

Figure 4.17 Comparison of modeling and measured values of the hot gas temperature at various heat transfer elements of the HRSG at the Neka CCPP.

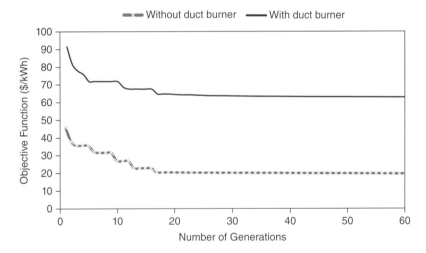

Figure 4.18 Convergence of objective functions for various numbers of generations in the genetic algorithm.

exergy increase in the water side to the exergy decrease in the gas side [13]. Usually, we observe that the higher the exergy unit price, the greater is the exergy efficiency. Optimum values of selected design parameters are listed in Table 4.14 for a range of values of C_E.

A smaller pinch temperature corresponds to a larger heat transfer surface area and a more costly system, as well as a higher exergy efficiency and lower operating cost. Figure 4.19 shows the variation of optimum values of the pinch temperature difference obtained from the developed program versus the unit exergy cost for the HRSG with and without the duct burner. The higher unit exergy cost increases the third term in Equation 4.54. To reduce the numerical value of the objective function, the exergy efficiency needs to increase to provide a lower exergy destruction rate $\dot{I}$ as well as a lower

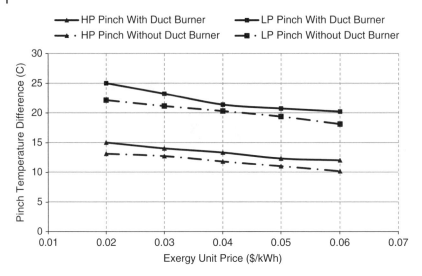

Figure 4.19 Variation of HP and LP pinch temperatures with unit exergy cost for HRSG with and without the duct burner.

operating cost. But note also that the capital cost increases when the pinch temperature difference decreases.

Figure 4.20 shows the variation of the optimum HP and LP drum pressures for the HRSG with and without a duct burner. An increase in the unit exergy price C_E results in an increase in both the HP and LP drum pressures, which in turn results in an increase in steam production and its exergy rate $(\dot{E}x_s)$. In this case, therefore, both the numerator and denominator of $(C_T/\dot{E}x_s)$ increase. However, the increase in $\dot{E}x_s$ dominates and causes the objective function $(C_T/\dot{E}x_s)$ to decrease. Furthermore, as shown in Figure 4.20, both the HP and LP drum pressures for the HRSG with the duct burner are greater than those for the HRSG without the duct burner. This is due to an increase in the hot gas temperature in the HRSG duct, which corresponds to greater values of drum and steam pressures and temperatures.

As Figure 4.21 shows, an increase in the unit exergy cost C_E leads to an increase in the HRSG exergy efficiency for both HRSG cases (with and without a duct burner). By increasing C_E, the third term in Equation 4.54, that is, $C_E \dot{I} t_{op}$, increases and the optimization scheme causes $\dot{I}$ to vary such that it decreases the objective function. By decreasing the exergy destruction rate $\dot{I}$, the exergy efficiency increases, as shown in Figures 4.21 and 4.22. This variation is observed in the HRSG both with and without the duct burner. In the former case, however, the results show a larger numerical value for the exergy destruction rate. Note that, at a fixed C_E, the exergy efficiency for the HRSG case with the duct burner is smaller than that for the case without the duct burner. This is due to the fact that, with the duct burner, the temperature difference between the hot gas and water rises, which increases the heat loss as well as the waste heat associated with the exhaust from the stack.

Figure 4.23 shows the variation of duct burner fuel consumption rate with unit exergy price. On increasing C_E and increasing the first term in Equation 4.54, the second term, that is, the operational cost, decreases to keep the objective function as low as possible.

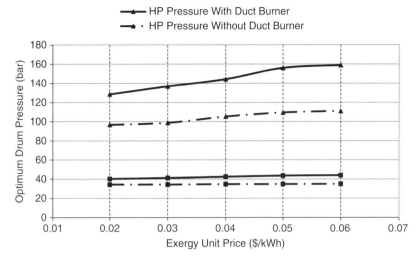

Figure 4.20 Variation of optimum values of drum pressures with unit exergy price, for HRSG (with and without duct burner).

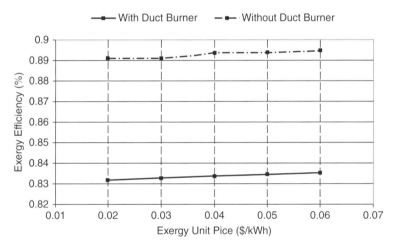

Figure 4.21 Variation of exergy efficiency with unit exergy price, for the HRSG (with and without duct burner).

This results in a decrease in Equation 4.54, which leads to a decrease in the exergy destruction rate and a decrease in the size of the duct burner and its corresponding capital cost.

To study the effect of exhaust gas temperature and mass flow rate for typical power plant gas turbines at various nominal power outputs, the range of variation of exhaust gas temperature and mass flow rate at various nominal power outputs are collected. Therefore the optimum design parameters are obtained for various inlet hot gas enthalpies into the HRSG. Note that the inlet water mass flow rate to the deaerator evaporator economizer also changes with the hot gas temperature, to avoid overheating of HRSG heating elements, such as the superheater and evaporators.

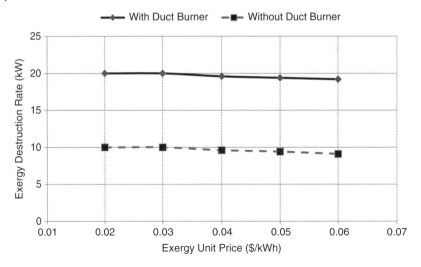

Figure 4.22 Variation of exergy destruction rate with unit exergy price, for the HRSG (with and without duct burner).

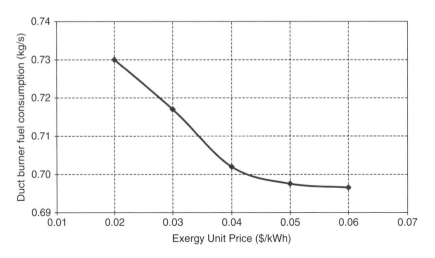

Figure 4.23 Variation of fuel consumption rate with unit exergy price for an HRSG with duct burner.

The results of varying inlet hot gas enthalpy are also presented in Table 4.15. The variation of optimum HP and LP pinch temperature differences with inlet gas enthalpy are also shown in Figure 4.24 for the HRSG, with and without the duct burner. The higher inlet gas enthalpy leads to a higher heat transfer surface area and the higher capital cost. This necessitates selecting higher LP and HP pinch temperatures to reduce the required heat transfer surface area, and to decrease the capital cost correspondingly. As shown in Figure 4.24, both HP and LP pinch temperature differences increase with a rise in inlet hot gas enthalpy. This enthalpy increase causes the heat transfer surface area and its capital cost to increase in order to be able to recover the hot gas enthalpy. Therefore the optimal design value for the pinch temperature difference increases in order to decrease the heat transfer surface area.

Table 4.15 Effects of inlet gas temperature on optimum design parameters for $C_E = 0.02$ $/kWh (with duct burner).

Inlet gas temperature T_g (°C)	High pressure level design parameters				Low pressure level design parameters			η_{ex}	C_T/Ex_s
	P (bar)	PP (K)	$\dot{m}_r$	$\dot{m}_{DB}$ (kg/s)	P (bar)	PP (K)	$\dot{m}_r$	(%)	($/kWh)
500	128.42	15	0.814	0.73	40.23	25	0.186	0.8317	62.92
510	143.44	15.23	0.800	0.72	43.106	26	0.2	0.8360	60.073
520	161	17.32	0.784	0.713	46.928	26.92	0.216	0.8372	56.97
530	164.03	20.03	0.7734	0.7076	48. 37	27.2	0.2266	0.8385	55.6
540	164.8	21.3	0.773	0.697	48.391	28.19	0.227	0.8387	53.42

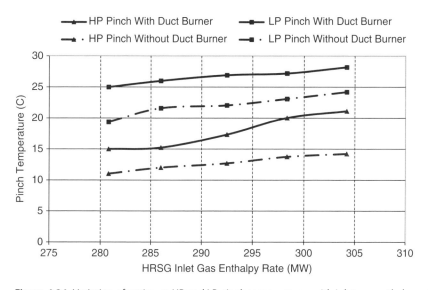

Figure 4.24 Variation of optimum HP and LP pinch temperatures with inlet gas enthalpy rate for $C_E = 0.02$ $/kWh (with and without duct burner).

The variation of HP and LP drum pressures with inlet hot gas enthalpy is shown in Figure 4.25. A higher drum pressures causes the steam temperature and mass flow rate to increase, providing a higher steam turbine power output. The higher inlet gas enthalpy causes a higher heat transfer surface area in the HP and LP sections to be required. Hence the higher HP and LP optimum pressure levels increase the steam exergy produced and decrease the objective function ($C_T/\dot{Ex}_s$).

Figures 4.26 and 4.27 show the variation of inlet hot gas enthalpy rate with HRSG exergy efficiency for the cases of the HRSG with and without duct burner. Between inlet hot gas enthalpy rates of 280 to 300 MW, the exergy efficiency increases noticeably (while the exergy destruction rate decreases in this range). The variation in HRSG exergy efficiency as the pinch temperature varies is shown in Figure 4.28. There, the HRSG exergy efficiency is observed to decrease with increasing pinch temperature difference due to

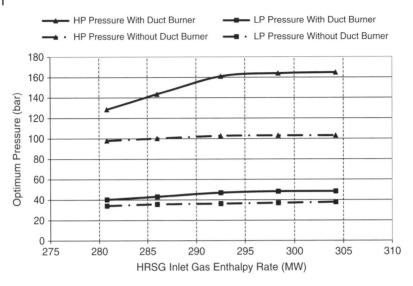

Figure 4.25 Variation of optimum HP and LP drum pressures with inlet gas enthalpy rate for $C_E = 0.02$ $/kWh (with and without duct burner).

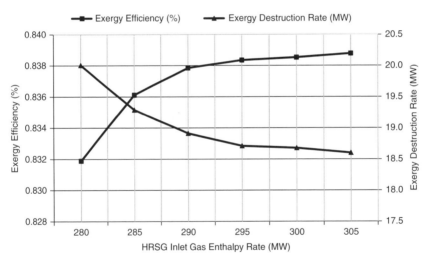

Figure 4.26 Variation of HRSG exergy efficiency and exergy destruction rate with inlet gas enthalpy rate for $C_E = 0.02$ $/kWh (with duct burner).

the fact that the higher difference values between the gas and water temperatures permit less energy recovery from the hot gases.

4.5.6 Sensitivity Analysis Results

To gain further insight into this analysis, a comprehensive sensitivity analysis is conducted. Table 4.16 shows the results of the sensitivity analysis, in terms of the variations in selected system design parameters when fuel price and investment cost vary. The highest values of variation for design parameters occur for the HP drum pressure and

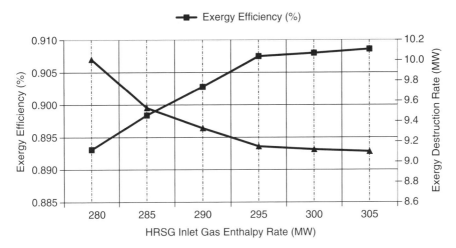

Figure 4.27 Variation of HRSG exergy efficiency and exergy destruction rate with inlet gas enthalpy rate for $C_E = 0.02$ \$/kWh (without duct burner).

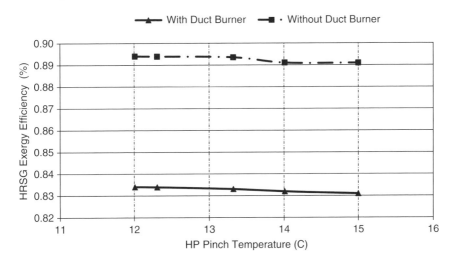

Figure 4.28 Variation of HRSG exergy efficiency with HP pinch temperature for HRSG (with and without duct burner).

HP pinch temperature difference. This provides the decision makers with useful information to guide the selection of the design parameters when fuel price is changing.

4.6 Concluding Remarks

Information about heat exchangers is provided along with comprehensive descriptions and illustrations of the modeling and optimization of various kinds of heat exchanger. The main heat exchangers used in a wide range of energy systems are considered. To ensure the accuracy of the model and corresponding simulation, each heat exchanger is validated with actual data from the industry. In addition, various new objective functions from technical, economic, and environmental points of view are defined, and then

Table 4.16 Sensitivity of selected design parameters to variations in fuel and investment costs (with duct burner).

Design parameter variation	Variation value (%)			
	Fuel Cost		Investment Cost	
	+25	−25	+25	−25
$\Delta T_{\text{PP-HP}}/T_{\text{PP-HP}}$	+7.1	−4.6	+13.1	−10
$\Delta T_{\text{PP-LP}}/T_{\text{PP-LP}}$	+8.2	−6.8	+6.1	−5.6
$\Delta P_{\text{HP}}/P_{\text{HP}}$	−6.1	+5.1	−2.2	+4.2
$\Delta P_{\text{LP}}/P_{\text{LP}}$	−1.7	+3.2	−4.0	+2.0
$\Delta \dot{m}_{\text{r-HP}}/\dot{m}_{\text{r-HP}}$	+0.25	−0.60	−0.18	+0.38
$\Delta \dot{m}_{\text{DB}}/\dot{m}_{\text{DB}}$	+1.2	−2.0	+0.85	−1.0

optimized using genetic algorithms while satisfying several reasonable constraints. This is done to make the optimization as realistic as possible, so as to provide useful information for manufacturers and designers.

Some key concluding remarks as extracted from this chapter follow:

- An increase in shell and tube heat exchanger exergy efficiency leads to an increase in heat exchanger cost. In this case, the lower the exergy destruction rate, the higher the total cost required. In the optimization of shell and tube heat exchangers, pressure drop and a large temperature difference between cold and hot stream cause higher irreversibility, which demonstrates why comprehensive optimization is merited.
- Interesting results flow from the optimization of a cross flow plate fin heat exchanger. The exergy analysis reveals that, when the exergy efficiency of the heat exchanger increases, the total cost of the heat exchanger also increases. Also, a decrease in exergy destruction rate leads to an increase in total annual cost. A set of Pareto optimal frontier points are shown, and a correlation between the optimal values of two objective functions is proposed. Fin pitch and fin height are found to be significant design parameters that lead to a conflict between two objective functions. In contrast, cold stream flow length, no-flow length, and hot stream flow length exhibit little or no effect on the conflict between the two optimized objective functions, and their maximum allowable values improve both objective functions simultaneously.
- The HRSG optimization also yields useful findings. As the exergy unit cost increases, the optimum values of design parameters tend to be selected so as to decrease the objective function. For example, pinch temperature can be decreased to reduce the exergy destruction rate. Furthermore, it is found that, at higher inlet gas enthalpies, the required heat transfer surface area (and its corresponding capital cost) increase.

References

1 Ahmadi, P., Hajabdollahi, H. and Dincer, I. (2011) Cost and entropy generation minimization of a cross-flow plate fin heat exchanger using multi-objective genetic algorithm. *Journal of Heat Transfer* **133**:021801.

2 Shah, R. K. and Sekulic, D. P. (2003) *Fundamentals of Heat Exchanger Design,* John Wiley & Sons, Inc., New York.

3 Taborek, J. (1991) *Industrial Heat Exchanger Design Practice,* John Wiley & Sons Inc., New York.

4 Kakac, S., Liu, H. and Pramuanjaroenkij, A. (2012) *Heat Exchangers: Selection, Rating, and Thermal Design,* CRC press.

5 Taal, M., Bulatov, I., Klemeš, J. and Stehlík, P. (2003) Cost estimation and energy price forecasts for economic evaluation of retrofit projects. *Applied Thermal Engineering* **23**:1819–1835.

6 Hajabdollahi, H., Ahmadi, P. and Dincer, I. (2012) Exergetic optimization of shell-and-tube heat exchangers using NSGA-II. *Heat Transfer Engineering* **33**:618–628.

7 Mishra, M., Das, P. and Sarangi, S. (2009) Second law based optimisation of cross-flow plate-fin heat exchanger design using genetic algorithm. *Applied Thermal Engineering* **29**:2983–2989.

8 Xie, G., Sunden, B. and Wang, Q. (2008) Optimization of compact heat exchangers by a genetic algorithm. *Applied Thermal Engineering* **28**:895–906.

9 Ahmadi, P., Dincer, I. and Rosen, M. A. (2011) Exergy, exergoeconomic and environmental analyses and evolutionary algorithm based multi-objective optimization of combined cycle power plants. *Energy* **36**:5886–5898.

10 Casarosa, C., Donatini, F. and Franco, A. (2004) Thermoeconomic optimization of heat recovery steam generators operating parameters for combined plants. *Energy* **29**:389–414.

11 Baukal Jr, C. E. (2004) *Industrial burners handbook.* CRC Press.

12 Ameri, M., Ahmadi, P. and Khanmohammadi, S. (2008) Exergy analysis of a 420 MW combined cycle power plant. *International Journal of Energy Research* **32**:175–183.

13 Hajabdollahi, H., Ahmadi, P. and Dincer, I. (2011) An exergy-based multi-objective optimization of a heat recovery steam generator (HRSG) in a combined cycle power plant (CCPP) using evolutionary algorithm. *International Journal of Green Energy* **8**:44–64.

Study Questions/Problems

1 Two fluids, with different properties, flow with equal free stream velocities parallel to a flat plate. What property of the fluid determines whether the velocity boundary layer of one is thicker than the other?

2 What is the most important reason for placing baffles in a shell and tube heat exchanger? How does the presence of baffles affect the heat transfer and the pumping power requirements?

3 What is the difference between pinch point temperature and approach point temperature in heat recovery heat exchangers, in combined cycle power plants? What is the effect of these temperatures on the size and cost of the heat exchangers?

4 Sketch a temperature profile of the dual pressure HRSG studied in this chapter and explain why the deaerator evaporator is used as an additional heating element.

5 Why are duct burners used in some combined cycle power plants?

6 What are the benefits of fins in heat exchangers? Provide a practical example where fins are widely used.

7 What are possible reasons for using a multi pass exchanger with cross-parallel flow instead of cross-counter flow?

8 Steam in the condenser of an advanced steam power plant is to be condensed at a temperature of 25°C using a wet cooling tower. This steam enters the tubes of the condenser at 15°C and exits at 23°C. The surface area of the tubes is 50 m², and the overall heat transfer coefficient is 2100 W/m².°C. Calculate the mass flow rate of cooling water required and the rate of condensation of the steam in the condenser.

9 Design a shell and tube type heat exchanger of TEMA E type with shell side fluid mixed. The heat exchanger is required to cool hot water with a mass flow rate of 20 kg/s and temperature of 100°C using cold water having a mass flow rate of 25 kg/s and a temperature of 25°C. What heat transfer rate is obtained using 220, 5 m long tubes (inner diameter = 16 mm and outer diameter = 19 mm, with $k_w = 380$ W/mK). The hot fluid flows in the tubes while the cold fluid flows in the shell. Determine the minimum shell diameter assuming that the tubes are arranged in a triangular pattern with a pitch of 38 mm. What are the tube side and shell side pressure drops?

10 Design, analyze, and optimize a triple pressure heat recovery steam generator (HRSG) for a large advanced combined cycle power plant whose net steam power output is 150 MW. The gas turbine exhaust gases can be treated as ideal gases and the outlet temperature and mass flow rate are 550°C and 500 kg/s respectively. Water enters a low pressure economizer for this HRSG at $P = 200$ kPa and $T = 40$°C. Make reasonable assumptions as required.

5

Modeling and Optimization of Refrigeration Systems

5.1 Introduction

Refrigeration plays an important role in daily life and is used for an extensive range of applications, including cooling of food, homes, and electronic devices. Although mainly considered a discipline within mechanical engineering, refrigeration is somewhat multi-disciplinary, drawing on other disciplines including chemical engineering, process engineering, food engineering, HVAC, and cryogenics. Thermodynamics is at the core of refrigeration, and optimization is an important tool for finding the best refrigeration system for a given application.

In general, refrigeration is a process involving the transfer of heat from a region at a lower temperature to one at a higher temperature. A refrigeration process occurs in a refrigerator and the cycle on which it operates is a refrigeration cycle. In such sectors as food, drink, and chemicals, refrigeration often represents a significant proportion of the energy costs for a site (up to 90% in the case of some cold storage facilities). Presently, the refrigeration industry is in need of and would benefit from enhanced procedures for energy and exergy analyses of refrigeration systems in conjunction with system design and optimization methods and increased applications of optimal refrigeration techniques. Since refrigeration requires electricity to transfer heat from lower to higher temperature areas, the optimization of such systems to reduce both electricity consumption and capital cost has been found to be of significant value. The growing concerns regarding climate change and global warming and the effects on humans and other species make efforts at reducing electricity and using more efficient and cost effective refrigeration increasingly important.

The primary objective of this chapter is to describe refrigeration cycles and their energy and exergy analyses, and to model comprehensively and optimize selected refrigeration systems. The latter objective involves defining objective function(s) and, usually, applying multi-objective optimization to determine the best optimal design parameters of the system. Relevant constraints need to be considered, so that proper optimization is carried out. In order to enhance understanding of the design criteria, a sensitivity analysis is conducted to assess how each objective function varies when design parameters are varied. We then provide some closing remarks related to the efficient design of refrigeration systems.

Optimization of Energy Systems, First Edition. Ibrahim Dincer, Marc A. Rosen, and Pouria Ahmadi.
© 2017 John Wiley & Sons Ltd. Published 2017 by John Wiley & Sons Ltd.

5.2 Vapor Compression Refrigeration Cycle

The vapor compression refrigeration (VCR) cycle is the most common refrigeration cycle, and forms the basis of most air conditioning systems and heat pumps. A VCR system consists of a compressor, a condenser, an expansion valve, and an evaporator. Such a system is shown in Figure 5.1, with corresponding *T-s* and *p-h* diagrams for the refrigeration cycle. The basic cycle, with some idealizations, has the following four main processes (step numners refer to Figure 5.1):

1) Constant pressure heating in an evaporator (steps 1–2)
2) Isentropic compression in a compressor (steps 2–3)
3) Constant pressure heat rejection in a condenser (steps 3–4)
4) Throttling in an expansion valve (steps 4–1)

In Figure 5.1, PPTD1 is the pinch point temperature difference for the condenser and PPTD2 is the pinch point temperature for the evaporator, both of which are important parameters for modeling. In a VCR cycle, refrigerant enters the compressor at point 2 as a saturated vapor and its pressure is increased. The temperature of the refrigerant increases with the isentropic compression to a temperature above the ambient temperature. Superheated vapor exits the compressor at point 3 and enters the condenser, where

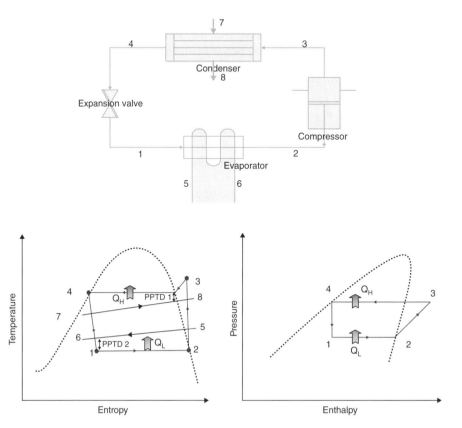

Figure 5.1 Schematic of vapor compression refrigeration cycle (top) and corresponding *T-s* and *P-h* diagrams.

heat is rejected to the environment and the refrigerant exits the condenser as a saturated liquid at point 4. The temperature at this point is still above the ambient temperature. To reduce the temperature to the desired temperature for the cooling application, the refrigerant temperature is decreased significantly by expanding (throttling) it to the evaporator pressure. This happens in an expansion valve, in which an isenthalpic process occurs that leads to a significant pressure drop and corresponding temperature decrease. The refrigerant enters the evaporator at point 1 as a low-quality saturated vapor and is evaporated by absorbing heat from the refrigerated space. The refrigerant exits the evaporator at point 2 and re-enters the compressor to complete the cycle.

A more detailed explanation of the essential components of a simple vapor compression refrigeration system as shown in Figure 5.1 follows:

- Evaporator. In this is device, heat exchange occurs from the cold space, providing refrigeration. The liquid refrigerant boils at low temperature as it absorbs heat.
- Suction line. This is the tube between the evaporator and the compressor. After the liquid has been vaporized in the evaporator, it flows through the suction line to the compressor.
- Compressor. This device separates the low pressure side of the system from the high pressure side and its main purpose is to compress the low temperature refrigerant vapor, creating a high temperature, high pressure, superheated vapor.
- Hot gas discharge line. This tube connects the compressor with the condenser. After the compressor has discharged the high pressure, high temperature, superheated refrigerant vapor, the vapour flows through the hot gas discharge line to the condenser.
- Condenser. This device is used for heat exchange, similar to the evaporator, except that its purpose is to reject heat rather than absorb it. The condenser changes the state of the superheated refrigerant vapor back to a liquid. This is done by removing enough heat to cause the refrigerant to condense.
- Liquid line. This line connects the condenser with the refrigerant control device, including the expansion valve. Only liquid refrigerant flows in this line, which is warm because the refrigerant is still under high pressure.
- Refrigerant control. This last control works as a metering device. It monitors the liquid refrigerant that enters the evaporator and ensures all the liquid is boiled off before the refrigerant enters the suction line. This avoids liquid refrigerant entering the suction line and then the compressor, where it can cause a malfunction.

5.2.1 Thermodynamic Analysis

To model/analyze a VCR system, we can consider a steady state flow and apply the first law of thermodynamics to each of the main components (Figure 5.2a). Energy and mass are conserved in each component and also by the overall system. Assuming changes in kinetic and potential energies are negligible, energy and mass rate balance equations for each component of the system can be expressed as follows:

Compressor

$$\dot{m}_{in} = \dot{m}_{out}$$
$$\dot{E}_{in} = \dot{E}_{out}$$
$$\dot{m}\,h_1 + \dot{W} = \dot{m}\,h_2 \rightarrow \dot{W} = \dot{m}(h_2 - h_1) \tag{5.1}$$

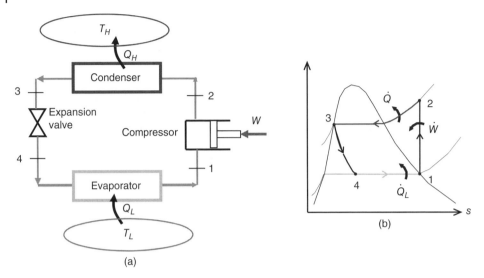

Figure 5.2 (a) An ideal vapor compression refrigeration system for analysis and (b) its temperature-entropy diagram.

where $\dot{m}$ denotes mass flow rate of refrigerant (in kg/s), h specific enthalpy (kJ/kg), and $\dot{W}$ compressor power input (kW).

Condenser

$$\dot{m}_2 = \dot{m}_3$$
$$\dot{m}\,h_2 = \dot{m}\,h_3 + \dot{Q}_H$$
$$\dot{Q}_H = \dot{m}(h_2 - h_3) \tag{5.2}$$

where $\dot{Q}_H$ is the heat rejection rate from the condenser to the high temperature environment.

Expansion valve

$$\dot{m}_3 = \dot{m}_4$$
$$\dot{m}h_3 = \dot{m}h_4$$
$$h_3 = h_4 \tag{5.3}$$

Evaporator

$$\dot{m}_4 = \dot{m}_1$$
$$\dot{m}h_4 + \dot{Q}_L = \dot{m}h_1 \rightarrow \dot{Q}_L = \dot{m}(h_1 - h_4) \tag{5.4}$$

where $\dot{Q}_L$ is the heat transferred from the low temperature environment to the evaporator.

For the overall refrigeration system, the energy rate balance can be written as

$$\dot{W} + \dot{Q}_L = \dot{Q}_H \tag{5.5}$$

The coefficient of performance (COP) of the refrigeration system can be expressed as:

$$\text{COP} = \frac{\dot{Q}_L}{\dot{W}} \tag{5.6}$$

and the isentropic efficiency of an adiabatic compressor as:

$$\eta_{comp} = \frac{\dot{W}_{isen}}{\dot{W}} = \frac{h_{2s} - h_1}{h_2 - h_1} \tag{5.7}$$

where h_{2s} is the specific enthalpy of the refrigerant at the turbine exit if the compression process is isentropic (i.e., reversible and adiabatic).

A *T-s* diagram of an ideal vapor compression refrigeration cycle is shown in Figure 5.2b. In this cycle, the refrigerant enters the compressor as a saturated vapor, is compressed isentropically in the compressor, and is cooled and condensed at constant pressure in the condenser by rejecting heat to a high temperature medium until it is a saturated vapor. The refrigerant is expanded in an expansion valve during which its specific enthalpy remains constant: it is evaporated in the evaporator at constant pressure by absorbing heat from the refrigerated space, and it exits the evaporator as a saturated vapor.

In energy analysis of this kind of vapor compression system, specific enthalpy values are required and can be obtained via three practical methods:

- using log *P-h* (pressure-enthalpy) diagrams, which provide the thermodynamic properties of the refrigerants,
- using tabulated numerical values of the thermodynamic properties of the refrigerants, or
- using known values of the latent heats and specific heats of the refrigerants and making use of the fact that areas on the *T-s* diagrams represent heat quantities.

Since modeling and optimization of such systems require many thermodynamic properties of various refrigerants, it is difficult to use tabulated data. To model such systems, coded thermodynamic properties are usually used, especially when performing optimization. For this, Engineering Equation Solver (EES), which was introduced in Chapter 1, is linked to the software package (Matlab) to facilitate the use of its thermophysical properties for various working fluids.

Example 5.1: Refrigerant-134a enters the compressor of a vapor compression refrigeration cycle at 100 kPa and −20°C at a rate of 0.5 m³/min and exits at 800 kPa. The isentropic efficiency of the compressor in 75%. The refrigerant enters the throttling valve at 750 kPa and 26°C. Sketch the cycle on a *T-s* diagram and calculate (a) the compressor consumption power, (b) the cooling load of the refrigerator, and (c) the pressure drop and rate of heat gain between the evaporator and compressor. Conduct a parametric study to investigate the effects of varying compressor isentropic efficiency and evaporator pressure on both compressor consumption power and COP.

Solution: We assume that all the devices operate at steady state conditions and that kinetic and potential energy changes are negligible.

The temperature-entropy diagram for the cycle, showing all main system points, is shown in Figure 5.3.

a) Using the properties of R-134a from thermodynamic tables:

$$P_1 = 100 \text{ kPa}, \ T_1 = -20°C \rightarrow \begin{cases} h_1 = 239.5 \frac{kJ}{kg} \\ s_1 = 0.97 \frac{kJ}{kgK} \\ v_1 = 0.198 \frac{m^3}{kg} \end{cases}$$

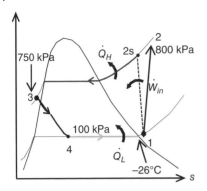

$$P_2 = 800 \text{ kPa}, \; s_{2s} = s_1 \rightarrow h_{2s} = 284.07 \text{ kJ/kg}$$
$$P_3 = 750 \text{ kPa}, \; T_3 = 26°C \rightarrow h_3 = h_{f@T_1=26°C} = 87.83 \text{ kJ/kg}$$
$$h_4 = h_3 = 87.83 \text{ kJ/kg}$$
$$T_5 = -26°C \text{ and saturated vapor} \rightarrow P_5 = 101 \text{ kPa and } h_5 = 234.6 \text{ kJ/kg}$$

Then, the mass flow rate and compressor consumption power can be calculated as:

$$\dot{m}_1 = \frac{\frac{0.5}{60} \frac{\text{m}^3/\text{min}}{\text{s/min}}}{0.198 \frac{\text{m}^3}{\text{kg}}} = 0.042 \frac{\text{kg}}{\text{s}}$$

$$\dot{W}_{comp} = \frac{\dot{m}_1(h_{2s} - h_1)}{\eta_{comp}} = 2.4 \text{ kW}$$

b) The cooling load of the refrigerator follows:

$$\dot{Q}_L = \dot{m}(h_5 - h_4) = 6.17 \text{ kW}$$

c) The pressure drop and rate of heat gain, respectively, can be calculated as follows:

$$\Delta P = P_5 - P_1 = 101.73 - 100 = 1.73 \text{ kPa}$$
$$\dot{Q}_{gain} = \dot{m}(h_1 - h_5) = 0.2 \text{ kW}$$

The results of the parametric study are presented in Figures 5.4 through 5.6. It is seen that increasing either compressor isentropic efficiency or evaporator efficiency causes the COP to rise. The higher is the compressor isentropic efficiency, the lower is the compressor consumption power, which leads to an increase in the COP of the refrigeration system. The parametric study also shows that an increase in evaporator pressure from $100 \times$ kPa to $130 \times$ kPa (corresponding to increasing the evaporator temperature from $-26.3°C$ to $-20.5°C$) increases the COP from 2.57 to 3.02—an increase of 15%. The results are in agreement with those in [1], where the COP of a VCR is observed to improve by 2–4% for each 1°C rise in evaporator temperature.

5.2.2 Exergy Analysis

Consider a vapor compression refrigeration cycle operating between a low temperature medium (T_L) and a high temperature medium (T_H) (i.e., the cycle in Figure 5.2).

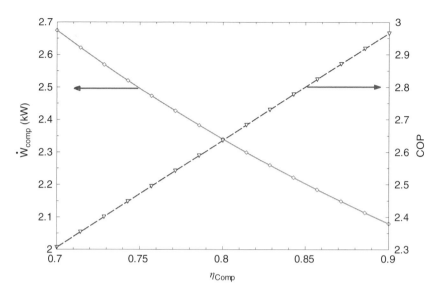

Figure 5.4 Effect of compressor isentropic efficiency on the compressor consumption power and system COP for $P_1 = 100$ kPa.

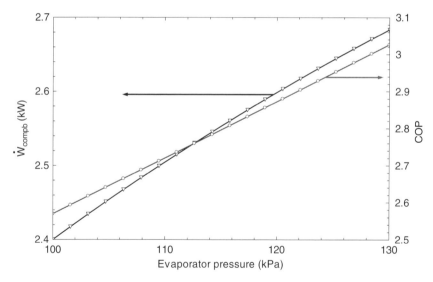

Figure 5.5 Effect of evaporator pressure on compressor power consumption and system COP for $\eta_{comp} = 0.78$.

The maximum COP of a refrigeration cycle operating between temperature limits of T_L and T_H based on the Carnot refrigeration cycle can be expressed as

$$\text{COP}_{\text{Carnot}} = \frac{T_L}{T_H - T_L} = \frac{1}{T_H/T_L - 1} \tag{5.8}$$

As reflected in their lower COPs, real refrigeration systems are not as efficient as the Carnot refrigerator, due to losses. Equation 5.8 suggests that a smaller temperature

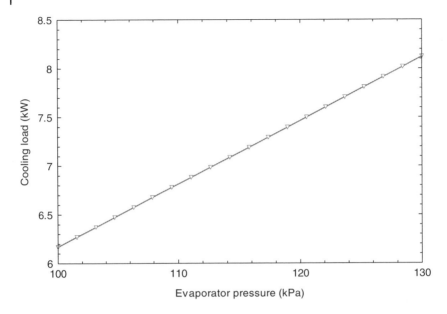

Figure 5.6 Effect of evaporator pressure on the system cooling load for $\eta_{comp} = 0.78$.

difference between the heat sink and the heat source $(T_H - T_L)$ provides a more efficient refrigeration system, that is, one with a greater COP.

Exergy analysis is a useful tool for determining the location and magnitude of exergy waste emissions and destructions for a refrigeration system and its components. It can thereby help determine components where improvements are necessary and how they should be prioritized. The exergy destruction rate in a component can be determined from an exergy rate balance for it. The exergy destruction rate can also be determined as a function of the entropy generation rate due to irreversibilities as follows:

$$\dot{Ex}_{dest} = T_0\dot{S}_{gen} \tag{5.9}$$

where T_0 is the dead state temperature or reference environment temperature. In a refrigerator, T_0 is usually equal to the temperature of the high temperature medium T_H, or the ambient temperature.

The exergy destruction rates and exergy efficiencies for the main components of a refrigeration cycle are described below.

Compressor:

$$\dot{Ex}_{in} - \dot{Ex}_{out} - \dot{Ex}_{dest,1-2} = 0$$
$$\dot{Ex}_{dest,1-2} = \dot{Ex}_{in} - \dot{Ex}_{out}$$
$$\dot{Ex}_{dest,1-2} = \dot{W} + \dot{Ex}_1 - \dot{Ex}_2$$
$$= \dot{W} - \Delta\dot{Ex}_{12} = \dot{W} - \dot{m}[h_2 - h_1 - T_0(s_2 - s_1)]$$
$$= \dot{W} - \dot{W}_{rev} \tag{5.10}$$

or

$$\dot{Ex}_{dest,1-2} = T_0\dot{S}_{gen,1-2} = \dot{m}T_0(s_2 - s_1) \tag{5.11}$$

$$\eta_{ex,comp} = \frac{\dot{W}_{rev}}{\dot{W}} = 1 - \frac{\dot{Ex}_{dest,1\text{-}2}}{\dot{W}} \tag{5.12}$$

Condenser:

$$\dot{Ex}_{dest} = \dot{Ex}_{in} - \dot{Ex}_{out}$$
$$\dot{Ex}_{dest} = (\dot{Ex}_2 - \dot{Ex}_3) - \dot{Ex}_{\dot{Q}_H}$$
$$= \dot{m}\left[h_2 - h_3 - T_0(s_2 - s_3)\right] - \dot{Q}_H\left(1 - \frac{T_0}{T_H}\right) \tag{5.13}$$

or

$$\dot{Ex}_{dest,2\text{-}3} = T_0 \dot{S}_{gen,2\text{-}3} = \dot{m}T_0\left(s_3 - s_2 + \frac{q_H}{T_H}\right) \tag{5.14}$$

$$\eta_{ex,cond} = \frac{\dot{Ex}_{\dot{Q}_H}}{\dot{Ex}_2 - \dot{Ex}_3} = \frac{\dot{Q}_H\left(1 - \frac{T_0}{T_H}\right)}{\dot{m}\left[h_2 - h_3 - T_0(s_2 - s_3)\right]} = 1 - \frac{\dot{Ex}_{dest}}{\dot{Ex}_2 - \dot{Ex}_3} \tag{5.15}$$

Expansion valve:

$$\dot{Ex}_{dest,3\text{-}4} = \dot{Ex}_{in} - \dot{Ex}_{out}$$
$$\dot{Ex}_{dest,3\text{-}4} = \dot{Ex}_3 - \dot{Ex}_4 = \dot{m}\left[h_3 - h_4 - T_0(s_3 - s_{43})\right] \tag{5.16}$$

or

$$\dot{Ex}_{dest,3\text{-}4} = T_0 \dot{S}_{gen,3\text{-}4} = \dot{m}T_0(s_4 - s_3) \tag{5.17}$$

$$\eta_{ex,exp\ valve} = 1 - \frac{\dot{Ex}_{dest,3\text{-}4}}{\dot{Ex}_3 - \dot{Ex}_4} = 1 - \frac{\dot{Ex}_3 - \dot{Ex}_4}{\dot{Ex}_3 - \dot{Ex}_4} \tag{5.18}$$

Evaporator:

$$\dot{Ex}_{dest} = \dot{Ex}_{in} - \dot{Ex}_{out}$$
$$\dot{Ex}_{dest} = -\dot{Ex}_{\dot{Q}_L} + \dot{Ex}_4 - \dot{Ex}_1$$
$$\dot{Ex}_{dest} = (\dot{Ex}_4 - \dot{Ex}_1) - \dot{Ex}_{\dot{Q}_L}$$
$$= \dot{m}[h_4 - h_1 - T_0(s_4 - s_1)] - \left[-\dot{Q}_L\left(1 - \frac{T_0}{T_L}\right)\right] \tag{5.19}$$

or

$$\dot{Ex}_{dest,4\text{-}1} = T_0 \dot{S}_{gen,4\text{-}1} = \dot{m}T_0\left(s_1 - s_4 - \frac{q_L}{T_L}\right) \tag{5.20}$$

$$\eta_{ex,evap} = \frac{\dot{Ex}_{\dot{Q}_L}}{\dot{Ex}_1 - \dot{Ex}_4} = \frac{-\dot{Q}_L\left(1 - \frac{T_0}{T_L}\right)}{\dot{m}[h_1 - h_4 - T_0(s_1 - s_4)]} = 1 - \frac{\dot{Ex}_{dest,4\text{-}1}}{\dot{Ex}_1 - \dot{Ex}_4} \tag{5.21}$$

The total exergy destruction rate for the cycle can be determined by adding exergy destruction rates for all components:

$$\dot{Ex}_{dest,\ total} = \dot{Ex}_{dest,1\text{-}2} + \dot{Ex}_{dest,2\text{-}3} + \dot{Ex}_{dest,3\text{-}4} + \dot{Ex}_{dest,4\text{-}1} \tag{5.22}$$

The second law efficiency (or exergy efficiency) of the cycle is defined as

$$\eta_{II} = \frac{\dot{Ex}_{\dot{Q}_L}}{\dot{W}} = \frac{\dot{W}_{min}}{\dot{W}} = 1 - \frac{\dot{Ex}_{dest,total}}{\dot{W}} \tag{5.23}$$

Substituting $\dot{W} = \frac{\dot{Q}_L}{\text{COP}}$ and $\dot{Ex}_{\dot{Q}_L} = -\dot{Q}_L\left(1 - \frac{T_0}{T_L}\right)$ into this efficiency expression yields:

$$\eta_{\text{II}} = \frac{\dot{Ex}_{\dot{Q}_L}}{\dot{W}} = \frac{-\dot{Q}_L\left(1 - \frac{T_0}{T_L}\right)}{\frac{\dot{Q}_L}{\text{COP}}} = -\dot{Q}_L\left(1 - \frac{T_0}{T_L}\right)\frac{\text{COP}}{\dot{Q}_L} = \frac{\text{COP}}{\frac{T_L}{T_0 - T_L}} = \frac{\text{COP}}{\text{COP}_{\text{Carnot}}}$$

(5.24)

Example 5.2: A refrigeration system operates on the ideal refrigeration cycle with R134-a as a working fluid. The evaporator and condenser pressures are 200 kPa and 2000 kPa, respectively. The temperatures of the low temperature and high temperature media are −9°C and 27°C, respectively. The heat rejection rate by the condenser is 18 kW. Determine (a) the mass flow rate of R134-a, (b) the compressor power consumption and the cycle COP, and (c) the exergy efficiency of the cycle and the total exergy destruction rate. Also, conduct a parametric study to determine the effect of varying evaporator and condenser pressure on COP, exergy efficiency and total exergy destruction rate.

Solution: A *T-s* diagram of the cycle follows:

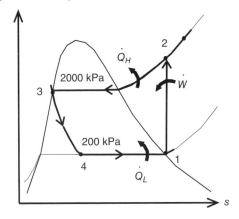

Using thermodynamic tables for R134-a, we find:

$$P_1 = 200 \text{ kPa}, \ x_1 = 1 \ \rightarrow \ h_1 = 244.5\frac{\text{kJ}}{\text{kg}}, \ s_1 = 0.937\frac{\text{kJ}}{\text{kg K}}, \ v_1 = 0.0998\frac{\text{m}^3}{\text{kg}}$$

Since the compressor is ideal, that is, isentropic, $s_2 = s_1$

$$P_2 = 2000 \text{ kPa}, \ s_2 = s_1 \ \rightarrow h_2 = 292.5\frac{\text{kJ}}{\text{kg}}$$

$$P_3 = P_2, \ x_3 = 0 \ \rightarrow h_3 = 151.8\frac{\text{kJ}}{\text{kg}}, \ s_3 = 0.5251\frac{\text{kJ}}{\text{kg K}}$$

$$h_4 = h_3 = 151.8\frac{\text{kJ}}{\text{kg}}, \ P_4 = P_1 \ \rightarrow s_4 = 0.5854\frac{\text{kJ}}{\text{kg K}}$$

a) $\qquad \dot{Q}_{\text{cond}} = \dot{m}(h_2 - h_3) \rightarrow \dot{m} = \frac{\dot{Q}_{\text{cond}}}{(h_2 - h_3)} = \frac{18 \text{ kW}}{292.5 - 151.8} = 0.128\frac{\text{kg}}{\text{s}}$

b)
$$\dot{W} = \dot{m}(h_2 - h_1) = 6.14 \text{ kW}$$

$$COP = \frac{\dot{Q}_L}{\dot{W}} = \frac{\dot{m}(h_1 - h_4)}{\dot{W}} = 1.93$$

c) The exergy rate associated with the heat transfer rate from the low temperature medium is:

$$\dot{Ex}_{Q_L} = -\dot{Q}_L \left(1 - \frac{T_L}{T_H}\right) = -\dot{m}(h_1 - h_4)\left(1 - \frac{T_L}{T_H}\right) = 1.61 \text{ kW} \tag{5.25}$$

The cycle exergy efficiency is

$$\psi = \frac{\dot{Ex}_{Q_L}}{\dot{W}} = 0.26 \tag{5.26}$$

The total exergy destruction rate in the cycle is the difference between the rate at which exergy is supplied, which is the compressor power in this case, and the exergy rate of the heat transfer rate from the low temperature medium. Thus:

$$\dot{Ex}_{des} = \dot{W} - \dot{Ex}_{Q_L} = 4.52 \text{ kW} \tag{5.27}$$

The parametric study results are shown below (see Figures 5.7–5.10). It is seen that the evaporator pressure has a positive effect on both cycle exergy efficiency and COP. That is, the higher is the evaporator pressure, the greater is the system performance. When the evaporator pressure increases, the cooling capacity of the refrigeration system increases while the compressor power consumption decreases, but the combination of these two effects leads to a net increase in COP.

However, raising the condenser pressure has a negative effect on system performance. That is, an increase in condenser pressure while other parameters are fixed leads to a decrease in both the cycle exergy efficiency and the COP. In this example, the condenser

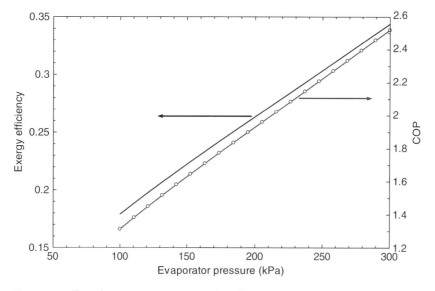

Figure 5.7 Effect of evaporator pressure on the refrigeration cycle exergy efficiency and COP.

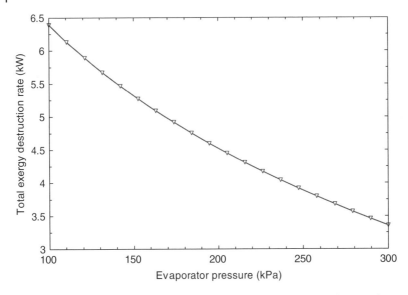

Figure 5.8 Effect of evaporator pressure on the refrigeration cycle total exergy destruction rate.

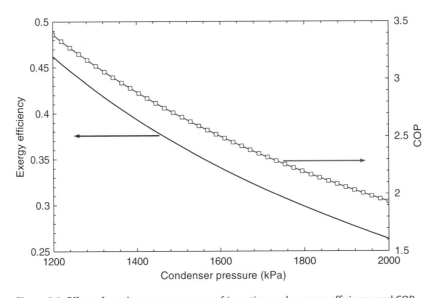

Figure 5.9 Effect of condenser pressure on refrigeration cycle exergy efficiency and COP.

heat rejection rate is fixed, so an increase in condenser pressure results in an increase in compressor pressure ratio, causing more power to be required. An increase in compressor power consumption reduces the COP of the system.

5.2.3 Optimization

Several parameters significantly affect the performance of vapor compression refrigeration cycles. Tools such as exergy analysis can help us develop ways to improve system performance, but that sometimes comes at an expense, for example, more efficient

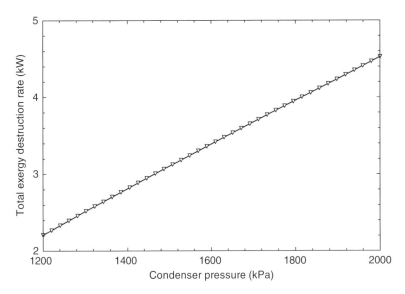

Figure 5.10 Effect of condenser pressure on total exergy destruction rate of the refrigeration cycle.

devices may have higher costs. Addressing such trade-offs rationally and beneficially is where multi-objective optimization is particularly important in that it allows technical and economic performance to be simultaneously addressed. In this section, we apply a genetic algorithm optimization to the VCR cycle introduced earlier. Two objective functions are considered for multi-objective optimization: system exergy efficiency and total cost rate. The exergy efficiency is maximized and the total cost rate minimized, while satisfying several practical constraints. The objective functions can be written as follows:

$$\text{OF I} = \text{Exergy efficiency} = \psi = \frac{\dot{Ex}_{Q_L}}{\dot{W}} = \frac{-\dot{Q}_L \left(1 - \frac{T_L}{T_H}\right)}{\dot{W}} \tag{5.28}$$

$$\text{OF II} = \dot{Z}_{\text{tot}} = \sum_k \dot{Z}_k \tag{5.29}$$

Here, $\dot{Z}_k$ can be expressed as follows:

$$\dot{Z}_k = \frac{Z_k CRF\varphi}{N \times 3600}$$

Also, Z denotes the equipment purchase cost and CRF the previously defined capital recovery factor, which is a function of interest rate i and the total operating period of the system in years n. Correlations for the purchase costs of the main components of the system are presented in Table 5.1.

5.2.3.1 Decision Variables

The decision variables (design parameters) in this optimization are evaporator temperature (T_{eva}), condenser temperature (T_{cond}), evaporator pinch point temperature (PPTD_1), condenser pinch point temperature (PPTD_2), and compressor isentropic efficiency (η_{comp}). Even though the decision variables may vary as part of the optimization procedure, each decision variable is normally restricted to a reasonable range. The constraints applied here and the reasons of their use are listed in Table 5.2, based on previous analyses [2, 3].

Table 5.1 Purchase cost correlations for selected components of refrigeration system.

Component	Purchase cost correlation
Evaporator	$Z_{eva}(\$) = 309.143\, A_{eva}(m^2) + 231.95$
Compressor	$Z_{comp}(\$) = \dfrac{573\, \dot{m}_{ref}\left(\dfrac{kg}{s}\right)}{0.8996 - \eta_{comp}}\left(\dfrac{P_{cond}}{P_{eva}}\right)\log\left(\dfrac{P_{cond}}{P_{eva}}\right)$
Condenser and expansion valve	$Z_{cond}(\$) = 516.621\, A_{cond}(m^2) + 268.45$

Table 5.2 Decision variables and their ranges for optimization of refrigeration system.

Decision variable	Lower Range	Upper Range	Reason
T_{eva} (K)	248	258	Commercial availability
T_{cond} (K)	313	325	Commercial availability
$PPTD_1$ (K)	5	10	Heat transfer limit
$PPTD_2$ (K)	5	10	Heat transfer limit
η_{comp} (−)	0.6	0.85	Commercial availability

5.2.3.2 Optimization Results

Figure 5.11 shows the Pareto frontier solution for the vapor compression refrigeration system in Figure 5.1. The objective functions for the multi-objective optimization are defined in Equations 5.28 and 5.29. It is observed in Figure 5.11 that, while the total exergy efficiency of the cycle increases to about 28%, the total cost rate increases slightly.

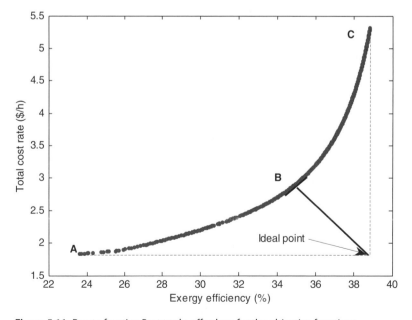

Figure 5.11 Pareto frontier: Best trade-off values for the objective functions.

An increase in total exergy efficiency from 28% to 34% corresponds to a larger increase in the cost rate of product. In addition, an increase in the exergy efficiency from 34% to a higher value leads to a significant increase in the total cost rate of the system. As shown in Figure 5.11, the maximum exergy efficiency exists at design point C (39%), while the total cost rate of products is the greatest at this point (5.4 $/hr).

The minimum value for the total product cost rate occurs at design point A in Figure 5.11 and is 1.84 $/hr. Design point A is the optimal situation when total cost rate of the product is the sole objective function, while design point C is the optimum point when exergy efficiency is the sole objective function. In multi-objective optimization, a process of decision-making for selection of the final optimal solution from the available solutions is required. Such decision-making is usually performed with the aid of a hypothetical point in Figure 5.11 (the ideal point), at which both objectives have their optimal values independent of the other objectives. It is clear that it is impossible to have both objectives at their optimum point simultaneously and, as shown in Figure 5.11, the ideal point is not a solution located on the Pareto frontier. The closest point to the ideal point that is on the Pareto frontier might be considered as a desirable final solution.

Nevertheless, in this case, the Pareto optimum frontier exhibits a weak equilibrium, that is, a small change in exergy efficiency by varying the operating parameters causes a large variation in the total product cost rate. Therefore, the ideal point cannot be utilized for decision-making in this problem. In selection of the final optimum point, it is desired to achieve a better magnitude for each objective than its initial value for the base case problem. Note that in multi-objective optimization and the Pareto solution, each point can be utilized as the optimized point. Therefore, the selection of the optimum solution depends on the preferences and criteria of decision makers, suggesting that each may select a different point as for the optimum solution depending on their needs. Table 5.3 shows all the design parameters for points A-C.

To better understand the variations of all design parameters, the scattered distribution of the design parameters are shown in Figures 5.12 to 5.16. The results demonstrate that the evaporator temperature tends to approach its higher values in the optimization, since this results in an increase in exergy efficiency. Similarly, as discussed in previous examples (e.g., Example 5.1), condenser temperature tends to approach its lower values. As explained earlier, lowering the condenser temperature leads to an increase in the exergy efficiency of the system. Unlike the evaporator and condenser temperatures, the pinch point temperatures have a scattered distribution between their allowable ranges. This is due to the fact that an increase in pinch point temperature leads in part to a decrease in the heat transfer mechanism in the heat exchanger, which results in a reduction in cooling load of the refrigeration system. However, an increase in this parameter simultaneously leads to a decrease in heat transfer surface area,

Table 5.3 Optimized values for design parameters of the system for three points on the Pareto frontier from multi-objective optimization.

Point	Ψ (%)	Total cost rate ($/h)	T_{eva} (K)	T_{cond} (K)	$PPTD_1$ (K)	$PPTD_2$ (K)	η_{comp} (−)
A	23.62	1.83	257.95	313.37	8.41	9.76	0.60
B	34.94	2.92	257.99	313.03	5.01	5.03	0.76
C	38.80	5.32	257.98	313.02	5.02	5.02	0.85

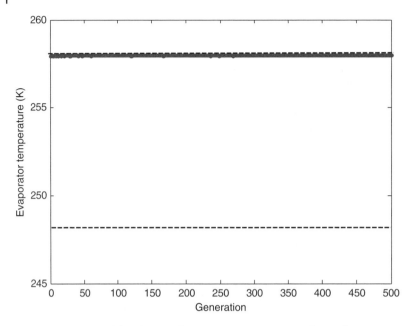

Figure 5.12 Scattered distribution of evaporator temperature with population in Pareto frontier.

which reduces the total cost of heat exchanger. These are competing effects on both objectives, which explains why a scattered distribution is obtained for these design parameters. Similar results are obtained for the compressor isentropic efficiency, with an increase in this design parameter improving the system performance by increasing the system COP while increasing the compressor purchase cost. When parameters exhibit scattered distributions in their allowable domains, these parameters are likely to have important effects on the trade-off between exergy efficiency and total cost rate. Having design parameters near their extreme values suggests that they do not exhibit a conflict between two objective functions, and that increasing those design parameters leads to an improvement of both objective functions.

For an enhanced understanding of the multi-objective optimization performed, a sensitivity analysis is carried out. The effects of each of the design parameters for points *A-C* on both objective functions are investigated. Figure 5.17 shows the effects of varying evaporator temperature on system exergy efficiency and total cost rate. An increase in evaporator temperature is seen to have a positive effect on both objective functions, that is, to decrease the total cost rate of the system while increasing its exergy efficiency. This observation is due in large part to the fact that an increase in this temperature is equivalent to an increase in evaporator pressure. Then, the higher the evaporator pressure, the lower the compressor power consumption, eventually resulting in an increase in system exergy efficiency. At the same time, a decrease in compressor power requirement leads to a decrease in the purchase cost of the compressor. The results for these parameters were already shown (see Figure 5.12), with the optimized values tending to reach their maximum values.

Figure 5.18 shows the effect of varying condenser temperature on both objective functions. An increase in condenser temperature is observed to lead to a decrease in exergy efficiency and an increase in total cost rate. As explained previously, an increase

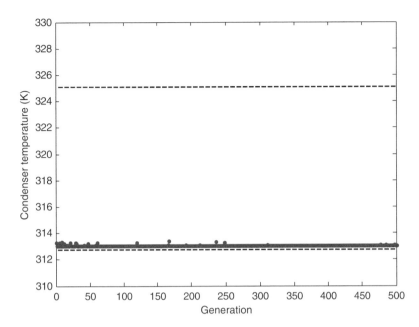

Figure 5.13 Scattered distribution of condenser temperature with population in Pareto frontier.

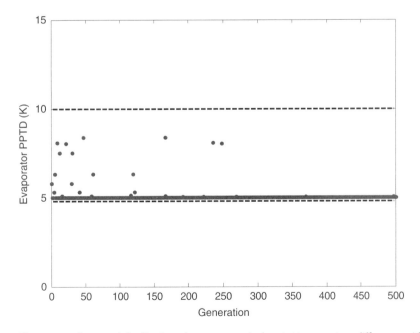

Figure 5.14 Scattered distribution of evaporator pinch point temperature difference with population in Pareto frontier.

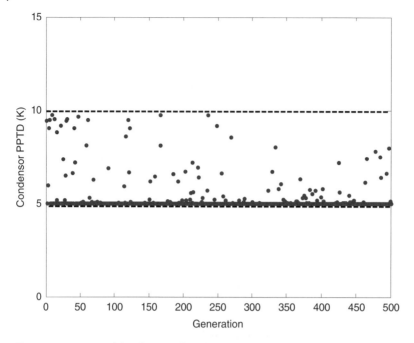

Figure 5.15 Scattered distribution of condenser pinch point temperature difference with population in Pareto frontier.

in condenser temperature is equivalent to an increase in condenser pressure, which requires more compressor power, decreasing the system exergy efficiency. Furthermore, an increase in condenser temperature results in an increase in surface heat transfer area for the heat exchanger in the condenser, ultimately raising the total cost level of the system. Hence, the scattered distribution of optimized points for this design parameter tend to reach their lower values. An increase in condenser and evaporator pinch point temperatures negatively affects the system exergy efficiency (see Figures 5.19 and 5.20). The higher the pinch point temperature, the lower the heat recovery, yielding a greater exergy destruction rate and lower system exergy efficiency. However, an increase in pinch point temperature results in a decrease in evaporator and condenser heat transfer surface area, and directly affects the purchase cost. Specifically, the lower the purchase cost, the lower the total cost rate of the system. Since an increase in pinch point temperatures has both positive and negative effects on the objective functions, the optimized values exhibit a scattered distribution in their domain as shown in Figures 5.14 and 5.15.

5.3 Cascade Refrigeration Systems

In some industrial applications, temperatures lower than a single vapor compression can provide are required. That is in part because a large temperature difference implies a large pressure range in the vapor compression cycle, which leads to poor compressor performance. One way to manage such situations is to use refrigeration in stages, where two or more refrigeration cycles operate in series. Such systems are called cascade refrigeration systems.

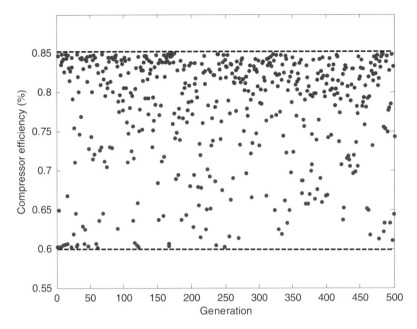

Figure 5.16 Scattered distribution of compressor isentropic efficiency with population in Pareto frontier.

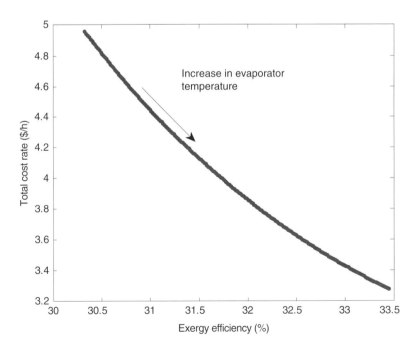

Figure 5.17 Effect of evaporator temperature on both objective functions.

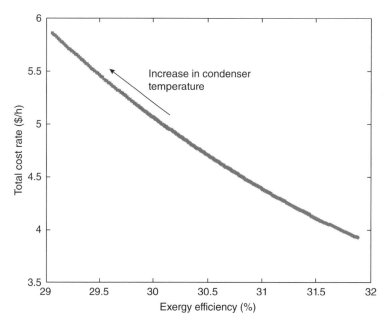

Figure 5.18 Effect of condenser temperature on both objective functions.

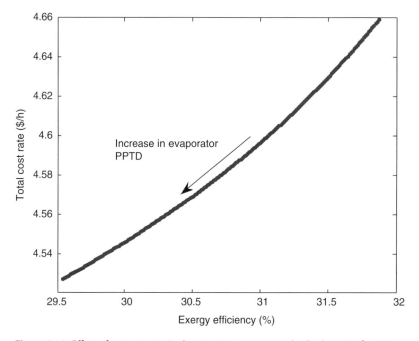

Figure 5.19 Effect of evaporator pinch point temperature on both objective functions.

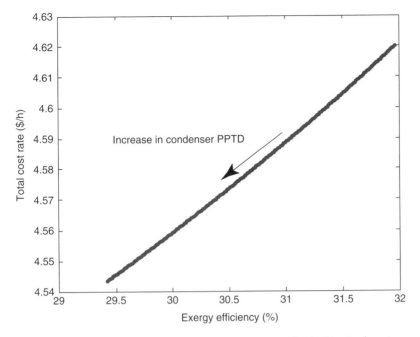

Figure 5.20 Effect of condenser pinch point temperature on both objective functions.

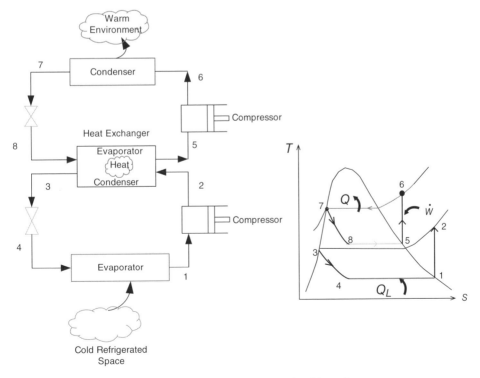

Figure 5.21 Schematic and *T-s* diagram of a two stage cascade refrigeration system.

Figure 5.21 illustrates a two stage cascade refrigeration system. Two refrigeration cycles are connected via a heat exchanger in the middle, which serves simultaneously as the evaporator for the topping cycle and the condenser for the bottoming cycle. Assuming a well-insulated heat exchanger and negligible kinetic and potential energies, the heat absorbed in the topping cycle is equal to the heat rejected from the bottoming cycle. Writing an energy balance for a control volume around the heat exchanger at steady state conditions yields:

$$\dot{m}_5(h_5 - h_8) = \dot{m}_2(h_2 - h_3) \quad \rightarrow \quad \frac{\dot{m}_5}{\dot{m}_2} = \frac{h_2 - h_3}{h_5 - h_8} \tag{5.30}$$

With this relation, the ratio of mass flow rates through each cycle can be calculated. The COP of this cascade system can be evaluated as:

$$COP = \frac{\dot{Q}_L}{\dot{W}_{net}} = \frac{\dot{m}_1(h_1 - h_4)}{\dot{m}_5(h_6 - h_5) + \dot{m}_2(h_2 - h_1)} \tag{5.31}$$

In the above cascade system, it is assumed that the refrigerant in both cycles is same and that no mixing occurs. However, different refrigerants can be used depending on the conditions and applications. In cascade refrigeration systems where the refrigerants are same, the heat exchanger is replaced by a flash chamber as it has better heat transfer performance. Such a system is called a multistage compression refrigeration system. A two stage cascade refrigeration system with a flash chamber is shown in Figure 5.22.

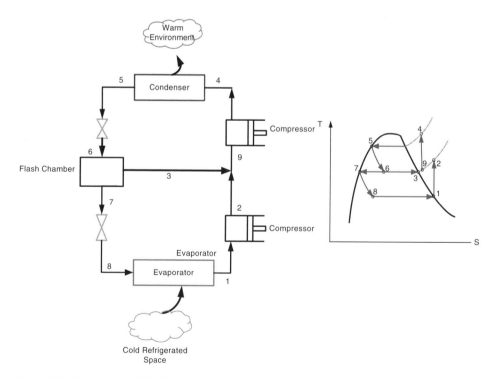

Figure 5.22 Schematic and *T-s* diagram of a two stage cascade refrigeration system with flash chamber.

Example 5.3: In the two stage compression refrigeration cycle with a flash chamber shown in Figure 5.22, the working fluid is R134a, and the minimum and maximum pressures of the cycle are 0.1 MPa and 1.4 MPa respectively. The refrigerant leaves the condenser as a saturated liquid and is throttled to a flash chamber operating pressure at 0.4 MPa. The refrigerant exiting the low pressure compressor at 0.4 MPa is also directed to the flash chamber. The vapor in the flash chamber is then compressed to the condenser pressure by a high pressure compressor, and the liquid is throttled to the evaporator pressure. The refrigerant exits the evaporator as a saturated vapor, and both compressors are assumed to be isentropic. Determine (a) the fraction of refrigerant that evaporates as it is throttled to the flash chamber, (b) the cooling load of the system if the mass flow rate through the condenser is 0.25 kg/s, and (c) the COP of the system. Also, (d) conduct a parametric study to determine the effect of flash chamber pressure on the system performance for various working fluids, (e) investigate the effect of the isentropic efficiencies of the compressors as well as the evaporator and condenser pressures on the system performance, and (f) optimize the COP of the system and determine the optimal design parameters.

Solution: Assumptions: Steady operating conditions exist. The changes in kinetic and potential energies are negligible. The flash chamber is adiabatic.

To calculate the thermodynamic properties at each state point (see Figure 5.23), Engineering Equation Solver (EES) is used. The values for the given points are:

$$h_1 = 234.4\frac{kJ}{kg}, \; h_2 = 262.6\frac{kJ}{kg}, \; h_3 = 255.5\frac{kJ}{kg}$$

$$h_5 = 127.2\frac{kJ}{kg}, \; h_6 = 127.2\frac{kJ}{kg}, \; h_7 = 63.9\frac{kJ}{kg}, \; h_8 = 63.9\frac{kJ}{kg}$$

a) The fraction of the refrigerant that evaporates is simply the quality at state 6, which can be calculated as:

$$x_6 = \frac{h_6 - h_f}{h_{fg}} = \frac{127.2 - 63.9}{191.6} = 0.33$$

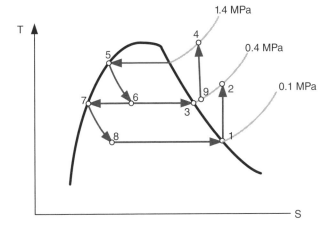

Figure 5.23 *T-s* diagram of the two stage compression refrigeration cycle with a flash chamber considered in the example.

b) To calculate the cooling load of the system, the specific enthalpy of point 9 is required. The relevant mass and energy balance equations are used for that calculation:

The mass rate balance equation: $\dot{m}_9 = \dot{m}_3 + \dot{m}_2$

The energy rate balance equation: $\dot{m}_3 h_3 + \dot{m}_2 h_2 = \dot{m}_9 h_9$

Dividing the energy rate balance equation by $\dot{m}_9$ yields:

$$h_9 = \frac{\dot{m}_3}{\dot{m}_9}h_3 + \frac{\dot{m}_2}{\dot{m}_9}h_2 \rightarrow h_9 = \frac{\dot{m}_3}{\dot{m}_3 + \dot{m}_2}h_3 + \frac{\dot{m}_2}{\dot{m}_3 + \dot{m}_2}h_2$$

Here, $\frac{\dot{m}_3}{\dot{m}_3 + \dot{m}_2}$ is equal to the fraction of the refrigerant evaporated in the flash chamber, which was already determined in part a, that is,

$$\rightarrow h_9 = x_6 h_3 + (1 - x_6)h_2 \rightarrow h_9 = 260.3\frac{kJ}{kg}$$

$$P_9 = 0.4 \text{ MPa} \,\&\, h_9 = 260.3\frac{kJ}{kg} \rightarrow s_9 = 0.94\frac{kJ}{kgK}$$

$$P_4 = 1.4 \text{ MPa and } s_4 = s_9 = 0.94\frac{kJ}{kgK} \rightarrow h_4 = 267\frac{kJ}{kg}$$

The mass flow rate through the condenser is $\dot{m}_6 = 0.25$ kg/s

The mass flow rate through the evaporator is $(1 - x_6)\dot{m}_6 = 0.167\frac{kg}{s}$

$$\dot{Q}_L = \dot{m}_1(h_1 - h_8) = 28.5 \text{ kW}$$

$$\dot{W}_{net} = \dot{W}_{compI} + \dot{W}_{compII} = \dot{m}_6(h_4 - h_9) + \dot{m}_1(h_2 - h_1) = 11.4 \text{ kW}$$

c) The COP of the system can be written as:

$$COP = \frac{\dot{Q}_L}{\dot{W}_{net}} = 2.5$$

d) EES is used to simulate the system and conduct a parametric study for different working fluids, and the results are shown in Figure 5.24. Here, R134a, R22 and R11 are considered as possible working fluids. Selecting proper working fluids has a significant effect on refrigeration systems. Two important parameters that need to be considered when selecting a working fluid are the temperature of the refrigerated space and the environment where the refrigerant exchanges heat. With the data given in this example, the results show that R134a is the best working fluid, among those considered, for the high flash chamber pressure. An increase in flash chamber pressure has a positive effect on COP of the system until it reaches its maximum value, and it has a negative effect above the optimal point. This result is due to the fact that an increase in flash chamber pressure leads to an increase in the cooling load of the system and a decrease in net compressor work rate. An increase in flash chamber pressure above a particular value results in an increase in compressor work rate, which eventually results in a decrease in the COP of the system. This is why an optimized value of this parameter is important.

e) To investigate the effects of varying other design parameters on system performance, EES computer code is developed with the ability of conducting a parametric study. Since the system using R134a has a higher COP compared the system with other working fluids, the parametric study is conducted for this working fluid only. However, it is noted that other working fluids exhibit similar behaviors. Figure 5.25 shows the effect of compressor isentropic efficiency on the COP of the system. It can be seen

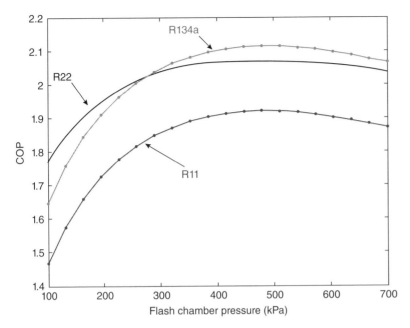

Figure 5.24 Effect of flash chamber pressure on COP of the system for three working fluids ($\eta_{\text{comp}} = 0.85$, $P_1 = 100$ kPa, $P_4 = 1400$ kPa).

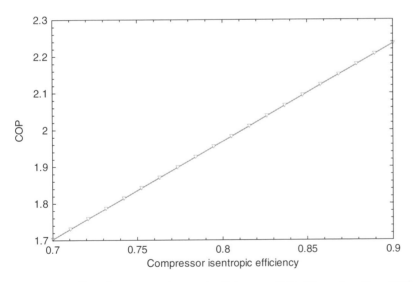

Figure 5.25 Effect of compressor isentropic efficiency on COP of the system ($P_1 = 100$ kPa, $P_4 = 1400$ kPa, $P_6 = 400$ kPa).

that an increase in compressor isentropic efficiency leads to an increase in the system COP. The higher the compressor isentropic efficiency, the lower the compressor work rate required, which results in an increase in system COP. Figure 5.26 shows the effect of evaporator pressure on the cooling load of the system and the net compressor work rate. The results show that an increase in evaporator pressure increases

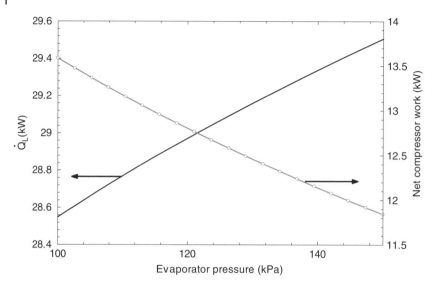

Figure 5.26 Effect of evaporator pressure on cooling load and net compressor work rate of the system ($\eta_{comp} = 0.85$, $P_4 = 1400$ kPa, $P_6 = 400$ kPa).

the system cooling load while decreasing the net compressor work rate. Note that a similar trend was obtained earlier for single stage vapor compression refrigeration systems. Figure 5.27 shows the effect of evaporator pressure on the COP of the system. The higher the evaporator pressure, the higher the system COP. Figures 5.28 and 5.29 display the effect of condenser pressure on the system cooling load, the net compressor work rate, and the COP of the system.

Figure 5.28 shows that an increase in condenser pressure increases the compressor work rate, because the higher pressure requires more work. Since the cooling load of the system decreases and the compressor work rate increases, the COP of the system decreases as shown in Figure 5.29. The parametric study demonstrates that an optimization will be likely to help in determining the optimal design parameters of the system.

f) **Optimization**

The parametric study shows that the compressor isentropic efficiency, evaporator and condenser pressure, and flash chamber pressure in the two stage vapor compression refrigeration system shown in Figure 5.22 all have major effects on system performance. But an increase in these parameters has differing effects on the system COP, so a genetic algorithm is applied to determine the optimal values for the system. The objective function here is:

$$\text{OF:} \; COP = \frac{\dot{Q}_L}{\dot{W}_{net}}$$

The design parameters for the optimization and their practical ranges are listed in Table 5.4.

The constraints listed in Table 5.4 are to be satisfied when applying the genetic algorithm. The optimal design parameters and other system performance data for the two stage compression refrigeration system with flash chamber are listed in Table 5.5.

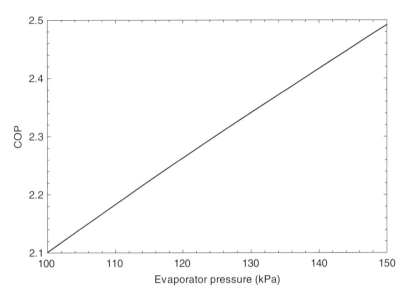

Figure 5.27 Effect of evaporator pressure on the COP of the system ($\eta_{comp} = 0.85$, $P_4 = 1400$ kPa, $P_6 = 400$ kPa).

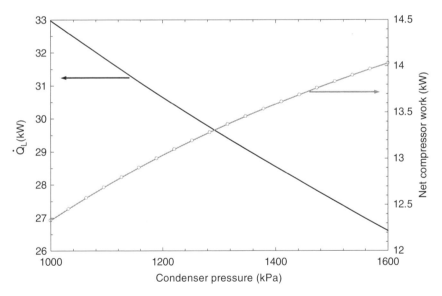

Figure 5.28 Effect of condenser pressure on cooling load and net compressor work rate of the system ($\eta_{comp} = 0.85$, $P_1 = 400$ kPa, $P_6 = 400$ kPa).

5.4 Absorption Chiller

An absorption chiller is a heat driven refrigeration machine that has several similarities to vapor compression refrigeration systems, but also some important differences. The main difference is the substitution of the compressor by a combination of an absorber, a solution pump, and a generator. When waste heat is available and cooling is required,

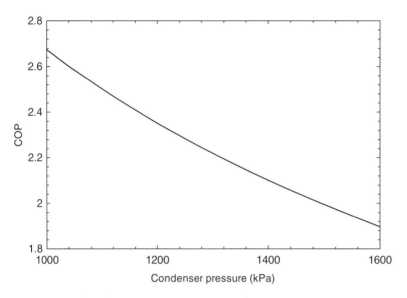

Figure 5.29 Effect of condenser pressure on COP of the system ($\eta_{comp} = 0.85$, $P_1 = 400$ kPa, $P_6 = 400$ kPa).

Table 5.4 Design parameters and their ranges for optimization study.

Design Parameter	Lower Bound	Upper Bound
Compressor isentropic efficiency (−)	0.75	0.9
Evaporator pressure (kPa)	100	150
Condenser pressure (kPa)	1000	1600
Flash chamber pressure (kPa)	100	700

Table 5.5 Optimized design parameters.

Design Parameter	Optimized Value
Evaporator pressure (kPa)	150
Condenser pressure (kPa)	1214
Flash chamber pressure (kPa)	470
Compressor isentropic efficiency	0.89
COP	3
Cooling load (kW)	31.58
Net compressor work rate (kW)	10.5
Condenser rejected heat (kW)	42.09

Figure 5.30 Single effect LiBr–water absorption chiller.

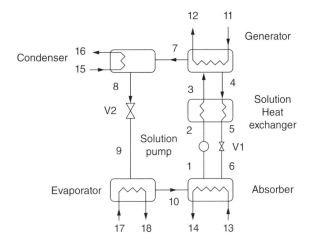

an absorption chiller can be an advantageous alternative to a vapor compression refrigerator.

Two fluids are used in an absorption chiller: the absorbent and the refrigerant. The refrigerant is chemically and physically absorbed by the absorber for the purpose of heat transfer. Three main flows are circulated inside the chiller, namely a strong solution, a weak solution, and the refrigerant.

Figure 5.30 shows a typical water lithium bromide (H_2O–LiBr) single effect absorption chiller, which consists of a generator, a condenser, an evaporator, an absorber, a solution heat exchanger, a solution pump, and two valves. Water is the refrigerant and lithium bromide the absorber. The strong solution of H_2O–LiBr separates from the absorber after being heated by the heat released by the weak solution of H_2O–LiBr in the solution heat exchanger. This facilitates the separation process in generator, after which the separated water is conveyed to the condenser and the H_2O–LiBr to the absorber. In the condenser, water in a saturated vapor state is converted to a saturated liquid or subcooled by releasing heat to the environment.

The water expands through a valve, decreasing its pressure to the evaporator pressure, and then enters the evaporator. There, the cooling effect occurs, with heat absorbed from the external water circuit; the internal water (refrigerant) becomes a saturated or superheated vapor. The produced vapor is transported to the absorber and mixed with the weak solution from the generator, while rejecting heat to the environment in order to obtain a strong solution of H_2O–LiBr. This strong solution passes through the solution heat exchanger and is input to the generator, completing the absorption cooling cycle.

5.4.1 Thermodynamic Analysis

An energy balance is applied to each component of the absorption chiller in order to determine the performance and values of the thermodynamic properties of the working fluid at all states. The fact that the refrigerant and the absorbent, and their compositions, differ in the generator and the absorber, is accommodated in the mass balance and energy balance equations for these two components.

Generator:

Mass rate balance: $\dot{m}_{ws}X_{ws} + \dot{m}_r = \dot{m}_{ss}X_{ss}$

Energy rate balance: $\dot{m}_3 h_3 + \dot{Q}_{gen} = \dot{m}_4 h_4 + \dot{m}_7 h_7$

Here, $\dot{Q}_{gen}$ is the heat input rate to the generator (in kW); X is the concentration; $\dot{m}_{ws} = \dot{m}_6$ is the mass flow rate of the weak solution (in kg/s); $\dot{m}_{ss} = \dot{m}_3$ is the mass flow rate of the strong solution (in kg/s); and $\dot{m}_r = \dot{m}_7$ is the mass flow rate of the refrigerant (in kg/s).

Absorber:

Mass rate balance: $\dot{m}_{ws} X_{ws} + \dot{m}_r = \dot{m}_{ss} X_{ss}$

Energy rate balance: $\dot{m}_6 h_6 + \dot{m}_{10} h_{10} = \dot{m}_1 h_1 + \dot{Q}_{abs}$

Here, $\dot{Q}_{abs}$ is the heat rejection rate by the absorber, X is the concentration, $\dot{m}_{ws} = \dot{m}_4$ is the mass flow rate of the weak solution, $\dot{m}_{ss} = \dot{m}_1$ is the mass flow rate of the strong solution, and $\dot{m}_r = \dot{m}_{10}$ is the mass flow rate of the refrigerant.

Solution pump:

Energy rate balance: $\dot{m}_1 h_1 + \dot{W}_P = \dot{m}_2 h_2$

Solution heat exchanger:

Energy rate balance: $\dot{m}_2 h_2 + \dot{m}_4 h_4 = \dot{m}_3 h_3 + \dot{m}_5 h_5$

Valve 1:

Energy rate balance: $\dot{m}_5 h_5 = \dot{m}_6 h_6 \Rightarrow h_5 = h_6$

Valve 2:

Energy rate balance: $\dot{m}_8 h_8 = \dot{m}_9 h_9 \Rightarrow h_8 = h_9$

Evaporator:

Energy rate balance: $\dot{m}_{10} h_{10} + \dot{Q}_{eva} = \dot{m}_9 h_9$

Condenser:

Energy rate balance: $\dot{m}_7 h_7 = \dot{m}_8 h_8 + \dot{Q}_{cond}$

5.4.2 Exergy Analysis

With the second law of thermodynamics and exergy principles, the following general exergy rate balance can be written:

$$\dot{Ex}_Q + \sum_i \dot{m}_i ex_i = \sum_e \dot{m}_e ex_e + \dot{Ex}_W + \dot{Ex}_D \tag{5.32}$$

where subscripts i and e denote the control volume inlet and outlet, respectively, $\dot{Ex}_D$ is the exergy destruction rate, and other terms are given as follows:

$$\dot{Ex}_Q = \left(1 - \frac{T_0}{T_i}\right) \dot{Q}_i \tag{5.33}$$

$$\dot{Ex}_W = \dot{W} \tag{5.34}$$

$$ex = ex_{ph} + ex_{ch} \tag{5.35}$$

Here, $\dot{Ex}_Q$ is the exergy rate of heat transfer crossing the boundary of the control volume at absolute temperature T, the subscript 0 refers to the reference environment conditions, and $\dot{Ex}_W$ is the exergy rate associated with shaft work. Also, ex_{ph} is defined as follows:

$$ex_{ph} = (h - h_0) - T_0(s - s_0) \tag{5.36}$$

The specific chemical exergy for a general gas mixture can be written as follows [4]:

$$ex_{mix}^{ch} = \left[\sum_{i=1}^{n} x_i ex_i^{ch} + RT_0 \sum_{i=1}^{n} x_i \ln x_i \right] \tag{5.37}$$

For the absorption cooling system, because water and the LiBr solution are not ideal fluids, the following expression is used for the molar chemical exergy calculation [5]:

$$\overline{ex}_{ch} = (1/\overline{M}_{sol}) \left[\sum_{i=1}^{n} y_i \, \overline{ex}_{ch}^{k} + \overline{R} \, T_0 \sum_{i=1}^{n} y_i \, \ln(a_i) \right] \tag{5.38}$$

Applying this equation to the LiBr-water solution we obtain:

$$\overline{ex}_{ch} = (1/\overline{M}_{sol}) \left[\begin{array}{c} y_{H_2O} \, \overline{ex}_{H_2O}^{0} + y_{LiBr} \, \overline{ex}_{LiBr}^{0} + \\ \overline{R} \, T_0 \, (y_{H_2O} \, \ln(a_{H_2O}) + y_{LiBr} \, \ln(a_{LiBr})) \end{array} \right] \tag{5.39}$$

Here, a_{H_2O} is the water activity, defined as the vapor pressure of water in the mixture divided by the vapor pressure of pure water, and a_{LiBr} is the LiBr activity, defined as the vapor pressure of LiBr in the mixture divided by the vapor pressure of LiBr. This equation consists of two parts: the standard chemical exergy of pure species and the exergy due to the dissolution process, defined as follows:

$$\overline{ex}_{ch}^{0} = \frac{1}{\overline{M}_{sol}} (y_{H_2O} \, \overline{ex}_{H_2O}^{0} + y_{LiBr} \, \overline{ex}_{LiBr}^{0}) \tag{5.40}$$

$$\overline{ex}_{ch}^{dis} = \frac{RT_0}{\overline{M}_{sol}} [y_{H_2O} \, \ln(a_{H_2O}) + y_{LiBr} \, \ln(a_{LiBr})] \tag{5.41}$$

where y_i is the molar fraction. The relevant molar fractions here are defined as

$$y_{H_2O} = \frac{(1 - x_{1w})\overline{M}_{LiBr}}{(1 - x_{1w})\overline{M}_{LiBr} + x_{1w}\overline{M}_{H_2O}} \tag{5.42}$$

$$y_{LiBr} = 1 - y_{H_2O} \tag{5.43}$$

Also, X_{1w} is defined as

$$x_{1w} = \frac{x_{LiBr}}{100} \tag{5.44}$$

where X_{LiBr} is the LiBr-water solution concentration (in %) and $\overline{M}_{LiBr}$ and $\overline{M}_{H_2O}$ are 86.85 kg/kmol and 18.02 kg/kmol, respectively.

To calculate the chemical exergy for constituents not listed in Table 5.6, we consider chemical reactions involving constituents for which the standard chemical exergy is already given. This approach allows us to calculate the chemical exergy for such constituents. Since the standard chemical exergy of LiBr is not listed in Table 5.5, the following reaction is considered in order to calculate its chemical exergy [6]:

$$\overline{ex}_{ch}^{0} = \Delta \overline{g}_f^{0} + \sum_{i=1}^{n} \overline{ex}_{ch,i}^{0} \tag{5.45}$$

Table 5.6 Standard chemical values for selected
substances at $T_0 = 298.15$ K and $P_0 = 1$ atm [6, 7].

Element	$\overline{ex}^0_{ch}$ (kJ/mol)	Element	$\overline{ex}^0_{ch}$ (kJ/mol)
Ag (s)	70.2	Kr (g)	34.36
Al (s)	888.4	Li (s)	393.0
Ar (s)	11.69	Mg (s)	633.8
As (s)	494.6	Mn (s_a)	482.3
Au (s)	15.4	Mo (s)	730.3
B (s)	628.8	N$_2$ (g)	0.72
Ba (s)	747.4	Na (s)	336.6
Bi (s)	274.5	Ne (g)	27.19
Br$_2$ (l)	101.2	Ni (s)	232.7
C (s, graphite)	410.26	O$_2$ (g)	3.97
Ca (s)	712.4	P (s, red)	863.6
Cd (s_a)	293.2	Pb (s)	232.8
Cl$_2$ (g)	123.6	Rb (s)	388.6
Co (s_a)	265.0	S (s, rhombic)	609.6
Cr (s)	544.3	Sb (s)	435.8
Cs (s)	404.4	Se (s, black)	346.5
Cu (s)	134.2	Si (s)	854.6
D$_2$ (g)	263.8	Sn (s, white)	544.8
F$_2$ (g)	466.3	Sn (s)	730.2
Fe(s_a)	376.4	Ti (s)	906.9
H$_2$ (g)	236.1	U (s)	1190.7
He (g)	30.37	V (s)	721.1
Hg (l)	115.9	W (s)	827.5
I$_2$ (s)	174.7	Xe (g)	40.33
K (s)	366.6	Zn (s)	339.2

$$Li + \frac{1}{2}Br_2 \rightarrow LiBr \tag{5.46}$$

$$\overline{ex}^0_{ch,LiBr} = \Delta\overline{g}^0_{f,LiBr} + \overline{ex}^0_{ch,Li} + \frac{1}{2}\overline{ex}^0_{ch,Br_2} \tag{5.47}$$

where $\Delta\overline{g}^0_{f,LiBr} = -324 \frac{kJ}{mol}$ [8].

Figure 5.31 shows the variation of specific chemical exergy as a function of LiBr concentration on a mass basis, following Equations 5.40 and 5.41. An increase in LiBr concentration is seen to raise the specific chemical exergy of the LiBr-water solution.

Based on LiBr concentration, therefore, the specific chemical exergy at each point of the single effect absorption chiller in Figure 5.30 can be calculated using code developed in Matlab software. By applying the exergy rate balance equation for each component in the single effect absorption chiller, the exergy destruction rates can be expressed.

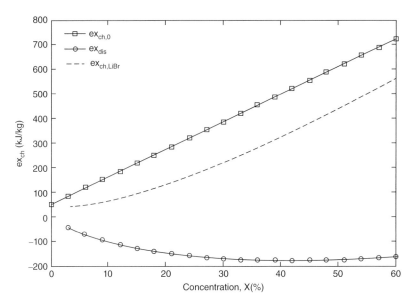

Figure 5.31 Variation of standard specific chemical exergy ($ex_{ch,0}$), specific chemical exergy due to dissolution (ex_{dis}) and specific chemical exergy, as a function of LiBr mass basis concentration at $T_0 = 25°C$.

Generator:

$$\dot{Ex}_{D,gen} = \dot{Ex}_{Q,gen} - (\dot{Ex}_4 + \dot{Ex}_7 - \dot{Ex}_3)$$

where $\dot{Ex}_{Q,gen}$ is the exergy rate of the heat input to the generator.

Absorber:

$$\dot{Ex}_{D,abs} = (\dot{Ex}_1 - \dot{Ex}_6 - \dot{Ex}_{10}) - \dot{Ex}_{Q,abs}$$

where $\dot{Ex}_{Q,abs}$ is the exergy rate of the heat rejected by the absorber.

Solution pump:

$$\dot{Ex}_{D,p} = \dot{Ex}_{W,p} - (\dot{Ex}_2 - \dot{Ex}_1)$$

where $\dot{Ex}_{W,p}$ is the exergy rate of the power input to the pump and is equal to the power consumed by the pump.

Solution heat exchanger:

$$\dot{Ex}_{D,shx} = (\dot{Ex}_4 - \dot{Ex}_5) - (\dot{Ex}_3 - \dot{Ex}_2)$$

Evaporator:

$$\dot{Ex}_{D,eva} = \dot{Ex}_{Q,eva} - (\dot{Ex}_{10} - \dot{Ex}_9)$$

where $\dot{Ex}_{Q,eva}$ is the exergy rate of the heat received by the evaporator, calculated as follows:

$$\dot{Ex}_{Q,eva} = -\left[\left(1 - \frac{T_0}{T_{eva}}\right) \times \dot{Q}_{eva}\right]$$

where

$$T_{eva} = \frac{T_9 + T_{10}}{2}$$

Condenser:

$$\dot{Ex}_{D,cond} = (\dot{Ex}_7 - \dot{Ex}_8) - \dot{Ex}_{Q,cond}$$

where $\dot{Ex}_{Q,cond}$ is the exergy rate of the heat rejected to the environment by the condenser.

Energy and exergy efficiencies Energy and exergy methods are used for assessing the system performance. However, it should be noted that deviations from ideal (i.e., reversible) conditions are not accurately identified with energy analysis but are with exergy analysis. Exergy analysis is therefore capable of overcoming a key shortcoming of energy analysis.

Energy and exergy measures of merit, in the form of coefficients of performance, for absorption chillers can be written as follows:

$$COP_{en} = \frac{\dot{Q}_{eva}}{\dot{Q}_{gen} + \dot{W}_p} \tag{5.48}$$

$$COP_{ex} = \frac{\dot{Ex}_{eva}}{\dot{Ex}_{gen} + \dot{Ex}_p} \tag{5.49}$$

Here,

$$\dot{Ex}_{eva} = \dot{Q}_{eva}\left(\frac{T_0 - T_{EVP}}{T_{EVP}}\right)$$

5.4.3 Exergoeconomic Analysis

Exergoeconomics is a combination of exergy analysis and economic principles that helps to understand how costs flow in a system and to optimize system performance. To identify the costs of inefficiencies and thereby to improve the cost effectiveness of the system, a cost balance is formulated for the absorption chiller as follows [9]:

$$\dot{C}_{Q,gen} + \dot{C}_p + \dot{Z}_{chiller} = \dot{C}_{Q,eva} \tag{5.50}$$

$$\dot{Z} = \frac{Z \times CRF \times \varphi}{N \times 3600} \tag{5.51}$$

where φ is the maintenance factor (often 1.06), N is the annual operational hours of the unit, i is the interest rate and n is the system life. CRF is the capital recovery factor which is expressible as follows:

$$CRF = \frac{i \times (1 + i)^n}{(1 + i)^n - 1} \tag{5.52}$$

The purchase cost of single effect absorption chiller Z can be expressed as a function of cooling capacity of the chiller ($\dot{Q}_{eva}$) (usually the cooling load in kW) as [5]

$$Z_{chiller} = 1144.3 \times (\dot{Q}_{eva})^{0.67} \tag{5.53}$$

5.4.4 Results and Discussion

Energy, exergy, and exergoeconomic analyses are used to evaluate the performance of the absorption chiller. Results are obtained by developing a simulation code in EES software and applying the input data in Table 5.7. Thermodynamic properties of each

Table 5.7 Input data for the simulation of a single effect LiBr–water absorption chiller.

Quantity	Value
Dead state temperature (°C)	25
Dead state pressure (kPa)	101.3
Generator inlet temperature (°C)	100
Evaporator inlet temperature (°C)	10
Solution heat exchanger efficiency (%)	64
Solution pump flow rate (kg/s)	0.05

Table 5.8 Thermodynamic properties at state points of the single effect absorption chiller.

State	$\dot{m}$ (kg/s)	T (°C)	P (kPa)	h (kJ/kg)	s (kJ/kgK)	ex_{ph} (kJ/kg)	ex_{ch} (kJ/kg)	ex (kJ/kg)
1	0.0500	32.905	0.680	85.8	0.1906	−1.2	522.3	521.1
2	0.0500	32.907	7.353	85.8	0.1906	−1.2	522.3	521.1
3	0.0500	63.198	7.353	147.0	0.379	3.9	522.3	526.2
4	0.0455	89.412	7.353	221.2	0.4872	7.5	609.2	616.7
5	0.0455	53.249	7.353	153.9	0.2919	−1.6	609.2	607.6
6	0.0455	44.683	0.680	153.9	0.2432	12.9	609.2	622.1
7	0.0045	76.824	7.353	2643.3	8.468	124.3	0	124.3
8	0.0045	39.928	7.353	167.2	0.5713	1.451	0	1.451
9	0.0045	1.465	0.680	167.2	0.6089	−9.752	0	−9.752
10	0.0045	1.465	0.680	2503.2	9.115	−208.7	0	−208.7
11	1.0000	100.00	200	419.1	1.307	34.11	0	34.11
12	1.0000	96.521	200	404.4	1.267	31.2	0	31.2
13	0.2800	25.000	200	104.8	0.367	0	0	0
14	0.2800	36.986	200	154.9	0.5318	0.9124	0	0.9124
15	0.2800	25.000	200	104.8	0.367	0	0	0
16	0.2800	34.569	200	144.8	0.4991	0.554	0	0.554
17	0.2800	10.000	200	42.0	0.151	1.503	0	1.503
18	0.2800	3.698	200	15.6	0.05654	3.203	0	3.203

state point of the system shown in Figure 5.30 are listed in Table 5.8. The thermodynamic properties obtained with the simulation code include specific enthalpy and entropy.

Using mass, energy and exergy balance rate equations, the cooling capacity, energy and exergy efficiencies, and total exergy destruction rate of the system are calculated (see Table 5.9).

Applying the exergy rate balance equation to each component of the absorption chiller allows the component exergy destruction rates to be calculated, as shown in Figure 5.32. It can be seen there that the evaporator and absorber account for the highest exergy destruction rates relative to the other components. This is mainly due to the fact that phase change with high temperature differences between the working

Table 5.9 Simulation results of the single effect absorption chiller system.

Quantity	Value
Cooling capacity (kW)	10.57
COP_{en}	0.72
Exergy efficiency (%)	31
Total exergy destruction rate (kW)	4
Total cost rate ($/h)	0.093

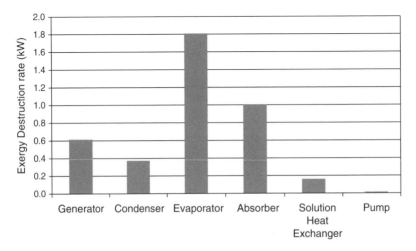

Figure 5.32 Exergy destruction rates of components of the absorption chiller system.

fluids and the refrigerated space occurs in the evaporator, which results in a significant entropy creation rate. The absorber incurs the second largest exergy destruction rate in the system, again because heat transfer occurs between fluids having significant temperature differences.

During absorption chiller operation, a change in an input variable in general influences other dependent variables. Here, to enhance understanding of system performance, a parametric study is conducted. The effects of varying some of the main design parameters on the energy and exergy efficiencies of the system and the total cost rate are examined. The generator inlet temperature (T_{11}), evaporator inlet temperature (T_{17}), solution heat exchanger efficiency (ε_{shx}), and solution pump flow rate (m_1) are considered as the main design parameters. Figure 5.31 shows the effect of generator inlet temperature on generator heat input rate and cooling capacity of the system. The temperature range selected here spans the practical range for a single effect absorption chiller. An increase in generator inlet temperature is observed in Figure 5.31 to increase both heat input rate to the generator and system cooling capacity. When the generator temperature increases, the inlet specific enthalpy increases accordingly, which results in an increase in the generator heat input rate. Considering energy and mass rate balance equations for the generator, an increase in generator temperature results in an increase in evaporator mass flow rate. An increase in evaporator mass flow rate, while keeping

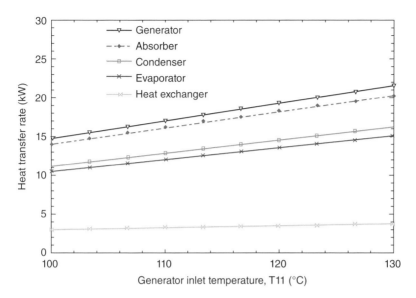

Figure 5.33 Effect of varying generator inlet temperature on evaporator and generator heat transfer rates.

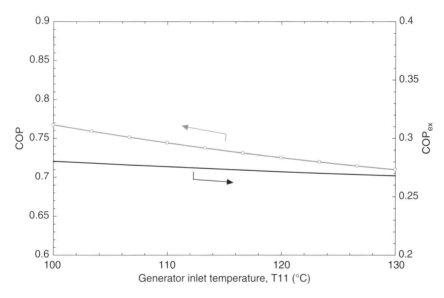

Figure 5.34 Effect of varying generator inlet temperature on energetic and exergetic COPs of the system.

other parameters fixed, leads to an increase in the evaporator load, which is the cooling capacity of the system. It is observed in Figure 5.33 that the slope of the line for the generator heat input rate is higher than the slope of the line for cooling capacity. This implies that an increase in the heat input rate to the generator is higher than the desired output of the system, which is the cooling capacity in this case. This observation is equivalent to a decrease in COP of the system, as shown in Figure 5.34.

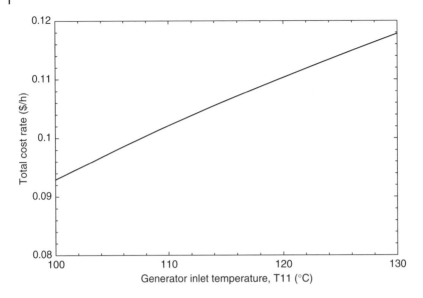

Figure 5.35 Effect of varying generator inlet temperature on the total cost rate of the chiller.

Although an increase in the generator inlet temperature increases the cooling capacity of the system, it decreases both the energetic and exergetic COPs of the system. Figure 5.35 illustrates the effect of this parameter on the total cost rate of the system, and shows that an increase in the generator inlet temperature results in an increase in the total cost rate of the system. This effect is due to the fact that the temperature increase raises the capacity of the system, but at an increased cost.

Another important parameter affecting system performance is the chiller inlet water temperature. Figure 5.36 shows the effect of a variation of chiller inlet water temperature on the COP of the system and cooling capacity. An increase in this temperature is seen to increase both system COP and cooling capacity. In part, this observation stems from the fact that the evaporator pressure increases as the chilled water inlet temperature increases, and the higher is the evaporator pressure, the higher is the system cooling capacity. In addition, the effect of evaporator pressure on the absorber is to decrease the mass fraction in the absorber. Figure 5.37 shows the effect of varying this temperature on the total cost rate of the system; an increase in chilled water inlet temperature is observed to increase the total cost rate of the system, due predominantly to an increase in cooling capacity of the system with higher cost.

The solution heat exchanger effectiveness also affects system performance notably. Figure 5.38 shows the effect of varying heat exchanger effectiveness on the energetic and exergetic COPs of the system. An increase in heat exchanger effectiveness is seen to increase both energetic and exergetic COPs of the system. To determine the reason for this behavior, we assess how this parameter affects the cooling capacity and input heat rate to the generator. Figure 5.39 shows the effect of varying heat exchanger effectiveness on the heat input rate to the generator and the cooling capacity of the system. An increase in heat exchanger effectiveness raises the cooling capacity of the system slightly but leads to a significant decrease in heat input rate to the generator. This is due to the fact that the higher the heat exchanger effectiveness, the greater the heat transfer that occurs between the two flows, which increases the temperature and specific enthalpy at

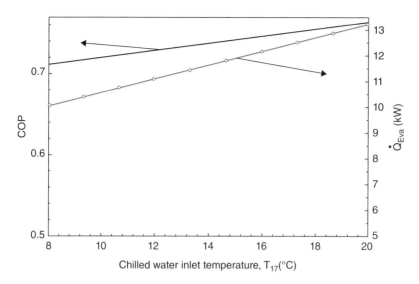

Figure 5.36 Effect of varying generator inlet temperature on COP and cooling capacity of the system.

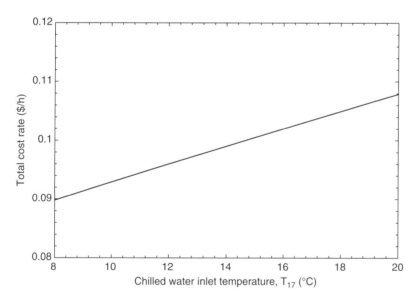

Figure 5.37 Variation of the total cost rate of the system with evaporator inlet temperature.

point 3. This increase in specific enthalpy leads to a decrease in heat input rate to the generator, based on an energy balance for the generator. Therefore, it is usually advisable to use a heat exchanger with a higher effectiveness, although the cost should also be considered. A higher heat exchanger effectiveness requires a higher heat transfer surface, with a higher cost. Consequently, there is a reasonable practical range for this effectiveness. Figure 5.40 shows the impact of varying heat transfer effectiveness on the total cost rate of the system, confirming that a higher heat exchanger effectiveness results in a higher total cost rate of the system.

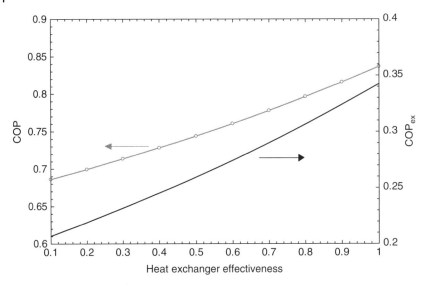

Figure 5.38 Variation of energetic and exergetic coefficients of performance of the system with solution heat exchanger efficiency.

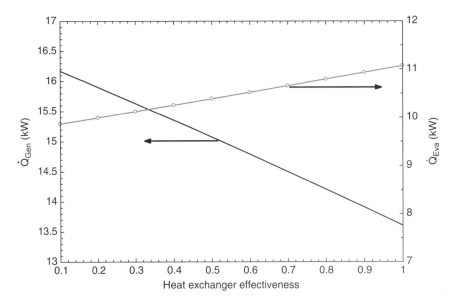

Figure 5.39 Variation of evaporator and generator heat transfer rates with solution heat exchanger effectiveness.

The effect of varying solution pump mass flow rate on system performance is shown in Figure 5.41. The COP of the system varies from 0.77 to 0.64 over the range of solution pump mass flow rates considered. To clarify the reason, the effect is considered of this increase on the load of the solution heat exchanger. As the mass flow rate increases, the heat exchanger load increases significantly, as more energy is available in the solution stream leaving the generator and more energy exits the absorber. The combination of these two effects results in an increase in system losses, leading to a system

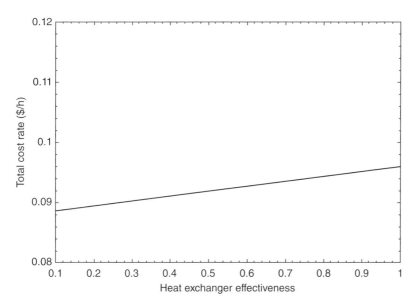

Figure 5.40 Variation of the total cost rate of chiller with solution heat exchanger efficiency.

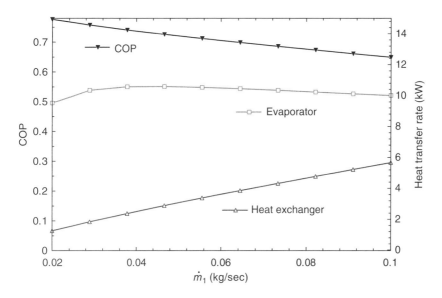

Figure 5.41 Variation with solution pump flow rate of evaporator and generator heat transfer rates and COP of the system.

COP decrease. Figure 5.41 also shows that an increase in solution pump mass flow rate increases the capacity of the system to a maximum value beyond which it decreases. The results show that the evaporator temperature exhibits a minimum temperature at maximum capacity. At the maximum capacity, the temperature profile in the absorber is well matched, with a lower pinch point temperature for better heat transfer. The effect of varying solution pump mass flow rate on the total cost rate of the system is shown

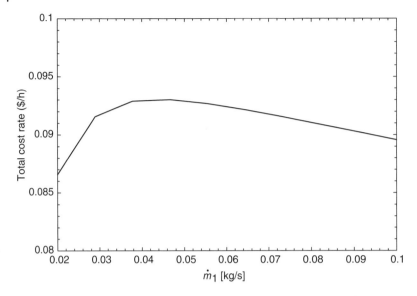

Figure 5.42 Variation with solution pump flow rate on the total cost rate of chiller.

in Figure 5.42. An increase in solution pump mass flow rate increases the total cost of the system as the capacity of the system increases accordingly to a maximum, beyond which it decreases.

5.4.4.1 Optimization

Optimization can help refrigeration engineers determine the best operation point of a refrigeration system subject to constraints, and is applied to the absorption chiller. We select the appropriate optimization method and formulate the optimization problem, objective functions, and constraints [7]. For optimizing the absorption chiller considered above, a multi-objective genetic algorithm is developed and applied. Two objective functions are considered here: COP of the system (to be maximized) and total cost rate of the system (to be minimized). These are expressed in Equations 5.48 and 5.53. Constraints that maintain decision variables within reasonable ranges are listed in Table 5.10 for the absorption chiller.

By applying a genetic algorithm via EES software with the above constraints and objective functions, an optimization is carried out. The results are shown in Table 5.11.

Table 5.10 Constraints on decision variables for the absorption chiller and their ranges.

Parameter	Lower bound	Upper bound
T_{11} (°C)	100	130
T_{17} (°C)	8	20
ε_{shx}	0.1	1
$\dot{m}_1$ (kg/s)	0.03	0.1

Table 5.11 Optimized values for design parameters of the system for two points on the Pareto frontier from multi-objective optimization.

Point	COP	$\dot{Z}_{chiller}\left(\frac{\$}{h}\right)$	$T_{11}(°C)$	$T_{17}(°C)$	ε_{shx}	$m_1(kg/s)$
A	0.74	0.06	101	8.03	0.60	0.08
B	0.75	0.09	90	16.35	0.80	0.07

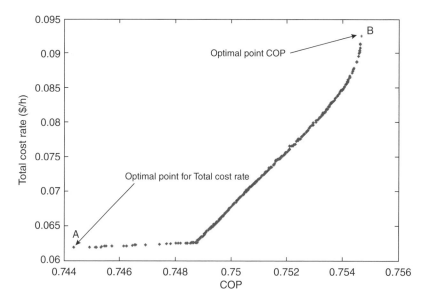

Figure 5.43 Pareto frontier, showing best trade-off values for the objective functions.

5.4.4.2 Optimization Results

Figure 5.43 exhibits the Pareto frontier solution for the absorption chiller system in Figure 5.30. The objective functions for the multi-objective optimization are indicated in Equations 5.48 and 5.53. The total cost rate is observed in Figure 5.43 to increase slightly, while the COP of the chiller increases to about 0.749. Note that increasing the COP further, from 0.749 to 0.755, increases the total cost rate significantly.

In this case, the Pareto optimum frontier exhibits a weak equilibrium, that is, a small change in COP caused by varying the operating parameters leads to a large variation in the total cost rate of the system. Therefore, the ideal point cannot be utilized for decision-making in this problem. When selecting the final optimum point, it is desired to achieve a better magnitude for each objective than its initial value for the base case problem. Note that in multi-objective optimization and the Pareto solution, each point can be utilized as the optimized point. Therefore engineering experience and the importance of each objective plays an important role in such decision-making, suggesting that each user may select a different point for the optimum solution depending on needs. Table 5.11 shows all the design parameters for points A and B.

To better understand the variations of all design parameters, the scattered distributions of the design parameters are shown in Figures 5.44 to 5.47. The results show that generator inlet temperature tends to reach its lower values, which results in an increase

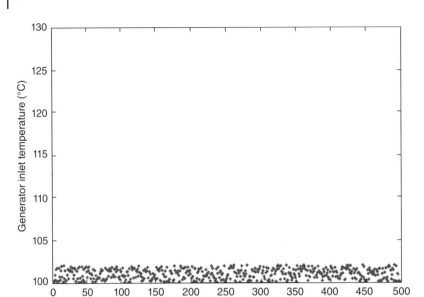

Figure 5.44 Scattered distribution of generator inlet temperature with population in the Pareto frontier.

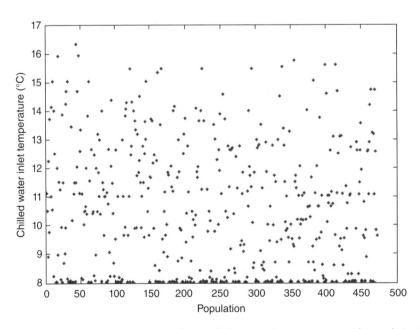

Figure 5.45 Scattered distribution of the chilled water inlet temperature with population in the Pareto frontier.

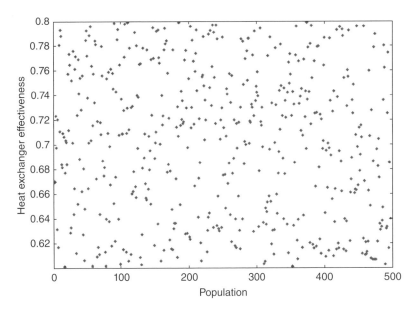

Figure 5.46 Scattered distribution of the heat exchanger efficiency with population in the Pareto frontier.

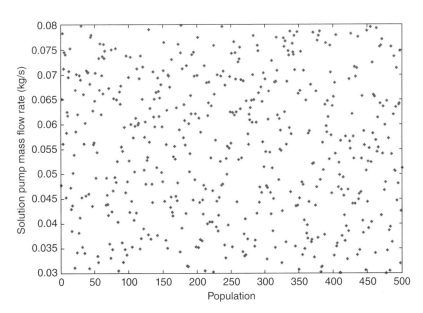

Figure 5.47 Scattered distribution of the solution pump mass flow rate with population in the Pareto frontier.

in COP because COP increases with decrement of the generator inlet temperature. But chilled water inlet temperatures exhibit a scattered distribution between their allowable ranges. This is due to the fact that an increase in chilled water inlet temperature increases the cooling load of the chiller which eventually results in an increment in COP of the system. Also, the cooling load of the chiller raises the total cost rate of the system. Similar results are obtained for the heat exchanger efficiency and solution pump mass flow rate, since increases in these design parameters improve the system performance and increase the COP while increasing the total cost rate.

5.5 Concluding Remarks

Refrigeration systems and their applications are described in this chapter, and modeling and optimization methods for them are explained and applied to the three main types of refrigeration systems.

The vapor compression refrigeration system is the most widely used refrigeration system. It is modeled analyzed, and a comprehensive parametric study is undertaken to determine how each design parameter affects the system performance. A multi-objective genetic algorithm is applied to determine the optimal design parameters for the vapor compression refrigeration system and a sensitivity analysis is carried out. The results show that evaporator pressure has a positive effect on both cycle exergy efficiency and COP, that is, the higher the evaporator pressure, the greater the system performance. Condenser pressure on the other hand has a negative effect on system performance; thus an increase in condenser pressure while other parameters are fixed leads to a decrease in both cycle exergy efficiency and COP.

Next, a cascade refrigeration system is described and modeled for applications where lower temperatures are required. A parametric study and optimization are also performed to enhance understanding of the system performance. The effect of changing working fluid on the cascade system performance is investigated and R134a is found to yield better performance for the application than the other working fluids considered. The results show that an increase in compressor isentropic efficiency leads to an increase in the COP of the system, while an increase in evaporator pressure increases the cooling load of the system while decreasing the net compressor work. An increase in flash chamber pressure has a positive effect on system COP up to a maximum values and then decreases beyond the optimal point.

Finally, an absorption refrigeration system is examined and a parametric study and optimization are conducted. The results demonstrate that an increase in generator inlet temperature increases both the heat input rate to the generator and the cooling capacity of the system. An increase in evaporator mass flow rate, while holding other parameters fixed, leads to an increase in evaporator load, that is, the cooling capacity of the system. In addition, an increase in chilled water inlet temperature increases both the COP and cooling capacity of the system.

References

1 Cengel, Y. A., Boles, M. A. and Kanoğlu, M. (2011) *Thermodynamics: An Engineering Approach*, McGraw-Hill, New York.

2 Ahmadi, P., Dincer, I. and Rosen, M. A. (2011) Exergy, exergoeconomic and environmental analyses and evolutionary algorithm based multi-objective optimization of combined cycle power plants. *Energy* **36**:5886–5898.

3 Ganjehkaviri, A., Jaafar, M. M., Ahmadi, P. and Barzegaravval, H. (2014) Modeling and optimization of combined cycle power plant based on exergoeconomic and environmental analyses. *Applied Thermal Engineering* **67**:566–578.

4 Bejan, A., Tsatsaronis, G. and Moran, M. (1995) *Thermal Design and Optimization*, John Wiley & Sons, Inc., New York.

5 Ahmadi, P. (2013) Modeling, analysis and optimization of integrated energy systems for multigeneration purposes. PhD thesis. Faculty of Engineering and Applied Science, University of Ontario Institute of Technology.

6 Kotas, T. J. (1985) *The Exergy Method of Thermal Plant Analysis*, Krieger, Malabar, Florida.

7 Dincer, I. and Rosen, M. A. (2013) *Exergy: Energy, Environment and Sustainable Development*, Second Edition, Elsevier.

8 Palacios-Bereche, R., Gonzales, R. and Nebra, S. A. (2012) Exergy calculation of lithium bromide–water solution and its application in the exergetic evaluation of absorption refrigeration systems LiBr-H_2O. *International Journal of Energy Research* **36**:166–181.

9 Bejan, A., Tsatsaronis, G. and Moran, M. J. (1996) *Thermal Design and Optimization*, John Wiley & Sons, Inc., New York.

Study Questions/Problems

1 Describe the four main processes that make up the simple vapor compression refrigeration cycle.

2 A refrigeration cycle is used to keep a food case at $-7°C$ in an environment at $25°C$. The total heat gain rate to the food case is estimated to be 8000 kJ/h and the heat rejection rate by the condenser is 1300 kJ/h. Determine (a) the power input to the compressor, (b) the COP of the refrigerator, and (c) the minimum power input to the compressor if a reversible refrigerator is used.

3 Consider the evaporator coils of a household refrigerator placed at the back of a freezer (see Figure 5.48). Refrigerant-134a enters the coils at 120 kPa with a quality of 0.25 and exits at 120 kPa and $-20°C$. If the compressor uses 600 W of power and the COP of the refrigerator is 1.3, determine (a) the mass flow rate of the refrigerant, and (b) the rate heat is rejected to the kitchen air. Considering the evaporator pressure, the compressor isentropic efficiency, and the compressor pressure ratio as design parameters, optimize the COP of the system and conduct a sensitivity analysis.

4 An automotive air conditioner operates on the vapor compression refrigeration cycle with refrigerant-134a as the working fluid. The refrigerant enters the compressor as a superheated vapor at pressure P_1 and temperature T_1 with a mass flow rate of 0.005 kg/s. The compressor isentropic efficiency is η and the working fluid exits the compressor at pressure P_2 and $60°C$. The R-134a is subcooled by $T_2°C$ at the exit of the condenser.

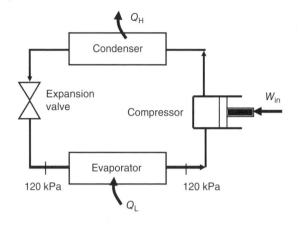

Figure 5.48 Schematic of the household refrigerator in problem 3.

a) Apply single objective optimization to determine the optimal values for P_1, T_1, P_2, T_2, and η satisfying reasonable constraints to maximize the COP of the system.

b) Apply multi-objective optimization considering COP and cost of the system as the two objective functions.

5 A refrigerated room is maintained at $-27°C$ by a vapor compression cycle using R-134a as the refrigerant. Heat is rejected to cooling water that enters the condenser at $16°C$ at a mass flow rate of 0.22 kg/s, and exits at $23°C$. The refrigerant enters the condenser at 1.2 MPa and $65°C$ and leaves at $42°C$. The inlet state of the compressor is 60 kPa and $-34°C$ and the compressor is estimated to gain a net heat of 150 W from the surroundings. Determine (a) the quality of the refrigerant at the evaporator inlet, (b) the refrigeration load, (c) the COP of the refrigerator, and (d) the theoretical maximum refrigeration load for the same power input to the compressor.

6 What is the main difference between a LiBr–water absorption chiller and an ammonia–water absorption chiller? Compare their COPs via suitable applications.

7 Repeat the optimization of the LiBr–water absorption chiller, but with ammonia as the working fluid. Make reasonable assumptions if needed.

8 Consider the double-effect LiBr–water absorption chiller system shown in Figure 5.49. High temperature water enters the high temperature generator at $T_{21} = 150°C$. The following mass flow rates are known: $\dot{m}_1 = 1$ kg/s, $\dot{m}_{21} = 8$ kg/s, $\dot{m}_{25} = 14$ kg/s, $\dot{m}_{27} = 20$ kg/s, and $\dot{m}_{23} = 12$ kg/s. The following vapor qualities are given: $X_1 = 0$, $X_4 = 0$, $X_{11} = 0$, $X_{14} = 0$, $X_{18} = 0$, $X_8 = 0$, and $X_{10} = 1$. The following temperatures are known: $T_{25} = 25°C$, $T_{27} = 12°C$, $T_{23} = 25°C$. Finally, the value of heat transfer coefficient products are known: $UA_{gen} = 25$ kW/K, $UA_{cond} = 65$ kW/K, $UA_{Eva} = 80$ kW/K, $UA_{ads} = 50$ kW/K, and $UA_{cd} = 10$ kW/K.

a) Determine all of the thermodynamic properties at each state point of the system

b) Calculate the energetic and exergetic COPs of the system

c) Evaluate the exergy destruction rate of each system component

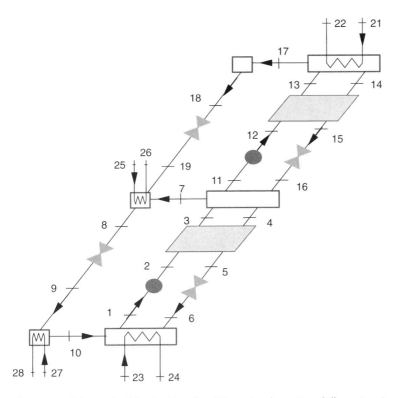

Figure 5.49 Schematic of the double-effect LiBr–water absorption chiller system in problem 8.

d) Apply a single objective optimization to calculate the maximum COP of the system.

9 Consider the triple effect ammonia–water absorption refrigeration cycle shown in Figure 5.50. Apply mass, energy, entropy, and exergy rate balance equations to each component to determine the thermodynamic properties of all state points, as well as the system exergy destruction rate and COP. Assume $T_0 = 20°C, P_0 = 101\,kPa, P_1 = 250\,kPa, \dot{m}_1 = 1\,kg/s, P_2 = 400\,kPa$, the condenser heating load is 240 kW, and the concentrations at point 29 and point 30 are 0.6 and 0.4, respectively.

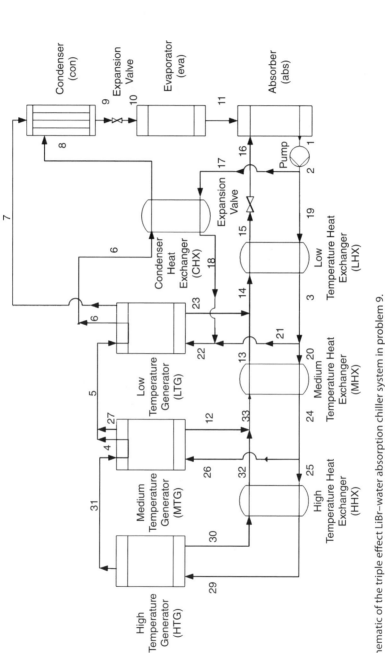

Figure 5.50 Schematic of the triple effect LiBr–water absorption chiller system in problem 9.

6

Modeling and Optimization of Heat Pump Systems

6.1 Introduction

A heat pump is a device that conveys heat from a source of heat to a heat sink at a higher temperature. Heat pumps are designed to move thermal energy in a direction opposite to that of spontaneous heat flow by absorbing heat from a cold space and releasing it to a warmer one. When a heat pump is used for heating, it employs the same basic refrigeration cycle used by an air conditioner or a refrigerator, but with an opposite intention—releasing heat into the conditioned space rather than the surrounding environment. The principle governing the operation of the heat pump was recognized before the start of the 1900s and is the basis of all refrigeration. The idea of using a heat engine in a reverse mode, as a heat pump, was proposed by Lord Kelvin in the nineteenth century, but it was only in the twentieth century that practical machines began to be used, mainly for refrigeration [1]. Heat pumps are also installation-cost competitive with central combustion furnace/central air conditioner combinations. Hence, heat pumps now routinely provide central air conditioning as well as heating. Today, heat pumps are widely used not only for air conditioning and heating, but also for cooling, producing hot water, and preheating feed water in various types of facilities including office buildings, computer centers, public buildings, restaurants, hotels, district heating and cooling systems, and industrial plants.

A heat pump is essentially a heat engine operating in reverse and can be defined as a device that moves heat from a region of low temperature to a region of higher temperature. The residential air-to-air heat pump, the type most commonly in use, extracts heat from low temperature outside air and delivers it indoors. To accomplish this and avoid violating the second law of thermodynamics, work is done on the heat pump working fluid (e.g., a refrigerant). Heat pump efficiency is determined by comparing the amount of energy delivered by the heat pump to the amount of energy it consumes. Note that efficiency measures are usually based on laboratory tests and do not necessarily measure how a heat pump performs in actual use.

The COP is the most common measure of heat pump efficiency. The COP is the ratio of the heat output of a heat pump to its driving energy input, expressible as

$$COP = Heat\ output/Energy\ input$$

Air-source heat pumps generally have COPs ranging from 2 to 4, implying that they deliver 2 to 4 times more energy than they consume, typically in the form of electricity.

Optimization of Energy Systems, First Edition. Ibrahim Dincer, Marc A. Rosen, and Pouria Ahmadi.
© 2017 John Wiley & Sons Ltd. Published 2017 by John Wiley & Sons Ltd.

Water- and ground-source heat pumps normally have COPs of 3 to 5 [1]. The COP of an air-source heat pump decreases as the outside temperature drops. Therefore, two COP ratings are usually given for a system: one at 8.3°C (47°F) and the other at −9.4°C (17°F). When comparing COPs, one must be sure that the ratings are based on the same outside air temperature to avoid inconsistencies. COPs for ground- and water-source heat pumps do not vary as widely because ground and water temperatures are more constant than air temperatures. Some heat pump units have an energy-saving feature that allows the unit to defrost only when necessary. Others enter a defrost cycle at set intervals whenever the unit is in the heating mode.

Another factor that lowers the overall efficiency of air-to-air heat pumps is their inability to provide sufficient heat during the coldest days of winter. This weakness causes a back-up heating system to be required. The back-up is often provided by electric resistance heating, which has a COP of only one. When the temperature drops to the −3.8°C to −1.1°C range, or a different system-specific balance point, this electric resistance heating engages and overall system efficiency decreases.

The primary objectives of this chapter are to describe the air/water heat pump system and its energy and exergy analyses, and to model comprehensively and optimize selected systems. The last objective involves defining objective function(s) and, usually, applying multi-objective optimization to determine the best optimal design parameters of the system. Relevant constraints need to be considered, so that appropriate optimization is carried out. In order to enhance understanding of the design criteria, a sensitivity analysis is conducted to assess how each objective function varies when design parameters are modified. We then provide some closing remarks related to the efficient design of heat pump systems.

6.2   Air/Water Heat Pump System

A schematic of an air/water heat pump system is shown in Figure 6.1. The system consists of two separate circuits: (1) a heat pump circuit (refrigerant circuit), and (2) a heat distribution circuit (water circuit). The refrigerant circuit consists of a compressor, a condenser, an expansion valve, and an evaporator. The refrigerant is R-134a. The heat distribution circuit consists of a storage tank and a circulating pump. Device I in Figure 6.1 is a fully hermetically sealed reciprocating piston compressor. The condenser (II) is of a coaxial pipe cluster heat exchanger construction that works on the counter flow principle. The refrigerant expands in an expansion valve (III). The evaporator (IV) is of a finned tube construction and has a large surface area. The refrigerant flows through the evaporator and draws heat from the ambient air over the large surface area. Heat transfer is enhanced by two fans that draw air through fins. During the operation period assessed, the control valve is adjusted so that the flow rate in the hot water circuit is approximately 0.020 m³/h. After the pressures on the suction and delivery sides of the working medium circuit have stabilized, data are recorded, including compressor power, hot water flow rate, and pressures and temperatures at various points of the unit.

Mass, energy, and exergy rate balances are employed to determine the heat input, the rate of exergy destruction, and energy and exergy efficiencies. Steady state, steady flow processes are assumed. A general mass rate balance can be expressed in rate form as

$$\sum \dot{m}_{in} = \sum \dot{m}_{out} \tag{6.1}$$

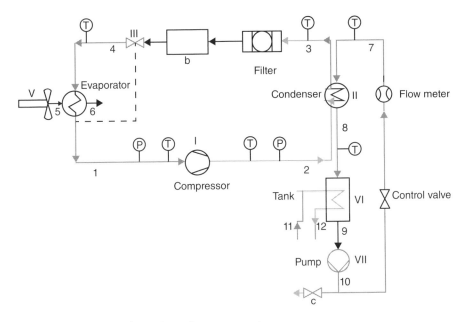

Figure 6.1 Schematic of an air/water heat pump system.

where ṁ is the mass flow rate, and the subscript *in* stands for inlet and *out* for outlet. Energy and exergy rate balances can be written respectively as

$$\dot{E}_{in} = \dot{E}_{out} \tag{6.2}$$

$$\dot{Ex}_{in} - \dot{Ex}_{out} = \dot{Ex}_D \tag{6.3}$$

The specific flow exergy of the refrigerant or water is evaluated as

$$ex = (h - h_0) - T_0(s - s_0) \tag{6.4}$$

where h denotes specific enthalpy, s denotes specific entropy, and the subscript zero indicates properties at the reference environment (dead) state (i.e., at P_0 and T_0).

The overall specific flow exergy of air is determined as [1]

$$ex_{air} = (C_{p,a} + \omega C_{p,v})T_0[(T/T_0) - 1 - \ln(T/T_0)] + (1 + 1.6078\omega)R_a T_0 \ln(P/P_0)$$
$$+ R_a T_0\{(1 + 1.6078\omega) \ln[(1 + 1.6078\omega_0)/(1 + 1.6078\omega)]$$
$$+ 1.6078\omega \ln(\omega/\omega_0)\} \tag{6.5}$$

where the specific humidity ratio is

$$\omega = \dot{m}_v/\dot{m}_a \tag{6.6}$$

The exergy rate of a flow is determined as

$$\dot{Ex} = \dot{m}(ex) \tag{6.7}$$

The exergy destruction rates in the heat exchanger (condenser or evaporator) and circulating pump respectively are evaluated as

$$\dot{Ex}_{D, HE} = \dot{Ex}_{in} - \dot{Ex}_{out} \tag{6.8}$$

$$\dot{Ex}_{D,pump} = \dot{W}_{pump} - (\dot{Ex}_{out} - \dot{Ex}_{in}) \tag{6.9}$$

where $\dot{W}_{pump}$ is the work rate of the pump.

The energy-based efficiency measure of the heat pump unit (COP_{HP}) and the overall heat pump system (COP_{sys}) can be defined as follows:

$$COP_{HP} = \frac{\dot{Q}_{cond}}{\dot{W}_{comp}} \tag{6.10}$$

and

$$COP_{sys} = \frac{\dot{Q}_{cond}}{\dot{W}_{comp} + \dot{W}_{pump} + \dot{W}_{fans}} \tag{6.11}$$

The exergy efficiency is generally expressed as the ratio of total exergy output to the total driving exergy input and can be written as

$$\psi = \frac{\dot{Ex}_{output}}{\dot{Ex}_{input}} \tag{6.12}$$

where "output" refers to "net output" or "product" or "desired value," and "input" refers to "driving input" or "fuel."

Here, the exergy efficiencies for the heap pump only and for the overall system can then be written as follows:

$$\psi_{HP} = \frac{\dot{Ex}^{Qc}}{\dot{W}_{comp}} \tag{6.13a}$$

and

$$\psi_{sys} = \frac{\dot{Ex}^{Qc}}{\dot{W}_{comp} + \dot{W}_{pump} + \dot{W}_{fans}} \tag{6.13b}$$

In addition, the exergy efficiency of the heat exchanger (condenser or evaporator) is determined as the increase in the exergy of the cold stream divided by the decrease in the exergy of the hot stream, on a rate basis, as follows:

$$\psi_{HE} = \frac{\dot{Ex}_{cold,out} - \dot{Ex}_{cold,in}}{\dot{Ex}_{hot,in} - \dot{Ex}_{hot,out}} = \frac{\dot{m}_{cold}(\psi_{cold,out} - \psi_{cold,in})}{\dot{m}_{hot}(\psi_{hot,in} - \psi_{hot,out})} \tag{6.14}$$

6.3 System Exergy Analysis

The following assumptions are made during the energy and exergy analyses:

- All processes are steady state and steady flow with negligible potential and kinetic energy effects and no chemical or nuclear reactions.
- Heat transfer to the system and work transfer from the system are positive.
- Air behaves as an ideal gas with a constant specific heat.
- Heat transfer and refrigerant pressure drops in the tubing connecting the components are negligible since their lengths are short.
- The compressor mechanical ($\eta_{comp,\,mech}$) and the compressor motor electrical ($\eta_{comp,\,elec}$) efficiencies are 68% and 69%, respectively. These values are based on actual data in which the power input to the compressor is 0.149 kW.
- The circulating pump mechanical ($\eta_{pump,\,mech}$) and the circulating pump motor electrical ($\eta_{pump,\,elec}$) efficiencies are 82% and 88%, respectively. These values are based on an electric power of 0.050 kW obtained from the pump characteristic curve.

- The fan mechanical ($\eta_{\text{fan, mech}}$) and the fan motor electrical ($\eta_{\text{fan, elec}}$) efficiencies are 40% and 80%, respectively.

The mass and energy rate balances as well as exergy destruction rates obtained from the exergy rate balances for each of the heat pump components illustrated in Figure 6.1 can be written as follows:

Compressor (I):

$$\dot{m}_1 = \dot{m}_{2,s} = \dot{m}_{\text{act},s} = \dot{m}_r \tag{6.15a}$$

$$\dot{W}_{\text{comp}} = \dot{m}_r(h_{2,\text{act}} - h_1) \tag{6.15b}$$

$$\dot{Ex}_{D,\text{comp}} = \dot{m}_r(ex_1 - ex_{2,\text{act}}) + \dot{W}_{\text{comp}} \tag{6.15c}$$

where heat interactions with the environment are neglected.

Condenser (II):

$$\dot{m}_2 = \dot{m}_3 = \dot{m}_r; \; \dot{m}_7 = \dot{m}_8 = \dot{m}_w \tag{6.16a}$$

$$\dot{Q}_{\text{cond}} = \dot{m}_r(h_{2,\text{act}} - h_3); \; \dot{Q}_{\text{cond}} = \dot{m}_w C_{p,w}(T_8 - T_7) \tag{6.16b}$$

$$\dot{Ex}_{D,\text{cond}} = \dot{m}_r(ex_{2,\text{act}} - ex_3) + \dot{m}_w(ex_7 - ex_8) \tag{6.16c}$$

Expansion (throttling) valve (III):

$$\dot{m}_3 = \dot{m}_4 = \dot{m}_r \tag{6.17a}$$

$$(h_3 = h_4) \tag{6.17b}$$

$$\dot{Ex}_{D,\text{exp}} = \dot{m}_r(ex_3 - ex_4) \tag{6.17c}$$

Evaporator (IV):

$$\dot{m}_4 = \dot{m}_1 = \dot{m}_r \tag{6.18a}$$

$$\dot{Q}_{\text{evap}} = \dot{m}_r(h_1 - h_4) \tag{6.18b}$$

$$\dot{Q}_{\text{evap}} = \dot{m}_{\text{air}} C_{p,\text{air}}(T_5 - T_6) \tag{6.18c}$$

$$\dot{Ex}_{D,\text{evap}} = \dot{W}_{\text{fan}} + \dot{m}_r(ex_4 - ex_1) + \dot{m}_{\text{air}}(ex_5 - ex_6) \tag{6.18d}$$

Fan (V):

$$\dot{m}_5 = \dot{m}_{5'} = \dot{m}_{\text{air}} \tag{6.19a}$$

$$\dot{W}_{\text{fan}} = \dot{m}_{\text{air}} \left[(h_5 - h_{5'}) + \frac{V^2_{\text{exit}}}{2} \right] \tag{6.19b}$$

$$\dot{Ex}_{D,\text{fan}} = \dot{W}_{\text{fan,elec}} + \dot{m}_{\text{air}}(ex_{5'} - ex_5) \tag{6.19c}$$

Storage tank (VI):

$$\dot{m}_8 = \dot{m}_9 = \dot{m}_w; \; \dot{m}_{11} = \dot{m}_{12} = \dot{m}_{tw} \tag{6.20a}$$

$$\dot{Q}_{st} = \dot{m}_w C_{p,w}(T_8 - T_9); \; \dot{Q}_{\text{tank}} = \dot{m}_{tw} C_{p,tw}(T_{12} - T_{11}) \tag{6.20b}$$

$$\dot{Ex}_{D,st} = \dot{m}_w(ex_8 - ex_9) + \dot{m}_{tw}(ex_{11} - ex_{12}) \tag{6.20c}$$

Circulating pump (VII):

$$\dot{m}_9 = \dot{m}_{10s} = \dot{m}_{10,\text{act}} = \dot{m}_w \tag{6.21a}$$

$$\dot{W}_{\text{pump}} = \dot{m}_w(h_{10,\text{act}} - h_9) \tag{6.21b}$$

$$\dot{Ex}_{D,\text{pump}} = \dot{m}_r(ex_9 - ex_{10,\text{act}}) + \dot{W}_{\text{pump}} \tag{6.21c}$$

Since the volume flow rate on the refrigerant side is not measured, COP_{act} is evaluated as follows:

$$COP_{act} = \frac{\dot{m}_w C_{p,w}(T_8 - T_7)}{\dot{W}_{comp,act}} = \frac{\dot{V}_w \rho_w C_{p,w}(T_8 - T_7)}{\dot{W}_{comp,act}} \qquad (6.22)$$

In addition to the energy and exergy efficiencies, energetic and exergetic coefficients of performance (*COP*s) can be developed for the heat pump and the overall system. These energetic and exergetic *COP*s help visualize actual system performance with variations in operating parameters. Energetic and exergetic *COP*s of the heat pump are defined respectively as

$$COP_{HP,en} = \frac{\dot{Q}_{con}}{\dot{W}_{comp}} \qquad (6.23)$$

$$COP_{HP,ex} = \frac{\dot{Ex}_2 - E}{\dot{x}_3 \dot{W}_{comp}} \qquad (6.24)$$

where $COP_{HP,en}$ denotes the energetic *COP* of the heat pump, $\dot{Q}_{con}$ the condenser load, $\dot{W}_{comp}$ the work input rate to the compressor, and $COP_{HP,ex}$ the exergetic *COP* of the heat pump, while $\dot{Ex}_2$ and $\dot{Ex}_3$ denote the exergy rates at states 2 and 3, respectively.

6.4 Energy and Exergy Results

Temperature, pressure, and mass flow rate data for the working fluid (R-134a), water, and air are given in Table 6.1 following the state numbers specified in Figure 6.1. Exergy rates evaluated for each state are also presented in Table 6.1. The reference environment state is taken to be the state of environment when the temperature and the atmospheric pressure were 2.2°C and 98.80 kPa, respectively [1]. The thermodynamic properties of water, air, and R-134a are found using the Engineering Equation Solver (EES) software package. In order to validate the simulation, the results of the experimental data acquired in the lab at University of Ontario Institute of Technology (UOIT) and the simulation code are compared and the results are shown in Figure 6.2. In this figure, the COP of the heat pump and the overall system are calculated and compared.

The comparison results show a good agreement between the experimental data and the simulation computer code. For evaluating the efficiency of heat pump systems, the most commonly used measure is the energy (or first law) efficiency, which is modified to a coefficient of performance (COP). However, for indicating the possibilities for thermodynamic improvement, energy analysis is inadequate and exergy analysis is needed. Exergy analysis of the air-source heat pump system presented in this chapter identifies improvement potential.

In order to enhance understanding of the system performance, a parametric study is conducted here to determine how each design parameter affects the performance. For the parametric study, the original system is scaled up by a factor of 10 in order to identify system performance variances more easily. The main benefit of such parametric analyses is that they help designers visualize how a system will perform under different operating conditions. Here the refrigerant mass flow rate, evaporator temperature, and condenser temperature are considered as the design parameters.

Table 6.1 Process data for flows in the heat pump system.

State No.	Description	Fluid	Phase	Temp., T (°C)	Pressure, P (kPa)	Specific humidity ratio, ω (kg$_{water}$/kg$_{air}$)	Specific enthalpy, h (kJ/kg)	Specific entropy, s (kJ/kg K)	Mass flow rate, $\dot{m}$ (kg/s)	Specific exergy, ex (kJ/kg)	Exergy rate, $\dot{Ex} = \dot{m}(ex)$ (kW)
0	–	Refrigerant	Dead state	2.2	98.80	–	257.4	1.041	–	0	0
0′	–	Water	Dead state	2.2	98.80	–	9.3	0.034	–	0	0
0″	–	Moist air	Dead state	2.2	98.80	0.002	–	–	–	0	0
1	Evaporator outlet/compressor inlet	Refrigerant	Superheated vapor	2.5	307		252.3	0.935	0.002	24.09	0.048
2,s	Condenser inlet/compressor outlet	Refrigerant	Superheated vapor	45.3	1011		256.2	0.935	0.002	27.98	0.056
2,act	Condenser inlet/compressor outlet	Refrigerant	Superheated vapor	54.6	1011		287.4	0.966	0.002	50.65	0.101
3	Condenser outlet/expansion valve inlet	Refrigerant	Compressed liquid	22.8	1011		83.4	0.313	0.002	26.45	0.053
4	Evaporator inlet	Refrigerant	Mixture	1.3	307		83.4	0.319	0.002	24.80	0.050
5	Fan air inlet to evaporator	Air	Gas	16		0.004	26.20		0.136	0.33	0.045
5′	Air inlet to fan	Air	Gas	15.9		0.004	26.09		0.136	0.23	0.032
6	Fan air outlet from evaporator	Air	Gas	14		0.004	24.17		0.136	0.27	0.037

(Continued)

Table 6.1 (Continued)

State No.	Description	Fluid	Phase	Temp., T (°C)	Pressure, P (kPa)	Specific humidity ratio, ω (kg_{water}/kg_{air})	Specific enthalpy, h (kJ/kg)	Specific entropy, s (kJ/kg K)	Mass flow rate, $\dot{m}$ (kg/s)	Specific exergy, ex (kJ/kg)	Exergy rate, $\dot{Ex} = \dot{m}(ex)$ (kW)
7	Water inlet to condenser	Water	Compressed liquid	16.9	230		71.1	0.252	0.011	1.77	0.020
8	Water outlet from condenser	Water	Compressed liquid	24.6	220		103.3	0.361	0.011	3.96	0.044
9	Water outlet from storage tank/circulating pump inlet	Water	Compressed liquid	16.0	200		67.3	0.239	0.011	1.55	0.017
10,s	Water outlet from tank/circulating pump outlet	Water	Compressed liquid	16.05	240		69.9	0.239	0.011	4.15	0.046
10,act	Water outlet from tank/circulating pump outlet	Water	Compressed liquid	16.9	240		71.0	0.251	0.011	1.96	0.022
11	Tap water inlet to tank	Water	Compressed liquid	9.1	500		38.7	0.138	0.024	0.76	0.018
12	Tap water outlet from tank	Water	Compressed liquid	13.1	500		54.9	0.196	0.024	1.00	0.024

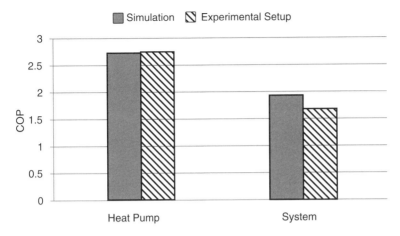

Figure 6.2 Comparison of simulation model and experimental measured data.

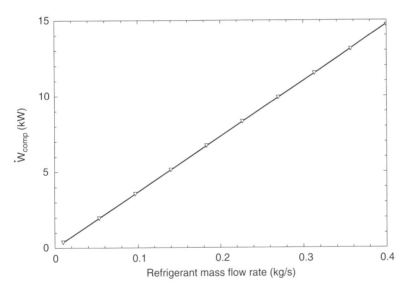

Figure 6.3 Effect of varying refrigerant mass flow rate on compressor work.

Figure 6.3 shows the effect of refrigerant mass flow rate on the compressor work. It is observed that an increase in refrigerant mass flow rate increases the compressor work rate, from 0.36 kW to 14.67 kW when $\dot{m}_r$ increases from 0.01 kg/s to 0.4 kg/s. Increasing the refrigerant mass flow rate implies that more refrigerant needs to be compressed, resulting in a higher energy requirement to drive the compressor.

The effect of changing refrigerant mass flow rate on cooling load and condenser load is shown in Figure 6.4. An increase in refrigerant mass flow rate is seen to result in an increase in both evaporator cooling load and condenser reject heat rate. It is clear that as the quantity of compressed refrigerant entering the condenser and the evaporator increases, the ability of the condenser and the evaporator to reject heat and to absorb heat, respectively, increases. Figure 6.5 shows the effect of evaporator temperature

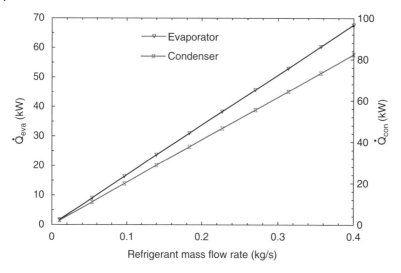

Figure 6.4 Effect of varying refrigerant mass flow rate on compressor work rate, condenser load, and cooling load.

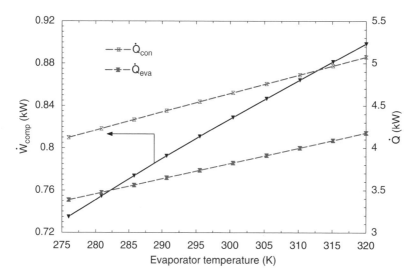

Figure 6.5 Effect of varying evaporator temperature on compressor work rate, condenser load, and cooling load.

on the compressor work rate and the evaporator and condenser loads. An increase in evaporator temperature is observed to lead to an increase in compressor work rate.

When the evaporator temperature increases from 276 K to 320 K, compressor work rate increases from 0.73 kW to 0.89 kW. An increase in evaporator temperature results in a higher energy stream entering the compressor. As the energy of the inlet stream increases, the work rate required to compress this stream also increases. It is also seen in Figure 6.5 that an increase in evaporator temperature increases both the evaporator and condenser loads. When the evaporator temperature increases, the outlet specific enthalpy of the evaporator increases which results in an increase in evaporator load.

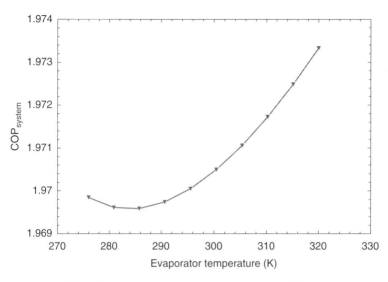

Figure 6.6 Effect of varying evaporator temperature on COP of the system

Also, an increase in evaporator temperature raises the amount of energy that is rejected through the condenser, leading to a higher condenser load. The effect of evaporator temperature on the system COP is shown in Figure 6.6. It is seen that an increase in evaporator temperature initially decreases the COP of the system to a minimum value and then causes the COP to increase. This is due to the fact that an increase in compressor work initially dominates compared to the desired heat, which makes the COP decrease; but a further increase in evaporator temperature increases the condenser load to a much greater extent than the increase in the compressor work.

Another important parameter is the condenser temperature (T_3). The effect of increasing T_3 on the condenser and evaporator loads can be seen in Figure 6.7. The condenser and evaporator loads are observed to decrease from 4.3 kW to 3.7 kW and 3.5 kW to 2.9 kW, respectively, as T_3 increases from 290 K to 310 K. This behavior is observed because an increase in T_3 results in a higher energy content of the stream exiting the condenser and entering the evaporator. As the temperature of the stream leaving the condenser increases for a fixed input temperature to the condenser, the heat rejection capacity of the condenser decreases. The effect of condenser temperature on the COP of the systems is shown in Figure 6.8. It is observed that an increase in this temperature decreases the COP of the system due to a decrease in condenser load, as previously explained.

Since different design parameters have different effects on the COP and evaporator and condenser loads of the system, optimization can be beneficial.

6.5 Optimization

For the optimization of the heat pump system, the compressor work rate and condenser load are considered as two objective functions. The compressor work rate should be minimized while the condenser rejected heat rate should be maximized to heat the water. A genetic algorithm optimization is performed for 450 generations, using a search

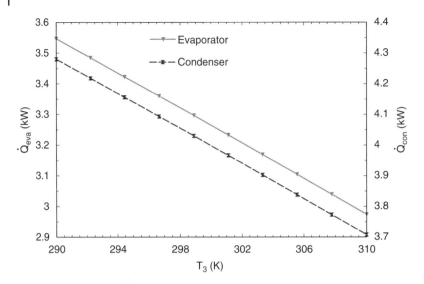

Figure 6.7 Effect of varying condenser temperature on condenser load and cooling load.

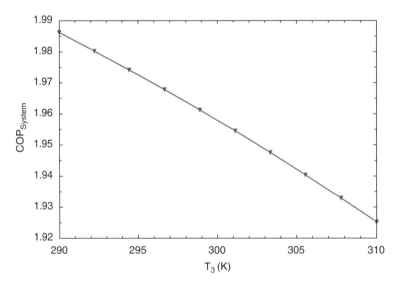

Figure 6.8 Effect of varying condenser temperature on COP of the system.

population size of $M = 100$ individuals, a crossover probability of $p_c = 0.9$, a gene mutation probability of $p_m = 0.035$, and a controlled elitism value $c = 0.55$.

Figure 6.9 shows the Pareto frontier solution for the multi-objective optimization of the heat pump system for the compressor work rate and condenser rejected heat rate as objective functions. The design parameters here are the evaporator temperature (T_1), the condenser temperature (T_3), and the refrigerant mass flow rate. The ranges of variation for these design parameters are listed in Table 6.2.

As shown in Figure 6.9, the maximum condenser load exists at design point C (112.2 kW), while the total compressor work rate is the greatest at this point (24.9 kW).

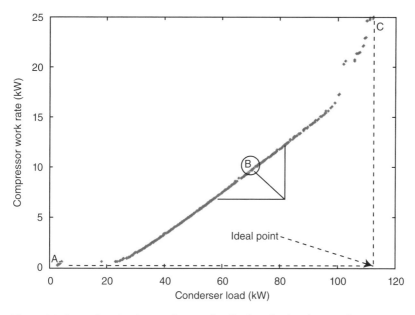

Figure 6.9 Pareto frontier showing best trade-off values for the objective functions.

Table 6.2 Optimization constraints.

Design parameter	Variation range
Evaporator temperature	$275\ \text{K} \leq T_1 \leq 320\ \text{K}$
Condenser temperature	$290\ \text{K} \leq T_1 \leq 310\ \text{K}$
Refrigerant mass flow rate	$0.01\ \text{kg/s} \leq \dot{m}_r \leq 0.4\ \text{kg/s}$

Also, the minimum value for the compressor work rate occurs at design point *A* and is about 0.2 kW. Design point *A* is the optimal situation when the compressor work rate is the sole objective function, while design point *C* is the optimum point when condenser load is the sole objective function.

In multi-objective optimization, a process of decision-making for selection of the final optimal solution from the available solutions is required. This process is usually performed with the aid of a hypothetical point in Figure 6.9 (the ideal point), at which both objectives have their optimal values independent of the other objectives. It is clear that it is impossible to have both objectives at their optimum point simultaneously and, as shown in Figure 6.9, the ideal point is not a solution located on the Pareto frontier.

The closest point to the ideal point on the Pareto frontier might be considered as a desirable final solution. Nevertheless, in this case, the Pareto optimum frontier exhibits weak equilibrium: a small change in condenser load from varying the operating parameters causes a large variation in the compressor work rate. Therefore, the ideal point cannot be utilized for decision-making in this problem. In selection of the final optimum point, it is desired to achieve a better magnitude for each objective than its initial value for the base case problem. Note that in multi-objective optimization and the Pareto solution, each point can be utilized as the optimized point. Therefore, the selection of

Table 6.3 Optimized values for design parameters of the system based on multi-objective optimization.

Design parameter	A	B	C
Evaporator temperature (K)	276	306	320
Condenser temperature (K)	301	299	290
Refrigerant mass flow rate (kg/s)	0.01	0.37	0.39

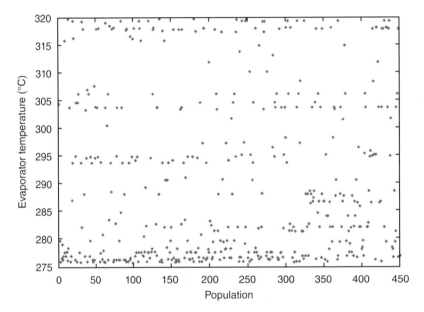

Figure 6.10 Scattered distribution of evaporator temperature with population in Pareto frontier.

the optimum solution depends on the preferences and criteria of the decision maker, suggesting that each may select a different point as the optimum solution depending on his/her needs. Table 6.3 shows all the design parameters for points *A-C*.

To better understand the variations of all design parameters, their scattered distributions are shown in Figures 6.10 to 6.12. The results show that refrigerant mass flow rate (Figure 6.12) tends to become as high as possible. This observation means that an increase in this parameters leads to better optimization results. The evaporator and condenser temperature have scattered distributions in their allowable domains, suggesting that these parameters have important effects on the trade-off between compressor work rate and condenser load.

6.6 Concluding Remarks

In this chapter, comprehensive energy and exergy analyses along with a multi-objective optimization are used for evaluating heat pump systems and their components. Actual data are utilized in the analysis and compared with the results of simulation code. A

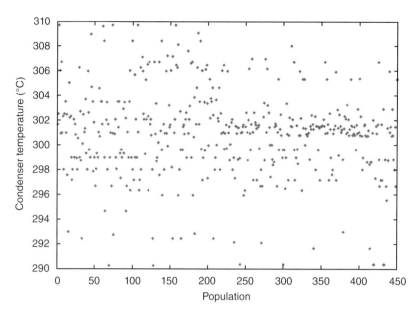

Figure 6.11 Scattered distribution of condenser temperature with population in Pareto frontier.

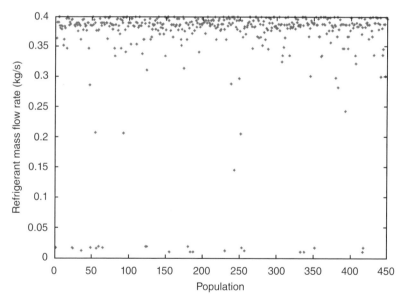

Figure 6.12 Scattered distribution of refrigerant mass flow rate with population in Pareto frontier.

good agreement is observed between the simulations and the experimentally measured data. The exergy analysis results show that the power required by the compressor, the condenser load, and the evaporator load all increase as the mass flow rate of refrigerant increases and as the evaporator temperature increases. In addition, an increase in the condenser temperature results in a reduction in COP, and the condenser and evaporator loads. Finally, a heat pump system is examined and a parametric study and optimization

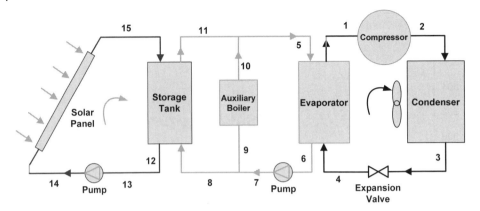

Figure 6.13 Schematic diagram of a solar assisted heat pump with storage tank and auxiliary boiler.

are conducted. A multi-objective optimization allows the optimal design parameters of the system to be determined.

Reference

1 Dincer, I. and Rosen, M. A. (2012) *Exergy: Energy, Environment and Sustainable Development*, Second edition, Elsevier.

Study Questions/Problems

1 Describe the four main processes that make up a simple vapor compression heat pump.

2 Compare air-source, water-source, and ground-source heat pump systems generally and from an exergetic point of view.

3 Determine the exergy efficiency of the following heat pump systems used to keep a house at 22°C and discuss the results: (a) an air-source heat pump with a COP of 1.8 that absorbs heat from outdoor air at 2°C, (b) a ground-source heat pump with a COP of 2.6 that absorbs heat from the ground at 12°C, and (c) a geothermal heat pump with a COP of 3.8 that absorbs heat from underground geothermal water at 60°C.

4 A schematic of a solar based heat pump is shown in Figure 6.13. The working fluid for this heat pump system is R134a. The solar panel surface area, evaporator pressure, condenser pressure, capacity of heat storage tank, and the values of superheating/subcooling in evaporator and condenser are selected as design parameters. Apply multi-objective optimization to determine the optimal value of the design parameters when exergy efficiency and total cost rate of the system are considered as the two objective functions.

7

Modeling and Optimization of Fuel Cell Systems

7.1 Introduction

Fuel cells are electrochemical energy conversion devices. They can be used in power generation, in vehicles, and even in laptop computers. Fuel cells are often considered zero-emission power sources, although they do emit water when fueled by hydrogen, and therefore are often viewed as environmentally benign. Although fuel cells are normally considered advanced energy systems and have attracted substantial interest only during the last decades, their operating principles were recognized long before that. Unlike the conventional heat engine, which converts fuel to heat and then uses that heat to generate electricity, fuel cells convert fuel to electricity directly using chemical reactions. An important advantage of fuel cells is their notably higher efficiencies compared to conventional power generation systems. The combination of high efficiency and low emissions permits fuel cells to be considered a clean energy technology. They are especially beneficial for transportation in locations where air quality is important.

A fuel cell is composed of three main parts: a fuel electrode (anode), an oxidant electrode (cathode), and an electrolyte between them. Figure 7.1 shows a single cell of a fuel cell. Fuel (H_2) enters the anode from the left and oxygen enters from the right to the cathode.

The overall chemical reaction is known as simple water splitting:

$$H_2 + \frac{1}{2}O_2 \rightarrow H_2O + \text{Work} + \text{Heat} \tag{7.1}$$

Although the overall reaction seems simple, water and thermal management in fuel cells is challenging. Issues need to be addressed such as water flooding and heat loss during operation. Fuel cells may be classified on a number of dimensions, such as operating temperature, type of electrolyte, and type of ion transferred through the electrolyte. Here we classify fuel cells based on their electrolyte, as follows:

- Alkaline fuel cell (AFC), which uses an alkaline water solution as an electrolyte
- Proton exchange membrane (PEM) or proton conducting membrane fuel cell
- Molten carbonate fuel cell (MCFC), which uses a molten carbonate salt as an electrolyte
- Solid oxide fuel cell (SOFC), which uses a solid oxide ion-conducting ceramic as an electrolyte

Further details about fuel cell classifications can be found elsewhere [1].

Optimization of Energy Systems, First Edition. Ibrahim Dincer, Marc A. Rosen, and Pouria Ahmadi.
© 2017 John Wiley & Sons Ltd. Published 2017 by John Wiley & Sons Ltd.

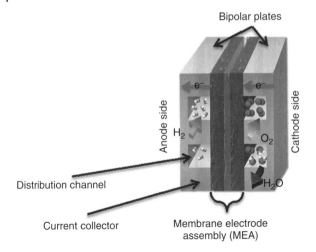

Bipolar plates

Anode side

Cathode side

H_2

O_2

e^-

e^-

H_2O

Distribution channel

Current collector

Membrane electrode
assembly (MEA)

Figure 7.1 Schematic of a single cell fuel cell.

In this chapter, we describe two common fuel cells: a PEM fuel cell, which is a low temperature fuel cell often employed for transportation applications, and an SOFC fuel cell, which is a high temperature suitable for stationary applications. Fundamental principles required for modeling are provided, and modeling of the two fuel cells is performed followed by parametric analyses. In addition, we comprehensively model and optimize some fuel cell systems for various objective functions and apply multi-objective optimization to determine the best optimal design parameters of the system. In order to enhance understanding of the design criteria, a sensitivity analysis is conducted to investigate how each objective function varies when a small change in a design parameter is made.

7.2 Thermodynamics of Fuel Cells

Some of the thermodynamic background required for the modeling of fuel cells is described in this section.

7.2.1 Gibbs Function

The Gibbs function is a thermodynamic property that is important when chemical reactions occur. The Gibbs energy is the energy that is available to do work, and can be defined as:

$$g = h - Ts \tag{7.2}$$

where s denotes specific entropy, h specific enthalpy and T absolute temperature. The Gibbs energy is a good indicator for predicting if the process is spontaneous or non-spontaneous. Similar to enthalpy of formation, the Gibbs function of a reaction is defined as the Gibbs function between the reaction products and reactants. If the products and reactants have the same pressure and temperature, the Gibbs function of the reaction can be expressed as:

$$\Delta g_{\text{reaction}} = g_P - g_R \tag{7.3}$$

If the formation reaction takes place at standard conditions ($T = 25\,°C$ and $P = 1$ atm), it is called the standard Gibbs function of formation, which is given in thermodynamic tables.

7.2.2 Reversible Cell Potential

The chemical energy of the fuel and the oxidant is converted to electricity in a fuel cell. This electricity is described in terms of cell potential and electrical current. The maximum electrical energy is obtained when the fuel cell operates reversibly from a thermodynamic perspective. In this case, the maximum cell potential is called the reversible cell potential. To determine the reversible cell potential, we apply fundamental thermodynamic principles. Considering the fuel cell in Figure 7.1 as a control volume, the overall reaction in a fuel cell can be written as

$$\text{Fuel (e.g., } H_2) + \text{Oxidant (e.g., } O_2) \rightarrow \dot{W} + \dot{Q} + \text{Product (e.g., } H_2O) \tag{7.4}$$

Here, the fuel, oxidant, and exhaust stream are at the same pressure and the same temperature, and it is assumed that the reaction is steady state. Writing the first and second laws of thermodynamics for a control volume around the fuel cell yields:

$$\frac{dE_{CV}}{dt} = \left[(\dot{N}h + KE + PE)_F + (\dot{N}h + KE + PE)_{\text{oxidant}} \right] - \left[(\dot{N}h + KE + PE)_P \right]$$
$$+ \dot{Q} - \dot{W} \tag{7.5}$$

$$\frac{dS_{CV}}{dt} = \left[(\dot{N}S)_F + (\dot{N}S)_{\text{oxidant}} \right] - \left[(\dot{N}S)_P \right] + \frac{\dot{Q}}{T} - \dot{S}_{\text{gen}} \tag{7.6}$$

where $\dot{N}$ denotes molar flow rate, h specific enthalpy, s specific entropy and $\dot{S}_{\text{gen}}$ the rate of entropy generation. The subscripts F and P denote the fuel and product, respectively, while KE and PE denote kinetic and potential energies, respectively. For a steady state process, there is no change in energy and entropy in the control volume, so the left hand sides of Equation 7.5 and Equation 7.6 become zero. It is also assumed that velocity and elevation changes are negligible, so that KE and PE can be neglected. Simplifying Equations 7.5 and 7.6 yields:

$$\left[(\dot{N}h)_F + (\dot{N}h)_{\text{oxidant}} \right] - \left[(\dot{N}h)_P \right] + \dot{Q} - \dot{W} = 0 \tag{7.7}$$

$$\left[(\dot{N}S)_F + (\dot{N}S)_{\text{oxidant}} \right] - \left[(\dot{N}S)_P \right] + \frac{\dot{Q}}{T} + \dot{S}_{\text{gen}} = 0 \tag{7.8}$$

For convenience, we denote the following:

$$h_{\text{in}} = h_F + \frac{\dot{N}_{\text{oxidant}}}{\dot{N}_F} h_{\text{oxidant}} \tag{7.9}$$

$$h_{\text{out}} = \frac{\dot{N}_P}{\dot{N}_F} h_P \tag{7.10}$$

$$s_{\text{in}} = s_F + \frac{\dot{N}_{\text{oxidant}}}{\dot{N}_F} s_{\text{oxidant}} \tag{7.11}$$

$$s_{\text{out}} = \frac{\dot{N}_P}{\dot{N}_F} s_P \tag{7.12}$$

Substituting Equations 7.9 to 7.12 into Equations 7.7 and 7.8 yields:

$$\dot{N}_F(h_{\text{in}} - h_{\text{out}}) + \dot{Q} - \dot{W} = 0 \tag{7.13}$$

$$\dot{Q} + T\dot{S}_{gen} + \dot{N}_F T(s_{in} - s_{out}) = 0 \tag{7.14}$$

Substituting Equation 7.14 into Equation 7.13, we have

$$\dot{W} = \dot{N}_F(h_{in} - h_{out}) - T\dot{S}_{gen} - \dot{N}_F T(s_{in} - s_{out})$$

Defining $w = \frac{\dot{W}}{\dot{N}_F}, q = \frac{\dot{Q}}{\dot{N}_F}$ and $s_{gen} = \frac{\dot{S}_{gen}}{\dot{N}_F}$, we have:

$$q = -Ts_{gen} - T(s_{in} - s_{out}) = T\Delta s - Ts_{gen}$$

$$w = (h_{in} - h_{out}) - Ts_{gen} - T(s_{in} - s_{out}) = (h_{in} - Ts_{in}) - (h_{out} - Ts_{out}) - Ts_{gen}$$

Using the Gibbs function in Equation 7.2 we have:

$$w = g_{in} - g_{out} - Ts_{gen} = -\Delta g - Ts_{gen}$$

According to the second law of thermodynamics, we know that entropy is always generated and never destroyed. Therefore, $s_{gen} \geq 0$, and the maximum possible work is obtained when $s_{gen} = 0$. Thus, the maximum available work for the fuel cell here is equal to the decrease in Gibbs function and can be expressed as:

$$w_{max} = -\Delta g \tag{7.15}$$

This equation is valid for all reversible processes regardless of the type of fuel cell. In fuel cell systems, the output electrical energy is expressed in terms of the potential difference between the cathode and anode. The potential is the potential energy per unit of charge (J/C), which is called a Volt. Potential energy is defined as the work done when charge is moved from one location to another in the electrical field. In the case of a fuel cell, the work is done by transferring one Coulomb of positive charge from a low to high potential. In this case the work done by a fuel cell can be expressed as follows:

$$w \text{ (J/mol fuel)} = E \times nF \tag{7.16}$$

where n is the number of moles of electrons transferred per mole of fuel consumed and F is the Faraday constant (96,487 C/mol electron).

Rearranging Equation 7.16, we have:

$$E = \frac{w}{nF} = \frac{-\Delta g - Ts_{gen}}{nF}$$

The maximum cell potential occurs when $s_{gen} = 0$ according to Equation 7.15 and can be called reversible cell potential and is expressed as:

$$E_r = \frac{-\Delta g}{nF}$$

Note that the Gibbs function is a thermodynamic property determined by state variables. Therefore the Gibbs function difference for a fuel cell is also a function of temperature and pressure:

$$\Delta g = \Delta h - T\Delta s$$

If the reaction occurs at standard temperature and pressure, this reversible cell potential is called standard cell potential and can be expressed as:

$$E_{r,0} = -\frac{\Delta g(T_{ref}, P_{ref})}{nF}$$

7.3 PEM Fuel Cell Modeling

As explained earlier, PEM fuel cells are used for various applications. Here we focus on vehicular applications where PEM fuel cells replace the internal combustion engines in conventional cars to reduce greenhouse gas emissions. For this investigation, Ballard's Xcellsis™ HY-80 fuel cell engine is used (Figure 7.2). Since the Xcellsis™ HY-80 fits beneath the floor of the vehicle, the size of the passenger compartment is not affected. The engine is a lightweight, 68 kW, hydrogen fueled, fuel cell system. The hydrogen is stored in a tank at 10 atm at 298 K. After pressure regulation, depending on the system pressure, the hydrogen is fed to the system. The fuel cell stack is composed of 97 cells each having a 900 cm^2 active surface area. A cooling system is employed to maintain a constant operating temperature inside the fuel cell stack.

The maximum voltage that can be produced by a cell without any irreversibility is called the reversible cell voltage. The following form of the Nernst equation [2] is used:

$$V_{rev} = 1.229 - 8.5 \cdot 10^{-4}(T_{FC} - 298.25)$$
$$+ 4.3085 \cdot 10^{-5} \cdot T_{FC} \left[\ln(p_{H_2}) + \frac{1}{2} \ln(p_{O_2}) \right]$$

as it allows investigation of the performance variations with system temperature, system pressure, and the partial pressures of both hydrogen and oxygen. The hydrogen and oxygen partial pressures can be written as

$$p_{H_2} = \frac{1 - x_{H_2O,A}}{1 + (x_A/2)(1 + \xi_A/(\xi_A - 1))} \cdot P_A$$
$$p_{O_2} = \frac{1 - x_{H_2O,C}}{1 + (x_C/2)(1 + \xi_C/(\xi_C - 1))} \cdot P_C$$

where x_{H_2O} is the water mole fraction, that is P_{sat}/P, for both the anode and cathode, x_A and x_C are the anode and cathode dry gas mole fractions respectively, and ξ_A and ξ_C are the anode and cathode stoichiometric ratios respectively. The operating cell voltage is less than the reversible cell voltage because of irreversibilities and over potentials.

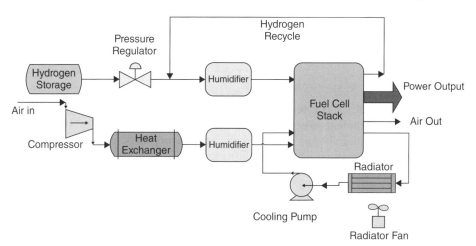

Figure 7.2 Flow diagram of Ballard Xcellsis™ HY-80 fuel cell engine. *Source*: Mert 2011. Reproduced with permission of Elsevier.

Since the over potentials depend on the system operating parameters, the operating cell voltage is not constant while the reversible cell voltage is constant for general cases. The operating cell voltage may be expressed by:

$$V_{operating} = V_{rev} - V_{irrev}$$

Activation losses, Ohmic losses, and mass transport or concentration losses are the over potentials that are taken to account in this study. That is,

$$V_{irrev} = V_{act} + V_{ohm} + V_{con} \qquad (7.17)$$

Here, V_{act} is the activation over potential, which arises from the slow rate of electro-chemical reactions and a portion of energy being lost on driving the rate of electro-chemical reactions in order to meet the rate required by the current demand. It can be expressed as:

$$V_{act} = \frac{\alpha_A + \alpha_C}{\alpha_A \alpha_C} \frac{RT_{FC}}{nF} \ln\left(\frac{J}{J_0}\right)$$

where α_A and α_C are the anode and cathode transfer coefficients respectively. Also, J_0 is the exchange current density, R is the universal gas constant and n is the number of electrons involved. In Equation 7.17, V_{ohm} is the Ohmic over potential, defined as:

$$V_{ohm} = Jt_{mem}\left[0.0051\lambda_{mem} - 0.0032\exp\left[1268\left(\frac{1}{303} - \frac{1}{T_{FC}}\right)\right]\right]^{-1}$$

Here, t_{mem} is the membrane thickness, λ_{mem} is the membrane humidity, and T_{FC} is the fuel cell operating temperature. Furthermore,

$$V_{con} = J\left(\beta_1 \frac{J}{J_{max}}\right)^{\beta_2}$$

where J_{max} is the limiting current density, and β_1 and β_2 are constants of the concentration over potentials. More details about fuel cell modeling can be found in reference [3].

After accounting for irreversibilities, the electrical power produced by the cells can expressed as:

$$\dot{W}_{Stack} = V_{operating} \cdot i \cdot A_{cell} \cdot n_{cell}$$

where n_{cell} is the number of fuel cells inside the stack, A_{cell} is the area of the each cell, and i is the current density.

7.3.1 Exergy and Exergoeconomic Analyses

Exergoeconomic analysis combines an economic assessment with exergy analysis, accounting for equipment costs and thermodynamic irreversibilities throughout the system. Exergy analysis was discussed in detail in previous chapters, so limited information on it is provided here. The following formulations are utilized for the net output power, and energy and exergy efficiencies for the overall system, respectively:

$$\dot{W}_{net} = \dot{W}_{FC} - \dot{W}_{comp} - \dot{W}_{cool,pump} - \dot{W}_{fan}$$

$$\eta_{sys,energy} = \frac{\dot{W}_{net}}{HHV_{H_2} \cdot F_{H_2,in}}$$

$$\eta_{sys,exergy} = \frac{\dot{W}_{net}}{\dot{E}_{in}}$$

where $\dot{E}_{in}$ is the exergy of the inlet streams of the system. An overall exergetic cost rate balance equation can be written as:

$$\sum(\dot{E}_{in,i} \cdot C_{in,i}) + \dot{Z}_{tot} = \sum(\dot{E}_{out,i} \cdot C_{out,i}) + \dot{W}_{net} \cdot C_W$$

Here, $\dot{E}_{in,i}$, $\dot{E}_{out,i}$, $C_{in,i}$, and $C_{out,i}$ are the exergies and exergy costs of the streams entering and leaving the control volume respectively, $\dot{Z}_{tot}$ is the annualized cost of the total system within the control volume, and C_W is the cost of the work or the power of the equipment. The cost balance is applied to the overall system to calculate the cost of the electrical power produced by the fuel cell system.

7.3.2 Multi-objective Optimization of a PEM Fuel Cell System

In order to apply a multi-objective optimization technique to the PEM fuel cell system considered, objective functions are required. In this study, a weighting method is used to form the objective function. That is, a weighting factor is applied to the objective function parameters, which are the produced work (W_{FC}), the energy efficiency $\eta_{sys,energy}$, the exergy efficiency $\eta_{sys,exergy}$, and the cost of produced work C_W. Note that C_W must be minimized, whereas the other parameters are to be maximized in the search for the optimum condition.

$$Z = w_1 W_{FC} + w_2 \eta_{energy} + w_3 \eta_{exergy} - w_4 C_w$$

The weighting factors are not selected; instead a parametric study for different weighting factors (Table 7.1) is carried out and the multi-objective problem is applied separately for each case.

The fitness function is formed depending on the objective function. The main problem for optimizing this objective function is the difference in the scales of the parameters. The value of $\dot{W}_{FC}$ is between 0 and 120 kW, while values of C_w lie between 2×10^{-3} and 6×10^{-3} \$/kW. Therefore, if the problem is solved without manipulating these conditions, the results will always seek the maximum work without attributing importance to the cost of production, regardless of the weight factor assigned to the

Table 7.1 Weighting factor sample for the parametric weighting.

Work (%)	Energy (%)	Exergy (%)	Cost (%)
0.900	0.033	0.033	0.033
0.800	0.067	0.067	0.067
0.700	0.100	0.100	0.100
0.600	0.133	0.133	0.133
0.500	0.167	0.167	0.167
0.400	0.200	0.200	0.200
0.300	0.233	0.233	0.233
0.200	0.267	0.267	0.267
0.100	0.300	0.300	0.300
0.000	0.333	0.333	0.333

Table 7.2 The ranges for parametric investigation.

Parameters	Range
Operating temperature	323–353 K
Operating pressure	2.5–4 atm
Membrane thickness	0.016–0.02 cm
Anode stoichiometry	1–3
Cathode stoichiometry	1.1–3
Reference temperature	253–323 K
Reference pressure	0.85–1 atm
Current density	0.01–2 A cm^{-2}

parameters. To avoid this situation, the parameters are normalized to form a fitness function between 0 and 1. The normalization is done by dividing each parameter by its maximum value, which is obtained by single optimization of that parameter with a computer program called MULOP, developed by the authors. Hence, the fitness function is modified as follows:

$$Z_{norm} = w_1 \frac{W_{FC}}{W_{FC,max}} + w_2 \frac{\eta_{energy}}{1} + w_3 \frac{\eta_{exergy}}{1} - w_4 \frac{C_w}{C_{w,max}}$$

The multi-objective optimization procedure involves a search for optimum values over the range of parameter values given in Table 7.2.

Figure 7.3 shows the change of work and energy and exergy efficiencies with respect to the weight of work in the objective function. It is seen that the work and efficiencies

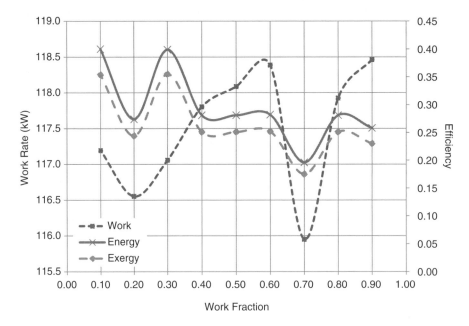

Figure 7.3 Variation of produced work and efficiencies with respect to weight of work in the objective function.

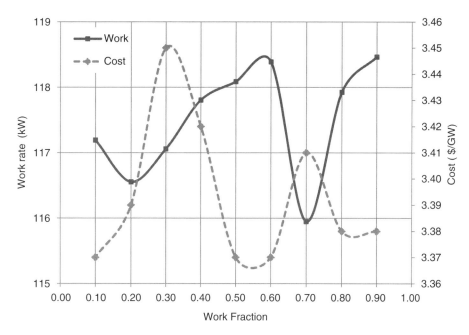

Figure 7.4 Variation of produced work and cost with respect to weight of work in the objective function.

are proportional to each other, which is an expected result since these values are all maximized. It is also seen that the best values for efficiencies lie toward the lower values of contribution, but for these values the produced work is low. Efficiency values between 0.5–0.6 are much more acceptable.

Figure 7.4 displays one maximization and one minimization objective; these are work and cost respectively. It is seen that when the work of the system increases, the cost decreases. This is an expected result since the optimum values in the figure are seen to be between 0.5–0.6, which is in harmony with the efficiency values.

Variations of the energy efficiency objective weight are investigated (see Figures 7.5 and 7.6). Similar to other figures, it is seen that when the weight of an objective function increases, a better result is not guaranteed for that objective, particularly because of the dynamics and complexity of the problem. The optimum results for these sets are for an energy efficiency weight of roughly 0.7, which has the minimum cost and meaningful efficiency and produced work results.

The results for the exergy efficiency objective are similar to the results for the energy efficiency objective, as seen in Figures 7.7 and 7.8. The result for exergy efficiency optimization is in accordance with the results for the energy efficiency values, and both optimized values depend on the performance of the fuel cell system tested.

The variation of electrical power, efficiencies, and cost with respect to the weight of cost in the objective function is shown in Figures 7.9 and 7.10. Surprisingly, it is seen that at a 0.6 value for the cost fraction, each of the work, the efficiencies, and the cost exhibit the lowest values. This is different to the other sets considered. The optimum results appear to lie between 0.8–0.9 in terms of cost and produced work. If the efficiencies are considered, the best optimum value seems to be for weights of 0.3–0.4.

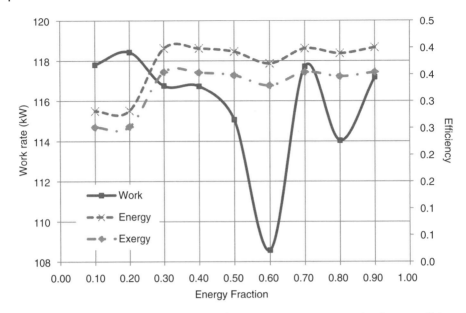

Figure 7.5 Variation of produced work and efficiencies with respect to weight of energy efficiency in the objective function.

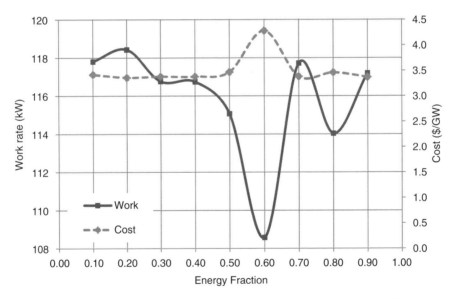

Figure 7.6 Variation of produced work and cost with respect to weight of energy efficiency in the objective function.

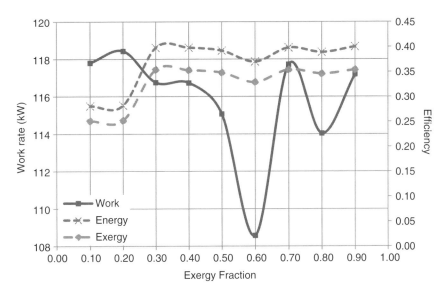

Figure 7.7 Variation of produced work and efficiencies with respect to weight of exergy efficiency in the objective function.

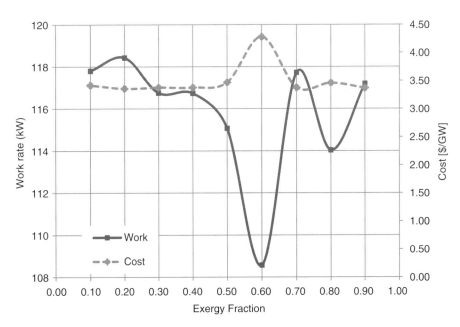

Figure 7.8 Variation of produced work and cost with respect to weight of exergy efficiency in the objective function.

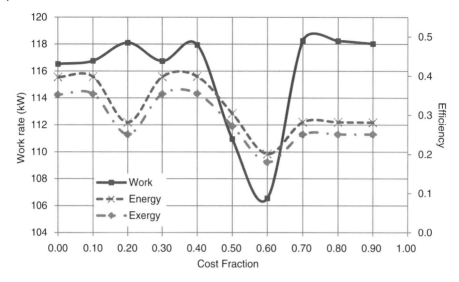

Figure 7.9 Variation of produced work and efficiencies with respect to weight of cost in the objective function.

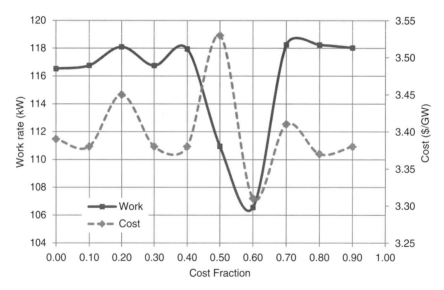

Figure 7.10 Variation of produced work and cost with respect to weight of cost in the objective function.

Table 7.3 presents the optimum cases for the investigation. The best values for the efficiencies and the electrical power values lie at high values of the work fraction as expected, roughly 0.9, which is taken as the maximum fraction given to the work in this study. The production cost is inversely proportional to the maximum produced work, and the selection of the optimum value depends on the needs of the system in which the fuel cell system is used, according to Figures 7.3 to 7.10. So, the best value from a cost point of view is 3.31 $/GW at a 0.6 work fraction.

Table 7.3 Selected results of the multi-objective optimization procedure.

Work fraction	Energy fraction	Exergy fraction	Cost fraction	Work rate (kW)	Cost ($/GW)	Exergy efficiency	Energy efficiency	Temp. (K)	Pressure (atm)	Membrane thickness (mm)	Current density (A/cm^2)
0.13	0.13	0.13	0.60	106	3.31	0.18	0.20	352	2.24	0.008	1.75
0.90	0.03	0.03	0.03	118	3.38	0.23	0.26	351	2.18	0.008	1.98
0.00	0.33	0.33	0.33	108	5.56	0.49	0.55	324	3.47	0.011	0.53
0.80	0.07	0.07	0.07	117	3.38	0.25	0.28	351	2.19	0.01	1.97
0.30	0.23	0.23	0.23	117	3.45	0.36	0.40	350	2.15	0.01	1.89
0.13	0.60	0.13	0.13	108	4.28	0.33	0.37	343	2.18	0.01	1.95

7.4 SOFC Modeling

The structure of a single SOFC is shown in Figure 7.11 and consists of six components: solid bipolar plate interconnection (of thickness x_1), fuel channels (x_2), anode electrode (x_3), electrolyte layer (x_4), cathode electrode (x_5) and oxidant channels (x_6). Fuel and oxidant are supplied at their respective channels and pass through the anode and cathode electrodes respectively, and react at the electronic conductor/gas interface/ionic conductor (TPB). The bipolar plate supplies a series of electrical connectors to the nearby cells or to the external circuit, and at the same time works as a gas barricade between the fuel channels and oxidant channels of nearby cells. In this analysis, pure hydrogen and air are regarded as the fuel and the oxidant, respectively. To produce more power, we consider 10 stacks of SOFCs, each consisting of 100 cells.

7.4.1 Mathematical Model

The technical model of an SOFC consists of two main parts: a mass balance model and an electrochemical model. The outputs of these models are output power and electrical efficiency.

In a solid oxide fuel cell, there are two electrochemical reactions, one occurring at the anode:

$$H_2 + O^{2-} \rightarrow H_2O + 2e^- \text{(anode)}$$

and the other at the cathode:

$$2e^- + \frac{1}{2}O_2 \rightarrow O^{2-} \text{(cathode)}$$

The fuel and oxidant channels provide hydrogen and oxygen, respectively. On the basis of electrochemical reaction equations, the molar flow rates of fuel at the inlet and outlet of fuel channels, and the hydrogen consumed by the electrochemical reaction, relations involving the operating current (I) can be written as in Equations 7.18 to 7.20. Here,

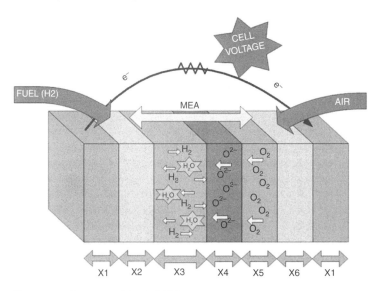

Figure 7.11 Structure of single SOFC.

n is the number of mole of electrons transferred per mole of reactant (H_2 or O_2). For the anode (hydrogen), $n = 2$, and for the cathode (oxygen), $n = 4$. In a similar manner, the molar flow rates of oxygen at the inlet and outlet of air channels and the oxygen absorbed by the electrochemical reactions are expressed as in Equations 7.21 to 7.23.

$$n_{H_2,in} = \frac{I}{nF\gamma_f} \tag{7.18}$$

$$n_{H_2,out} = \frac{I}{nF}\left(\frac{1}{\gamma_f} - 1\right) \tag{7.19}$$

$$n_{H_2} = \frac{I}{nF} \tag{7.20}$$

$$n_{O_2,in} = \frac{I}{nF}\lambda_{air} \tag{7.21}$$

$$n_{O_2,out} = (\lambda_{air} - 1)\frac{I}{nF} \tag{7.22}$$

$$n_{O_2} = \frac{I}{nF} \tag{7.23}$$

In this analysis, the oxidant provided in the cathode channels is air, which is assumed to consist of oxygen and nitrogen. The molar flow rate of nitrogen at the cathode inlet and outlet is given by Equation 7.24. In the fuel channels, the average mole fractions are approximated by Equation 7.25. In the oxidant channels, the average molar fractions are given by Equations 7.26 and 7.27.

$$n_{N_2,in} = n_{N_2,out} = \frac{I\lambda_{air}X_{N_2,in}}{nFX_{O_2,in}} \tag{7.24}$$

$$X_{H_2O} = \frac{\gamma_f}{2}, \quad X_{H_2} = 1 - \frac{\gamma_f}{2} \tag{7.25}$$

$$X_{O_2} = \frac{2\lambda_{air}X_{O_2,in} - X_{O_2,in} - X_{O_2,in}^2}{2(\lambda_{air} - X_{O_2,in})} \tag{7.26}$$

$$X_{N_2} = \frac{2\lambda_{air}X_{N_2,in} - X_{O_2,in}X_{N_2,in}}{2(\lambda_{air} - X_{O_2,in})} \tag{7.27}$$

One of the most important parameters in fuel cells is operating voltage. The operating cell voltage is determined by removing all over potentials (activation, Ohmic, and diffusion) from the standard Nernst voltage at the given temperature and pressure. Losses are functions of the operating current, pressure, and temperature as well as the SOFC structure. Thus, the operating voltage can be expressed as follows:

$$V = V_{OCV} - \left(\sum_j \eta_{ohm,j} + \eta_{diff,3} + \eta_{diff,5} + \eta_{act,3} + \eta_{act,5}\right), j = 1, 3, 4, 5 \tag{7.28}$$

Here, $\eta_{ohm,j}$ is the Ohmic loss in all solid compartments of the fuel cell. Also, V_{OCV} denotes the open circuit voltage, which can be written based on the assumption that both reactions occur at the fuel cell temperature as:

$$V_{OCV} = -\frac{1}{2F}\left(H_{H_2O} - H_{H_2} - \frac{1}{2}H_{O_2}\right) + \frac{T}{2F}\left(S^0_{H_2O} - S^0_{H_2} - \frac{1}{2}S^0_{O_2}\right)$$
$$+ \frac{RT}{2F}\ln\left(\frac{P_{H_2O}}{P_{H_2}(P_{O_2})^{\frac{1}{2}}}\right) \tag{7.29}$$

The interface of the electronic conductor phase, reacting gases, and ionic conductor phase is called the three-phase boundary (TPB), and corresponds to where the electrochemical reactions can take place. Also, $L_{j,\text{opt}}$ refers to the active TPB sites where the electrochemical reactions actually occur. In this analysis, the composite electrodes of the SOFC are considered as a combination of electronic and ionic conductors. The Ohmic loss is achieved on the basis of Ohm's law as written in Equations 7.30 and 7.31. The Ohmic losses in active TPB sites where the ionic resistance is dominant are accounted for in the first term on the right hand side of Equation 7.31. The second term refers to the remaining sites of the anode and cathode electrodes where only electronic conducting resistance exists. Additionally, β_1 and β_4 are the electrical resistances of the bipolar plate and electrolyte, which are dependent on material and geometric features as in Equations 7.32 and 7.33, $A_{j,\text{opt}}$ is applied to the area of the active TPB regions in each electrode, as written in Equation 7.34, and D_{Pj} denotes the average diameter of the parallel pores in the two electrodes (see Equation 7.35).

$$\eta_{\text{ohm},j} = I\beta_j, \qquad j = 1,4 \tag{7.30}$$

$$\eta_{\text{ohm},j} = \frac{I(A_{j,\text{opt}} - L_y L_z)L_{j,\text{opt}}}{2A_{j,\text{opt}}L_y L_z(1 - \phi_j)\sigma_4} + \frac{I(L_j - L_{j,\text{opt}})}{L_y L_z(1 - 1.8\phi_j)\sigma_i}, \quad j = 3,5 \tag{7.31}$$

$$\beta_1 = \frac{L_1}{L_y L_z \sigma_1} + \frac{L_2}{(1 + n_{ch})L_t L_y \sigma_1} + \frac{L_6}{(1 + n_{ch})L_t L_z \sigma_1} \tag{7.32}$$

$$\beta_4 = \frac{L_4}{L_y L_z \sigma_4} \tag{7.33}$$

$$A_{j,\text{opt}} = (1 + 2L_{j,\text{opt}}\phi_j/D_{Pj})L_y L_z \tag{7.34}$$

$$D_{Pj} = D_{W,j}\phi_j/(1 - \phi_j) \tag{7.35}$$

Here, P_{TPB} is applied to the partial pressures of reactants and products at the TPB regions accordingly:

$$P_{\text{H}_2\text{O},TPB} = p_f X_{\text{H}_2\text{O}} + \frac{RT\psi_3}{2FD_3\Phi_3}\frac{L_3 I}{L_y L_z} \tag{7.36}$$

$$P_{\text{H}_2,TPB} = p_f X_{\text{H}_2} - \frac{RT\psi_3}{2FD_3\phi_3}\frac{L_3 I}{L_y L_z} \tag{7.37}$$

$$P_{\text{O}_2,TPB} = p_{\text{air}} - (p_{\text{air}} - X_{\text{O}_2}p_{\text{air})\exp\left(\frac{RT\psi_5 L_5 I}{4FD_5\phi_5 p_{\text{air}}L_y L_z}\right) \tag{7.38}$$

$$D_3 = \frac{1.43 \times 10^{-7}T^{1.75}(M_{\text{H}_2} + M_{\text{H}_2\text{O}})^{\frac{1}{2}}}{p_f(2M_{\text{H}_2}M_{\text{H}_2\text{O}})^{\frac{1}{2}}\left(v_{\text{H}_2}^{\frac{1}{3}} + v_{\text{H}_2\text{O}}^{\frac{1}{3}}\right)^2} \tag{7.39}$$

$$D_5 = \frac{1.43 \times 10^{-7}T^{1.75}(M_{\text{O}_2} + M_{\text{N}_2})^{1/2}}{p_{\text{air}}(2M_{\text{O}_2}M_{\text{N}_2})^{1/2}(v_{\text{O}_2}^{1/3} + v_{\text{N}_2}^{1/3})^2} \tag{7.40}$$

$$\eta_{\text{diff},c} = \frac{RT}{4F}\ln\left(\frac{P_{\text{O}_2}}{P_{\text{O}_2,TPB}}\right) \tag{7.41}$$

$$\eta_{\text{diff},a} = \frac{RT}{4F}\ln\left(\frac{P_{\text{H}_2}P_{\text{H}_2\text{O},TPB}}{P_{\text{H}_2,TPB}P_{\text{H}_2\text{O}}}\right) \tag{7.42}$$

To calculate activation losses, the Butler–Volmer equation is used [1], expressed as follows:

$$i = i_{o,j} \left[\exp\left(\frac{\alpha n F}{RT} \eta_{act,j} \right) - \exp\left(-\frac{(1-\alpha)nF}{RT} \eta_{act,j} \right) \right], j = 3,5 \tag{7.43}$$

$$\eta_{act,j} = \frac{RT}{\alpha n F} \sinh^{-1}\left(\frac{i}{i_{0,j}} \right), \quad j = 3,5 \tag{7.44}$$

where n has the value of 2 and α represents the charge transfer coefficient, which is 0.5 here. Also, $i_{0,j}$ denotes the exchange current density and can be expressed as follows:

$$i_{0,j} = \frac{RT}{nF} \delta_j \exp\left(-\frac{E_j}{RT} \right), j = 3,5 \tag{7.45}$$

The electrical power density of a single SOFC is

$$P_{elec,cell} = V \times i \tag{7.46}$$

and the overall output power generated by 10 SOFC stacks can be written as

$$P_{total} = V \times i \times N \times A \tag{7.47}$$

where N is total number of single cells. The electrical efficiency can be expressed as

$$\eta_{elec} = \frac{V \times i \times N \times A}{n_{H_2} \times LHV_f} \tag{7.48}$$

Values of the relevant physical properties and model constants are listed in Table 7.4.

7.4.2 Cost Analysis

Generally in this analysis, the SOFC cost model consists of three main parts: capital cost, fuel cost, and maintenance cost. That is,

$$C = C_{capital} + C_{fuel} + C_{maintenance} \tag{7.49}$$

The output of this model is the cost of electricity produced from the SOFC stacks. For the capital cost, the cost model of the SOFC developed by Palazzi et al [4] is used. The SOFC capital cost includes two parts. The first is cost of the pressurized enclosure (insulation), expressible as:

$$C_{VP} = 10^{K_1 + K_2 \log(ThermalLoad)} \tag{7.50}$$

$$C_{VBM} = C_{VP} \cdot f_{BM} \cdot f_{pressure} \tag{7.51}$$

The second part of the capital cost includes the fuel cell cost and the stack housing cost, which can be written respectively as follows:

$$C_{cell} = A_{cell} C_{spec} \tag{7.52}$$

$$C_{FCstack} = f_{BM} \cdot (C_{cell} \cdot N_{cells} + 2 \cdot N_{stack} \cdot A_{cell} \cdot f_{hs} \cdot C_{h,spec}) \tag{7.53}$$

Where $N_{stack} = N_{cells}/N_{cells,max}$

The present value of the capital cost can be expressed as

$$C_{capital} = f_a \cdot (C_{VBM} + C_{FCstack}) \tag{7.54}$$

Here, f_a is the annual recovery factor for bringing the cost to a present value, defined as $f_a = \frac{r(1+r)^n}{(1+r)^n - 1}$.

Table 7.4 Parameter values for various properties and constants used in modeling of the SOFC.

Parameter	Value	Parameter	Value
Physical property parameters		A	100 cm^2
$D_{W,3}$	14×10^{-6} m	L_1	0.05 cm
$D_{W,5}$	14×10^{-6} m	L_2	0.1 cm
E_3	140×10^3 Jmol^{-1}	L_3	0.05 cm
E_5	137×10^3 Jmol^{-1}	L_4	0.002 cm
n_{ch}	10	L_5	0.005 cm
α	0.5	L_6	0.1 cm
δ_3	6.54×10^{11} Ω^{-1}m^{-2}	L_{ch}	0.2 cm
δ_5	2.35×10^{11} Ω^{-1}m^{-2}	L_t	0.8 cm
v_{O_2}	16.3 cm^3	L_y	10 cm
v_{H_2}	6.12 cm^3	L_z	10 cm
v_{H_2O}	13.1 cm^3	$L_{3,opt}$	3.75×10^{-4}
v_{N_2}	18.5 cm^3	$L_{5,opt}$	7.20×10^{-4}
σ_1	1.5×10^6 Ω^{-1}m^{-1}	Operation constants	
σ_3	8.0×10^4 Ω^{-1}m^{-1}	$P_f = P_{air}$	0.1 MPa
σ_4	$33.4 \times 10^3 \exp(-10.3 \times 10^3/\mathrm{T})$ Ω^{-1}m^{-1}	P_{ref}	0.1 MPa
σ_5	$8.4 \times 10^3 \Omega^{-1}$m^{-1}	T	1073 °K
Ψ_3	9.5	Global constants	
Ψ_5	7.2	F g	96 500 C mol^{-1}
Sizing parameters		R	8.314 J mol^{-1} K^{-1}
N	1000		

The capital cost and annual cost are utilized to determine the electricity cost ($/kWh):

$$C_{capital,annual} = C_{capital} \times \frac{r \times (1+r)^n}{(1+r)^n - 1} \tag{7.55}$$

The fuel cost is calculated based on the amount of methane that is consumed during the system life cycle and the heating value of hydrogen (kJ/kg). The cost analysis assumes hydrogen is produced from natural gas with a 10-year lifetime with a 9% interest rate [6]. Also, the annual maintenance cost of the system is assumed to be 3% of its capital cost. Hence, the fuel cost can be written as

$$C_{fuel} = C_{H_2} \times LHV_{H_2} \times n_{H_2} \times M_{H_2} \times PF \tag{7.56}$$

Here, PF is the power factor, which is equal to 6000 hr/year. By dividing the total cost given in Equation 7.49 by the output electrical power of the stacks, the electricity cost (in $/kWh) is obtained.

7.4.3 Optimization

In order to achieve optimum performance of the SOFC stacks, decision variables need to be determined using optimization, including objective functions, decision variables, and physical constraints. Using Equations 7.16 to 7.46, the output electrical power and electrical efficiency are determined. The electricity cost is also an important objective function, and can be determined using Equations 7.49 to 7.56. Here, three objective functions are considered as three strategies for creating optimal points.

The constraints and their variation ranges for decision variables are as follows:

$$1000 \leq N \leq 18000 \tag{7.57}$$

$$0 \leq i \leq 4.5 \tag{7.58}$$

$$900 \leq T \leq 1100 \tag{7.59}$$

$$1 \leq P \leq 4 \tag{7.60}$$

Figure 7.12 shows the variation of efficiency and output electrical power of the fuel cell with respect to current density. It is seen that initially, an increase in current density results in an increase in SOFC output power, up to a maximum value; as current density increases further, the output power decreases due to the fact that beyond a specific current density the voltage losses dominate and the power decreases. The efficiency of the system, however, decreases with increasing SOFC current density. This is mainly due to the fact that an increase in current density necessitates an increase in the hydrogen flow rate as a fuel to the fuel cell. Since a higher current density requires more fuel, an increase in this parameter increases the fuel cost and the cost of the electricity production. This observation indicates that, along with SOFC capacity, current density plays a significant role in the determination of the objective function and should be considered as one of the decision variables.

In order to optimize the SOFC fuel cell, three objective functions are considered: output power to be maximized, fuel cell efficiency to be maximized, and cost of electricity

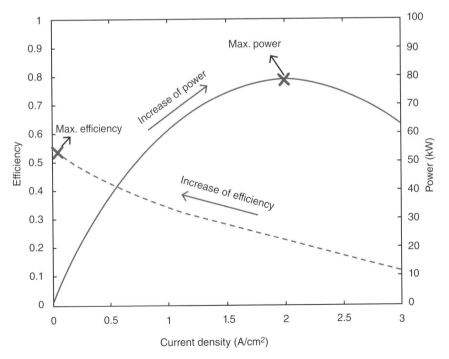

Figure 7.12 Variation of SOFC efficiency and output power with current density.

Table 7.5 Optimization results for SOFC fuel cell.

	First strategy: max. power (kW)	Second strategy: max. electrical efficiency	Third strategy: min. COE ($/kWh)
Objective function	222.95	0.635	0.222
		Decision variables	
Number of SOFC cells	1787	1587	1748
Current density (A/cm²)	2.7	0.1	0.4
SOFC temperature	1136.8	1088.7	1137.1
SOFC pressure	4.0	3.8	3.9

to be minimized. A genetic algorithm is applied in order to determine the optimal values of the design parameters for each objective function. The results are presented in Table 7.5.

An increase in SOFC fuel cell capacity, which corresponds to an increase in the number of fuel cell stacks, results in an increase in the electrical power output of the system. But, an increase in fuel cell current density initially increases the output power to an optimal value. This is where optimization is important for determining optimal values. When the output power of the fuel cell is considered in optimization, higher values for the number of stacks results in an increase in current density and fuel cell temperature, which are considered as constraints for the output power above a certain value. For the second objective function, which is the maximization of fuel cell efficiency, an increase in the number of fuel cell stacks at a fixed fuel mass flow rate to the system results in an increase in the fuel cell output power, which eventually increases the efficiency of the system. In addition, an increase in the number of fuel cell stacks at a fixed fuel mass flow rate causes the fuel cell current density to decrease, which increases the efficiency of the system. This observation is expected according to Figure 7.12. Note that an increase in the number of fuel cell stacks initially has more of an effect on the increase in system efficiency; after a certain number of fuel cell stacks, an increase in this parameter results in a decrease in the fuel cell operating temperature, affecting system efficiency. This explains why an optimized value for the number of fuel cell stacks exists for this objective function.

Similarly, an increase in fuel cell capacity initially results in a decrease in the cost of electricity production to a minimum value; beyond this value, an increase in the fuel cell capacity increases the cost of electricity. The reason is that an increase in the number of fuel cell stacks initially has a pronounced effect on the output power, which results in a decrease in the system investment cost per output power. But above an optimized value for the number of fuel cell stacks, any increase in the total investment cost is greater than the increase in fuel cell output power. This is why above a certain value for the number of fuel cell stacks, the cost of electricity production increases. This observation indicates that the number of fuel cell stacks, the fuel cell operating temperature, and the fuel cell current density are the most appropriate design parameters when the cost of electricity production is an objective function.

7.5   Concluding Remarks

This chapter starts by introducing fuel cells and their applications. The thermochemical equations for a fuel cell are formulated and explained. Two main fuel cells types, namely SOFC and PEM, are considered further and detailed modeling is provided. A genetic algorithm is applied to determine the optimal design parameters of each fuel cell as is a sensitivity analysis. The optimization of the PEM shows that the electricity cost is inversely proportional to the maximum produced electricity, and the selection of the optimum value depends on the needs of the system in which the fuel cell system is employed. The best value from a cost point of view is 3.31 \$/GW at a 0.6 cost fraction, from a work point of view is 118 kW at a 0.9 work fraction, and from an energy and exergy efficiency point of view are 0.49 and 0.55 at a 0.33 efficiency fraction. For the SOFC optimization, three objective functions are considered and a genetic algorithm applied. The results show that the number of fuel cell stacks, the current density, and the fuel cell operating temperature are the most important design parameters for SOFC optimization.

References

1 Li, X., Fields, L. and Way, G. (2006) Principles of fuel cells. *Platinum Metals Rev* **50**:200–201.
2 Mert, S. O., Özçelik, Z., Özçelik, Y. and Dinçer, I. (2011) Multi-objective optimization of a vehicular PEM fuel cell system. *Applied Thermal Engineering* **31**:2171–2176.
3 Ay, M., Midilli, A. and Dincer, I. (2006) Exergetic performance analysis of a PEM fuel cell. *International Journal of Energy Research* **30**:307–321.
4 Palazzi, F., Autissier, N., Marechal, F. M. and Favrat, D. (2007) A methodology for thermo-economic modeling and optimization of solid oxide fuel cell systems. *Applied Thermal Engineering* **27**:2703–2712.

Study Questions/Problems

1   A schematic of an integrated combined heat and power (CHP) system with an SOFC is shown in Figure 7.13. Consider compressor pressure ratio, compressor isentropic efficiency, number of fuel cell stacks, gas turbine inlet temperature, gas turbine isentropic efficiency, and HRSG pinch point temperature as design parameters and apply a genetic algorithm to determine optimal values to maximize the exergy efficiency of the system. You can assume that air and combustion gases can be treated as ideal gases and that the net output electrical power of the system is 2 MW.

2   A PEM fuel cell can be considered for low-grade thermal power production. A schematic of a CHP system based on a PEM fuel cell is shown in Figure 7.14. Use reasonable assumptions for modeling of the system and considering appropriate design parameters, apply multi-objective optimization with exergy efficiency and total cost rate considered as the two objective functions. Conduct sensitivity analyses for different working fluids for the Organic Rankine cycle (ORC) part of the system and compare the results.

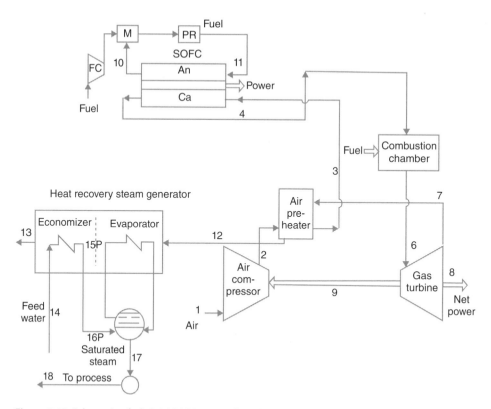

Figure 7.13 Schematic of a hybrid SOFC-GT combined heat and power system.

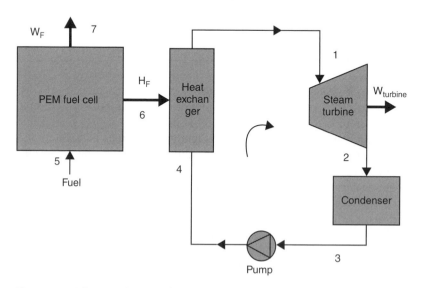

Figure 7.14 Schematic diagram of an organic Rankine cycle integrated with a PEM fuel cell for fuel cell heat recovery.

8

Modeling and Optimization of Renewable Energy Based Systems

8.1 Introduction

Renewable energies have attracted ample interest in recent years for various reasons, including the significant concerns about climate change. According to the Intergovernmental Panel on Climate Change (IPCC), most of the increase in global average temperatures since the mid-20th century is linked to the observed increase in atmospheric greenhouse gas (GHG) concentrations due to anthropogenic emissions. A greenhouse gas absorbs and emits radiation within the thermal infrared range [1, 2]. This process is the fundamental cause of the greenhouse effect. The primary greenhouse gases in the earth's atmosphere are water vapor, carbon dioxide, methane, nitrous oxide, and ozone. The greenhouse effect is a process by which thermal radiation from a planetary surface is absorbed by atmospheric greenhouse gases, and re-radiated in all directions. Since part of this re-radiation is back toward the surface and the lower atmosphere, it results in an elevation of the average surface temperature above what it would be in the absence of the gases [1]. It is generally agreed that one way to mitigate global warming is to use more renewable energies.

Renewable energy is derived from natural resources such as sunlight, wind, rain, tides, waves, geothermal heat, and biomass. These are naturally replenished when used. About 16% of global final energy consumption currently is from renewable resources, with 10% from traditional biomass, mainly for heating, and 3.4% from hydroelectricity [1]. New renewables (small hydro, modern biomass, wind, solar, geothermal, and biofuels) account for another 3% and are growing rapidly. Biomass, as a renewable energy source, is biological material from living, or recently living, organisms [2].

In this chapter, we describe various renewable energy systems and discuss their applications. An evolutionary algorithm based optimization is applied to determine the optimal design parameters of each system depending on different objective functions ranging from exergy efficiency to total cost rate. In order to enhance understanding of the design criteria, a sensitivity analysis is conducted to determine how each objective function varies when small changes in some design parameters are applied. Finally, some closing remarks are provided for efficient design of renewable energy systems, followed by some fundamental and practical questions.

Optimization of Energy Systems, First Edition. Ibrahim Dincer, Marc A. Rosen, and Pouria Ahmadi.
© 2017 John Wiley & Sons Ltd. Published 2017 by John Wiley & Sons Ltd.

8.2   Ocean Thermal Energy Conversion (OTEC)

A large amount of solar energy is stored as heat in the surface waters of the world's oceans, providing a source of renewable energy. Ocean thermal energy conversion (OTEC) is a process for harnessing this renewable energy in which a heat engine operates between the relatively warm ocean surface, which is exposed to the sun, and the colder (by about 5°C) water deeper in the ocean, in order to produce electricity. OTEC usually incorporates a low temperature Rankine cycle engine which boils a working fluid such as ammonia to generate a vapor which turns the turbine to generate electricity, and then is condensed back into a liquid in a continuous process. About 80% of the energy that is received from the sun by the earth is stored in the world's oceans [3, 4], and many regions of the world have access to this OTEC resource. OTEC can indirectly produce fuels, for example, by using OTEC-generated electricity to produce hydrogen that can be used in hydrogen fueled cars as well as in the development of synthetic fuels. For a small city, millions of tons of CO_2 are typically generated annually through fossil fuel use during the operation of devices, while with OTEC the value is almost zero. OTEC has the potential to replace some fossil fuel use.

An OTEC system utilizes low-grade energy and typically has a low energy efficiency (approximately 3–5%). Therefore, achieving a high electricity generating capacity with OTEC requires the use of large quantities of seawater and, correspondingly, large amounts of pumping power. These factors detract from the cost effectiveness of this technology. OTEC is not commercially viable today. In order to improve the effectiveness and economics of OTEC cycles, it is proposed to integrate them with industrial operations so that, apart from generating electricity, they could be used for fresh water production, air conditioning and refrigeration, cold water agriculture, aquaculture and mariculture, and hydrogen production [3]. Potential locations for OTEC have been identified, most of which are in the Pacific Ocean, and about 50 countries are examining its implementation as a sustainable source of energy and fresh water, including India, Korea, Palau, Philippines, the U.S.A., and Papua New Guinea [5]. In 2001, as a result of cooperation between Japan and India, a 1-MW OTEC plant was built in India [5], and others are planned for the near future [6].

Figure 8.1 shows an integrated OTEC system equipped with a flat plate solar collector and PEM electrolyzer. This integrated system uses the warm surface seawater to evaporate a working fluid such as ammonia or a Freon refrigerant, which drives a turbine to produce electricity, which in turn is used to drive a PEM electrolyzer to produce hydrogen. After passing through the turbine, the vapor is condensed in a heat exchanger that is cooled by cold deep seawater. The working fluid is then pumped back through the warm seawater heat exchanger, and the cycle is repeated continuously.

8.2.1   Thermodynamic Modeling of OTEC

For thermodynamic modeling purposes, the integrated OTEC system for hydrogen production considered here (Figure 8.1) is divided into three parts: flat plate solar collector, ocean thermal energy conversion (OTEC) unit, and PEM electrolyzer. Energy and exergy analyses are used to determine the temperature profile in the plant, input and output enthalpy and exergy flows, exergy destructions rates, and energy and exergy efficiencies. The relevant energy balances and governing equations for the main sections of the plant shown in Figure 8.1 are described in the following subsections.

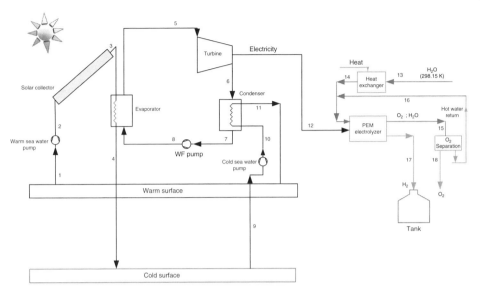

Figure 8.1 Schematic of ocean thermal energy conversion (OTEC) with a flat plate solar collector and PEM electrolyzer for hydrogen production.

8.2.1.1 Flat Plate Solar Collector

As shown in Figure 8.1, water enters the solar collector at point 2 and is heated by the collector. The useful heat gain rate by the working fluid can be written as:

$$\dot{Q}_u = \dot{m}C_p(T_3 - T_2) \tag{8.1}$$

where T_3, T_2, C_p and $\dot{m}$ are the water outlet temperature, inlet temperature, specific heat at constant pressure, and mass flow rate. The Hottel-Whillier equation for the heat gain rate by the flat plate collector considering heat losses from the collector is used here [7]:

$$\dot{Q}_u = A_P F_R [S - U_l(T_{in} - T_0)] \tag{8.2}$$

where T_0 is the ambient temperature and the F_R is heat removal factor, defined as:

$$F_R = \frac{\dot{m}C_p}{U_l A_P}\left[1 - e^{\left\{-\frac{F' U_l A_P}{\dot{m}C_p}\right\}}\right] \tag{8.3}$$

Here, F' is the collector efficiency factor which is around 0.914 for this case [7] and U_l is the overall collector loss coefficient, obtained from [7]. In Equation 8.2, the radiation flux absorbed by the absorber is calculated as:

$$S = (\tau\alpha)I \tag{8.4}$$

where $(\tau\alpha)$ is optical efficiency and I is solar radiation intensity. The energy efficiency of the solar flat plate collector is expressible as:

$$\eta = \frac{\dot{Q}_u}{IA_P} \tag{8.5}$$

More information about solar flat plate collector modeling is detailed in reference [8].

8.2.1.2 Organic Rankine Cycle (ORC)

As shown in Figure 8.1, the electricity production unit is based on an organic Rankine cycle, which is suitable for low-grade heat. Figure 8.2 shows the corresponding temperature-entropy (*T-s*) diagram of the ORC.

The net power output of the system is written as

$$\dot{W}_{net} = \dot{W}_G - (\dot{W}_{WS} + \dot{W}_{CS} + \dot{W}_{WF}) \tag{8.6}$$

where $\dot{W}_G$ is the turbine generator power, $\dot{W}_{WS}$ and $\dot{W}_{CS}$ are the warm and cold seawater pumping power, and $\dot{W}_{WF}$ is working fluid working power.

Turbine Generator Power The turbine power output is calculated by writing the energy balance for a control volume around a turbine, and is the product of working fluid mass flow rate $\dot{m}_F$ and the adiabatic heat loss between the evaporator and the condenser. The turbine power output can be expressed as follows:

$$\dot{W}_G = \dot{m}_F \eta_T \eta_G (h_5 - h_6) \tag{8.7}$$

Here, η_T is the turbine isentropic efficiency and η_G the generator mechanical efficiency.

Warm Seawater Pumping Power The warm seawater pumping power can be written as:

$$\dot{W}_{WS} = \frac{\dot{m}_{WS} \, \Delta H_{WS} \, g}{\eta_{WSP}} \tag{8.8}$$

where ΔH_{WS} is the total pump head difference of the warm seawater piping:

$$\Delta H_{WS} = (\Delta H_{WS})_P + (+H_{WS})_E \tag{8.9}$$

Here, $(eH_{WS})_P$ is the pump head of the warm seawater pipe, which can be expressed as [9]:

$$(HH_{WS})_P = (\Delta H_{WS})_{SP} + (\Delta H_{WS})_B \tag{8.10}$$

where $(hH_{WS})_{SP}$ is the friction loss of the straight pipe and $(sH_{WS})_B$ is the bending loss on the warm seawater pipe. These can be written as [9]:

$$(HH_{WS})_{SP} = 6.82 \frac{L_{WS}}{d_{WS}^{1.17}} \times \left(\frac{V_{WS}}{100}\right)^{1.85} \tag{8.11}$$

$$(H_{WS})_B = \sum \lambda_m \frac{V_{WS}^2}{2g} \tag{8.12}$$

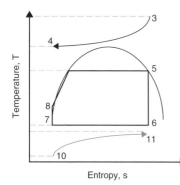

Point number	Specification
3	Warm water entering ORC evaporator
4	War water exiting the ORC evaporator to the ocean
7	Saturated liquid ammonia entering ORC pump
8	Liquid ammonia exiting ORC pump
5	Saturated vapour entering turbine
6	Saturated ammonia entering condenser
10	Cold water entering the condenser
11	Cold water outlet from condenser to the ocean

Figure 8.2 *T-s* diagram of the ocean thermal energy conversion (OTEC) for hydrogen production.

Here, L_{WS} is the length of the warm seawater pipe, d_{WS} is the warm seawater inner pipe diameter and V_{WS} is the velocity of warm seawater inside the pipe. Also, $(iH_{WS})_E$ is the pressure difference of warm seawater in the evaporator, expressible as:

$$(iH_{WS})_E = \lambda_E \frac{V_{WS}^2}{2g} \frac{L_E}{(D_{eq})_W} \tag{8.13}$$

Here, L_E is the length of the evaporator plate and λ_E is taken from reference [9]. Also, D_{eq} is the equivalent diameter, which is calculated as:

$$D_{eq} = 2\delta \tag{8.14}$$

where δ is the clearance.

Cold Seawater Pumping Power   The cold seawater pumping power is expressed as:

$$\dot{W}_{CS} = \frac{\dot{m}_{CS} \, \Delta H_{CW} \, g}{\eta_{CSP}} \tag{8.15}$$

where η_{CSP} is the cold seawater pump efficiency and ΔH_{CW} is the total pump head of the cold seawater piping, given as:

$$\Delta H_{CS} = (\Delta H_{CS})_P \; + (+H_{CS})_C + (\Delta H_{CS})_d \tag{8.16}$$

Here, $(eH_{CS})_P$ is the pump head of the cold seawater pipe [9]:

$$(HH_{CS})_P = (\Delta H_{CS})_{SP} \; + (\Delta H_{CS})_B \tag{8.17}$$

These two terms are similar to the terms in Equations 8.11 and 8.12. Also, $(hH_{CS})_C$ is the cold seawater pressure difference in the condenser, which is defined as

$$(iH_{CS})_C = \lambda_C \frac{V_{CS}^2}{2g} \frac{L_{EC}}{(D_{eq})_C} \tag{8.18}$$

where L_C is the length of the evaporator plate and λ_C is taken from reference [9]. Also, D_{eq} is the equivalent diameter given in Equation 8.14 and $(sH_{CS})_d$ is the pressure difference caused by the density difference between the warm seawater surface and cold deeper seawater, calculated as [9]:

$$(HH_{CS})_d = L_{CS} - \frac{1}{\rho_{CS}} \left(\frac{1}{2}(\rho_{WS} + \rho_{CS}) L_{CS} \right) \tag{8.19}$$

8.2.1.3   PEM Electrolyzer

The schematic diagram of the PEM electrolyzer for H_2 production is shown on the right side of Figure 8.1. During electrolysis, the required electricity and heat are both supplied to the electrolyzer to drive the electrochemical reactions. As shown in Figure 8.1, liquid water is fed to the PEM electrolyzer at ambient temperature. Liquid water enters a heat exchanger that brings it to the PEM electrolyzer temperature after which it then enters the electrolyzer. Leaving the cathode, the H_2 produced dissipates heat to the environment and cools to the reference environment temperature. The oxygen gas produced at the anode is separated from the water and oxygen mixture and then cooled to the reference environment temperature. The remaining water is then returned to the water supply stream for the next hydrogen production cycle. The overall PEM electrolysis reaction is simply water splitting, that is, electricity and heat are used to separate water into hydrogen and oxygen. The hydrogen is stored in a tank for later usage.

8.2.2 Thermochemical Modeling of a PEM Electrolyzer

Energy and exergy analyses of a PEM electrolyzer can be performed in conjunction with electrochemical modeling. The total energy needed by the electrolyzer can be obtained noting:

$$\Delta H = \Delta G + T\Delta S \tag{8.20}$$

where G is the Gibbs free energy and $T\Delta S$ is the thermal energy requirement. The values of G, S, and H for hydrogen, oxygen, and water can be obtained from thermodynamic tables [10]. The total energy need is the theoretical energy required for H_2O electrolysis without any losses. The catalyst used in PEM electrolysis provides an alternative path for the reaction with lower activation energy. The mass flow rate of hydrogen is determined by [11]:

$$\dot{N}_{H_2,out} = \frac{J}{2F} = \dot{N}_{H_2O,reacted} \tag{8.21}$$

where J is the current density and F is the Faraday constant. The electric energy input rate to the electrolyzer can be expressed as:

$$E_{electric} = JV \tag{8.22}$$

where $E_{electric}$ is the electric energy input and $Ex_{electric}$ the electric exergy input. Also, V is given as:

$$V = V_0 + V_{act,a} + V_{act,c} + V_{Ohm} \tag{8.23}$$

where V_0 is the reversible potential, which is related to the difference in free energy between reactants and products, and V_0 can be obtained by the Nernst equation as follows:

$$V_0 = 1.299 - 8.5 \times 10^{-4}(T_{PEM} - 298) \tag{8.24}$$

Also, $V_{act,a}$, $V_{act,c}$ and $V_{o=Ohm}$ are the activation overpotential of the anode, the activation overpotential of the cathode, and the Ohmic overpotential of the electrolyte, respectively. Ohmic overpotential in the proton exchange membrane is caused by the resistance of the membrane to the transport of hydrogen ions through it. The ionic resistance of the membrane depends on the degree of humidification and thickness of the membrane as well as the membrane temperature [12]. The local ionic conductivity $\sigma(x)$ of the proton exchange membrane can be expressed as [11, 12]:

$$\sigma_{PEM}[\lambda(x)] = [0.5139\lambda(x) - 0.326] \exp\left[1268\left(\frac{1}{303} - \frac{1}{T}\right)\right] \tag{8.25}$$

where x is the distance in the membrane measured from the cathode–membrane interface and $\lambda(x)$ is the water content at a location x in the membrane. The value of $\lambda(x)$ can be calculated in terms of the water content at the membrane–electrode edges:

$$\lambda(x) = \frac{\lambda_a - \lambda_c}{D} x + \lambda_c \tag{8.26}$$

where D is the membrane thickness, and λ_a and λ_c are the water contents at the anode–membrane and cathode–membrane interfaces, respectively. The overall Ohmic resistance can thus be expressed as [11]:

$$R_{PEM} = \int_0^D \frac{dx}{\sigma_{PEM}[\lambda(x)]} \tag{8.27}$$

Based on Ohm's law, the following equation can be written for the Ohmic overpotential:

$$V_{Ohm,PEM} = J R_{PEM} \tag{8.28}$$

The activation overpotential, V_{act}, caused by a deviation of net current from its equilibrium, and also an electron transfer reaction, must be differentiated from the concentration of the oxidized and reduced species [11, 12]. Then,

$$V_{act,i} = \frac{RT}{F} sinh^{-1}\left(\frac{J}{2J_{0,i}}\right), \ i = a, c \tag{8.29}$$

Here, J_o is the exchange current density, which is an important parameter in calculating the activation overpotential. It characterizes the electrode's capabilities in the electrochemical reaction. A high exchange current density implies a high reactivity of the electrode, which results in a lower overpotential. The exchange current density for electrolysis can be expressed as [12]:

$$J_{0,i} = J_i^{ref} exp\left(-\frac{E_{act,i}}{RT}\right), \ i = a, c \tag{8.30}$$

where J_i^{ref} is the pre-exponential factor and $E_{act,i}$ is the activation energy for the anode and cathode. More details about PEM electrolysis modeling can be found elsewhere [12, 13].

8.2.3 Exergy Analysis

Exergy analysis can help to develop strategies and guidelines for more effective use of energy, and has been applied to various thermal processes, including power generation, CHP, and trigeneration. Exergy can be divided into four components: physical, chemical, kinetic, and potential. For the processes involved in this analysis, the latter two are neglected since changes in elevation and speed are negligible [14, 15]. Physical exergy is defined as the maximum work obtainable as a system interacts with a reference environment at an equilibrium state [7, 8]. Chemical exergy is associated with the departure of the chemical composition of a system from the chemical equilibrium of a reference environment. Chemical exergy is important in combustion evaluation. Applying the first and second laws of thermodynamics, the following general exergy balance, which can be applied to each component, is obtained:

$$\dot{Ex}_Q + \sum_i \dot{m}_i ex_i = \sum_e \dot{m}_e ex_e + \dot{Ex}_W + \dot{Ex}_D \tag{8.31}$$

where the subscripts e and i denote the inlet and outlet of a control volume, $\dot{Ex}_D$ is the exergy destruction rate, and the other terms are as follows [14–17]:

$$\dot{Ex}_Q = \left(1 - T_0/T_i\right) \times \dot{Q}_i \tag{8.32}$$

$$\dot{Ex}_w = \dot{W} \tag{8.33}$$

$$ex = ex_{ph} + ex_{ch} \tag{8.34}$$

$$ex_{ph} = (h - h_0) - T_0(s - s_0) \tag{8.35}$$

Here, $\dot{Ex}_Q$ and $\dot{Ex}_w$ respectively are the corresponding exergy rates associated with heat transfer and work across the boundary of a control volume, T is the absolute temperature and the subscript o refers to the reference environment conditions. The reference environment considered here has a temperature $T_0 = 20°C$ and a pressure $P_0 = 1$ bar. The specific chemical exergy for gas mixtures is defined as follows [14]:

$$ex_{ch} = \sum x_k ex_{ch}^k + RT_0 \sum x_k \ln(x_k) \tag{8.36}$$

Further details about each term in this equation are given in references [14–17].

To improve environmental sustainability, it is necessary not only to use sustainable or renewable sources of energy, but also to utilize non-renewable sources like natural gas more efficiently, while minimizing environmental damage. In this way, society can reduce its use of limited resources and extend their lifetimes. Here, a sustainability index *SI* is used to relate exergy with environmental impact [18]:

$$SI = \frac{1}{D_P} \tag{8.37}$$

where D_P is a depletion number, which is defined as the exergy destruction/input exergy [7]. This relation demonstrates how reducing a system's environmental impact can be achieved by reducing its exergy destruction.

8.2.4 Efficiencies

The energy efficiency of the OTEC system is defined as the net power output of the system divided by input energy at the evaporator, which can be expressed as:

$$\eta = \frac{\dot{W}_{net}}{\dot{Q}_{EVP}} \tag{8.38}$$

where $\dot{W}_{net}$ is given in Equation 8.6, and $\dot{Q}_{EVP}$ is expressible as

$$\dot{Q}_{EVP} = \dot{m}_{WF}(h_1 - h_4) \tag{8.39}$$

8.2.4.1 Exergy Efficiency

The exergy efficiency is defined as the product exergy output divided by the exergy input [10]. According to Nag and Dai [19], the exergy efficiency of the ORC power generation cycle in an OTEC system is given as:

$$\Psi = \frac{\dot{W}_{net}}{\dot{Ex}_{in}} = \frac{\dot{W}_{net}}{\dot{Ex}_{in,WS} + \dot{Ex}_{in,CS}} \tag{8.40}$$

$$\dot{Ex}_{in,WS} = \dot{m}_{in,WS}[(h_{WS,in} - h_0) - T_0(s_{WS,in} - s_0)] \tag{8.41}$$

$$\dot{Ex}_{in,CS} = \dot{m}_{in,CS}[(h_{CS,in} - h_0) - T_0(s_{CS,in} - s_0)] \tag{8.42}$$

8.2.5 Exergoeconomic Analysis

The renewable energy-based multigeneration system in Figure 8.1 has various components. In this section, the cost functions of each subsystem are described.

8.2.5.1 Flat Plate Solar Collector in OTEC Cycle

As shown in Figure 8.1, a flat plate solar collector is used with the OTEC system. The purchase cost of the solar flat plate collector can be expressed as [20]:

$$Z_{FPC}(\$) = 235\, A_{FPC} \tag{8.43}$$

Here, A_{FPC} is the flat plate collector area, in m².

8.2.5.2 OTEC Cycle

The OTEC cycle has four major components, for each of which the cost is described as follows:

OTEC Turbine and Generator The cost function of the ORC turbine can be calculated as follows [20]:

$$Z_T(\$) = 4750 \, (\dot{W}_T)^{0.75} + 60(\dot{W}_T)^{0.95} \tag{8.44}$$

Here, $\dot{W}_T$ is the power generated by the turbine in kW.

OTEC Evaporator The purchase cost of OTEC evaporator can be calculated as follows [20]:

$$Z_{EVP}(\$) = 276\left(\frac{\dot{Q}_{EVP}}{U_{EVP}\Delta T_{\ln}}\right)^{0.88} \tag{8.45}$$

Here, U_{EVP} is the overall heat transfer coefficient for the evaporator, which is taken to be $4.39 \, \frac{kW}{m^2 K}$ [20].

Condenser The purchase cost of an OTEC condenser can be calculated as follows [20, 21]:

$$Z_{cond}(\$) = 150\left(\frac{\dot{Q}_{cond}}{U_{cond}\Delta T_{\ln}}\right)^{0.8} \tag{8.46}$$

Here, U_{cond} is the overall heat transfer coefficient for the evaporator ($4.69 \, \frac{kW}{m^2 K}$) [21].

Pumps The following expression is used to calculate the cost of the OTEC pumps [22]:

$$Z_{pump}(\$) = 3500(\dot{W}_P)^{0.41} \tag{8.47}$$

Here, $\dot{W}_P$ is the pump work rate in kW.

8.2.6 Results and Discussion

8.2.6.1 Modeling Validation and Simulation Code Results

To model the integrated OTEC system for hydrogen production, simulation code using Matlab software is developed. Three main parts are first individually modeled including each exergy flow rate. Engineering equation solver (EES) is linked to Matlab to calculate the properties of the different working fluids (i.e., water and ammonia), such as pressure, temperature, enthalpy, and entropy. Several simplifying assumptions are made here to render the analysis more tractable, while retaining adequate accuracy to illustrate the principal points of the work:

- All processes operate at steady state.
- The thermodynamic cycle of the integrated system in Figure 8.1 is an ideal saturated Rankine cycle using pure ammonia as the working fluid.
- All the components are adiabatic.
- Pressure drops in the ORC cycle are negligible.
- State 5 is saturated vapor.
- Heat losses from piping and other auxiliary components are negligible.

Table 8.1 Input data for the system simulation.

Parameter	Value	Parameter	Value
Turbine isentropic efficiency, η_T	0.80	Warm seawater mass flow rate (kg/s)	150
Generator mechanical efficiency, η_G	0.90	Cold seawater mass flow rate (kg/s)	150
Working fluid pump isentropic efficiency, η_{WFP}	0.78	Cold seawater pipe length (m)	1000
Seawater pumps isentropic efficiency, η_P	0.80	Cold seawater pipe inner diameter (m)	0.70
Ambient temperature ($^{\circ}$C)	25	Warm seawater pipe length (m)	50
Solar radiation incident on collector surface, I (W/m^2)	500	Warm seawater pipe length (m)	0.70
Warm seawater temperature, T_{WSI} ($^{\circ}$C)	22	Solar collector effective area (m^2)	5000
Cold seawater temperature at depth of 1000 m, T_{CSI} ($^{\circ}$C)	4	Electrolyzer working temperature ($^{\circ}$C)	80

Table 8.2 Input parameters used to model PEM electrolysis [10, 11].

Parameter	Value
P_{O_2} (atm)	1.0
P_{H_2} (atm)	1.0
T_{PEM} ($^{\circ}$C)	80
$E_{act,a}$ (kJ/mol)	76
$E_{act,c}$ (kJ/mol)	18
λ_a	14
λ_c	10
D (μm)	100
J_a^{ref} (A/m^2)	$1.7/10^5$
J_c^{ref} (A/m^2)	$4.6/10^3$
F (C/mol)	96 486

To conduct the simulation, reliable input data are required. For each subsystem, some data are input to the simulation code in order to determine the outputs. Table 8.1 lists the input parameters for the OTEC system simulation. Also, Table 8.2 lists the parameters used to simulate the PEM electrolyzer.

To ensure the accuracy and validity of the developed computer simulation code, the PEM electrolyzer is validated with experimental data from the literature. Specifically, the electrochemical model is used to simulate experiments published in the literature and the modeling results and experimental data are compared. The electrolyte used in the experiments [19] is Nafion, a polymer widely used as electrolyte in fuel cells and electrolyzers. The thicknesses of the electrolytes tested by Ioroi *et al.* [23] and Millet

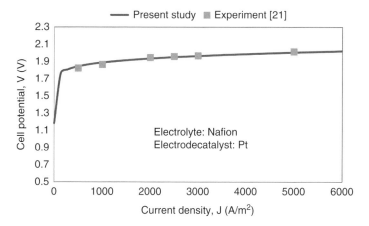

Figure 8.3 Comparison of present model with experimental data.

et al. [19] were 50 μm and 178 μm, respectively. Platinum was used as the electrode catalyst. The simulation code for the *J–V* characteristics for PEM electrolysis are compared with experimental data of Ioroi *et al.* [23] in Figure 8.3. The modeling results agree well with the experimental data, supporting the accuracy of the present model. It is found that the cell potential increases rapidly when the current density is less than 300 A/m². When *J* exceeds 300 A/m², the cell potential increases slightly with *J*. To enhance the understanding of the electrochemical performance of the PEM electrolyzer, Ohmic and activation overpotentials are examined and shown individually in Figure 8.4. This figure shows that the Ohmic overpotential is very small and increases slightly with current density. This observation is attributable to the fact that the membrane ionic conductivity (σ_{PEM}) is significantly higher for typical values of λ as well as the operating

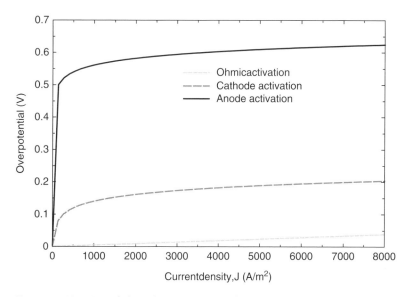

Figure 8.4 Variation of electrolyzer overpotentials at various current densities.

temperature, which leads to a lower R_{PEM} (see Equations 8.25 and 8.27), which in turn lowers the overall Ohmic resistance and the Ohmic overpotential.

8.2.6.2 Exergy Analysis Results

Results of the thermodynamic modeling and exergy analysis are presented here, including assessments of the effects of varying several design parameters on the cycle performance. As already discussed, the inputs of the simulation program are transferred to the developed code in order to calculate the outputs. The results of the simulation program are listed in Table 8.3. The net power output is seen to be around 101 kW, which leads to a hydrogen production rate of about 1.2 kg/hr. In addition, the exergy efficiency of the integrated OTEC system is much higher than the energy efficiency, mainly due to the fact that the work is produced using a low-grade (in terms of a temperature near to that of the reference environment) heat at the ocean surface.

The exergy analysis results are presented in Figure 8.5 and show that the highest exergy destruction occurs in the condenser, which is mainly due to the temperature difference between the two fluid streams passing through it, but also due to the pressure drop across the device. Another important result relates to the dimensionless exergy ratio, which is useful for prioritizing exergy losses in an intuitive manner. Both the exergy destruction rate and the dimensionless exergy destruction ratio are higher in the condenser than in other components, suggesting that it would likely be worthwhile to focus improvement efforts on this component.

Since the solar radiation intensity changes during the day, the variation of system performance with solar radiation intensity is investigated. Figure 8.6 shows the variation of exergy efficiency and exergy destruction rate of the OTEC system for various values of solar radiation intensity. Increasing the solar radiation intensity results in an increase in the exergy efficiency of the OTEC system, suggesting that during such a day the solar intensity increases the inlet temperature entering the evaporator, which leads to an increase in enthalpy at point 3 entering the turbine to produce work. The higher the temperature difference between cold and hot surfaces the more work is

Table 8.3 Parameter values resulting from energy and exergy analyses of the system.

Parameter	Value
Net power output, $\dot{W}_{net}$ (kW)	101.96
Exergy efficiency, Ψ (%)	22.70
Energy efficiency, η (%)	3.60
Sustainability index, SI	1.29
Total exergy destruction rate, $\dot{Ex}_{D,tot}$ (kW)	42.12
Hydrogen production rate, $\dot{m}_{H_2}$ (kg/hr)	1.20
PEM electrolyzer exergy efficiency, Ψ_{PEM} (%)	56.34
Warm surface pump power, $\dot{W}_{WS}$ (kW)	1.30
Cold surface pump power, $\dot{W}_{CS}$ (kW)	3.13
Working fluid pump power, $\dot{W}_{WF}$ (kW)	0.88

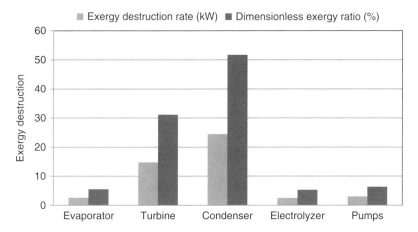

Figure 8.5 Exergy destruction rate and dimensionless exergy destruction ratio for each component of the ocean thermal energy conversion (OTEC) system for hydrogen production.

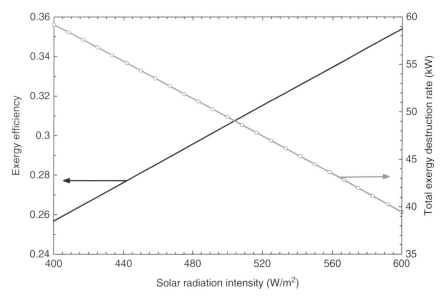

Figure 8.6 Variation with solar radiation intensity of the OTEC exergy efficiency and total exergy destruction rate.

produced. Since the cold surface temperature is almost constant in the deep ocean, an increase in evaporator inlet temperature results in an increase in the output power. Also, Figure 8.6 shows that an increase in solar radiation leads to a decrease in total exergy destruction rate.

The effect of solar radiation intensity on exergy destruction rate for each component is shown in Figure 8.7. An increase in solar radiation intensity is observed to have major effects on the exergy destruction rates of the turbine and the condenser while the changes in exergy destruction rate for the evaporator and the pump are not drastic.

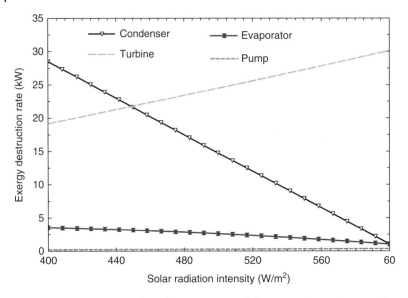

Figure 8.7 Variation with solar radiation intensity of the exergy destruction rate of components.

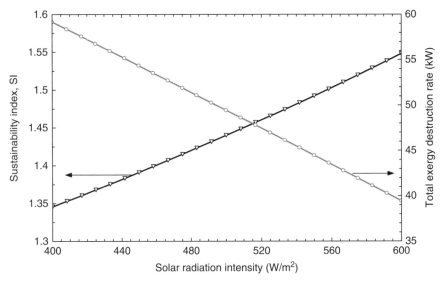

Figure 8.8 Variation with solar radiation intensity of the sustainability index and total exergy destruction rate.

So, since the decrease in the exergy destruction rate of the condenser is much higher than that for the turbine, the total exergy destruction rate decreases with an increase in solar radiation intensity. Figure 8.8 shows the effect of solar radiation intensity on both sustainability index (SI) and total exergy destruction rate. The results are similar to those in Figure 8.6, that is, the overall exergy destruction of the cycle decreases and the sustainability index increases with increasing solar radiation intensity. The exergy efficiency, exergy destruction, and sustainability are thus observed to be linked in such systems, supporting the utility of exergy and environmental analyses.

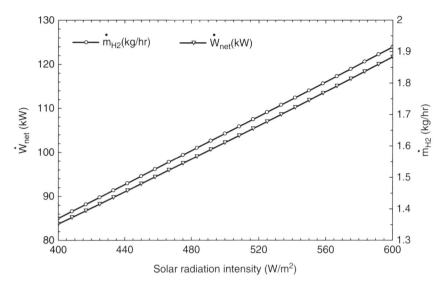

Figure 8.9 Variation with solar radiation intensity of the OTEC net power output and hydrogen production rate.

Figure 8.9 shows the effect of solar radiation intensity on hydrogen production mass flow rate. As the solar intensity increases the hydrogen production increases since the turbine work increases and correspondingly the electrical input to PEM electrolysis increases.

Another major parameter in an OTEC plant is ambient temperature because it affects the surface temperature and also the value for inlet exergy at each point of the plant. Hence, this temperature is an important parameter. Figure 8.10 shows the variation of exergy efficiency and sustainability index with ambient temperature. As can be seen,

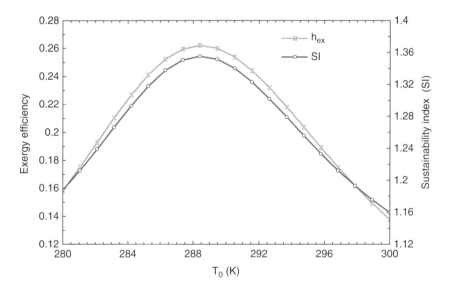

Figure 8.10 Variation with ambient temperature of the exergy efficiency and sustainability index.

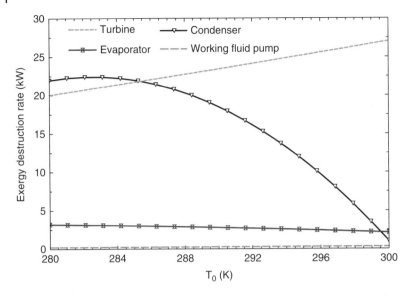

Figure 8.11 Variation with ambient temperature of the exergy destruction rate of each component.

an increase in ambient temperature first leads to an increase in exergy efficiency and sustainability index, due to an increase in the fluid temperature entering the evaporator which increases the work of the turbine. Above 285 K, the exergy efficiency decreases because the exergy input to the system, which is the denominator of Equation 8.40, increases. Hence two different effects occur and the effect of an increase in inlet exergy values dominates, so the exergy efficiency of the OTEC system decreases. The variation of exergy destruction rate of each component of the OTEC system with ambient temperature is shown in Figure 8.11, which demonstrates the same trend as Figure 8.10. An increase in T_0 has major effects on the exergy destruction rate for the turbine and the condenser, while for the evaporator and working fluid pump it is almost constant. An increase in temperature for the condenser first leads to an increase in exergy destruction rate and then above 285 K the exergy destruction rate decreases. However, the exergy destruction rate always increases for the turbine. The summation of these two effects results in an initial increase in the total exergy destruction rate, then it leads to a decrease in the total exergy destruction as shown in Figure 8.12.

To enhance the analysis, the effect of condenser temperature on the exergy efficiency of the OTEC system is also investigated (see Figure 8.13). An increase in condenser temperature is seen to reduce the exergy efficiency of the system. Increasing condenser temperature leads to an increase in enthalpy at point 2 and results in a reduction in turbine work, which lowers the exergy efficiency. Since the exergy destruction rate in the condenser is much higher than in other components, any changes in the input exergy flow to the condenser lead to an increase in its exergy destruction, so an increase in condenser temperature leads to an increase in exergy flow at point 2 keeping other parameters constant.

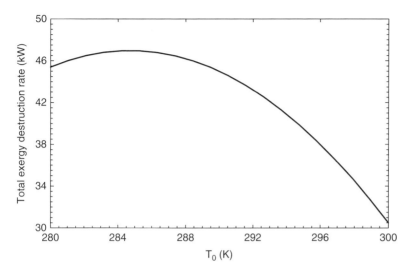

Figure 8.12 Variation with ambient temperature of the total exergy destruction rate.

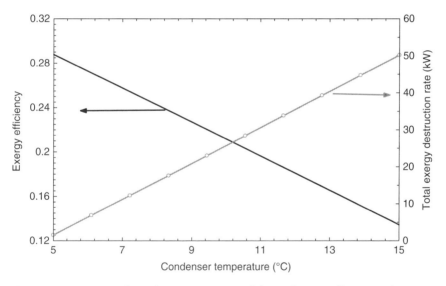

Figure 8.13 Variation with condenser temperature of the total exergy efficiency and exergy destruction rate.

8.2.7 Multi-objective Optimization

A multi-objective optimization method based on an evolutionary algorithm is applied to the OTEC system for hydrogen production to determine the best design parameters for the system. Objective functions, design parameters and constraints, and overall optimization are described in this section.

8.2.7.1 Objectives

Two objective functions are considered here for multi-objective optimization: exergy efficiency (to be maximized) and total cost rate of product (to be minimized). Consequently, the objective functions in this analysis can be expressed as explained below.

Exergy Efficiency

$$\Psi = \frac{\dot{W}_{net}}{\dot{Ex}_{in}} = \frac{\dot{W}_{net}}{\dot{Ex}_{in,WS} + \dot{Ex}_{in,CS}} \tag{8.48}$$

Total Cost Rate

$$\dot{C}_{tot} = \sum_k \dot{Z}_k + \dot{C}_{env} \tag{8.49}$$

where the cost rates of environmental impact and fuel are expressed as

$$\dot{C}_{env} = C_{Co_2}\dot{m}_{Co_2} \text{ and } \dot{C}_f = c_f \dot{m}_f \, LHV \tag{8.50}$$

Here, $\dot{Z}_K$ is the purchase cost of each component, as explained in the previous section.

8.2.7.2 Decision Variables

The following decision variables (design parameters) are selected for this study: warm surface mass flow rate ($\dot{m}_{WS}$), OTEC evaporator pinch point temperature (PP) difference, OTEC turbine isentropic efficiency (η_T), pump isentropic efficiency (η_p), OTEC condenser temperature (T_{cond}) and flat plate collector area (A_{FP}). Although the decision variables may be varied in the optimization procedure, each is normally required to be within a reasonable range. Such constraints, based on earlier reports, are listed in Table 8.4.

8.2.8 Optimization Results

The genetic algorithm optimization is performed for 300 generations, using a search population size of $M = 100$ individuals, a crossover probability of $p_c = 0.9$, a gene mutation probability of $p_m = 0.035$, and a controlled elitism value $c = 0.55$. The results of the optimization are described below. Figure 8.14 shows the Pareto frontier solution for this system with objective functions indicated in Equations 8.48 and 8.49 for the multi-objective optimization. It can be seen in this figure that the total cost rate of products increases moderately as the total exergy efficiency of the cycle increases to about

Table 8.4 Optimization constraints and their rationales.

Constraint	Rationale
$70 < \dot{m}_{WS} < 150$	Required pump work limitation
$3 < PP < 6°C$	Heat transfer limit
$\eta_T < 0.9$	Commercial availability
$\eta_p < 0.9$	Commercial availability
$400 \text{ m}^2 < A_{FP} < 1000 \text{ m}^2$	Commercial availability
$5°C < T_{cond} < 9°C$	Heat transfer limit

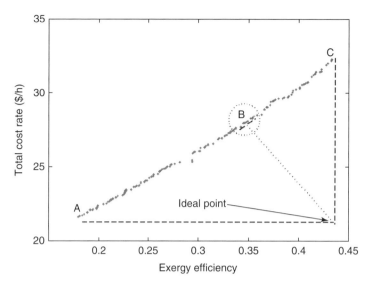

Figure 8.14 Pareto frontier, showing best trade-off values for the objective functions.

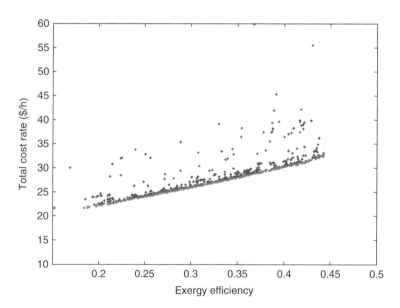

Figure 8.15 Results of all evaluations during 300 generations using genetic algorithm. A clear approximation of the Pareto frontier is visible on the lower part of the figure.

35%. Increasing the total exergy efficiency from 35% to 45% increases the cost rate of the product significantly. The results of optimum exergy efficiency and total cost rate for all points evaluated over 300 generations are shown in Figure 8.15.

As shown in Figure 8.14, the optimized values for exergy efficiency on the Pareto frontier range between 17% and 45%. To provide a good relation between exergy efficiency and total cost rate, a curve is fitted on the optimized points obtained from the evolutionary algorithm. This fitted curve is shown in Figure 8.15 and the expression for this fitted curve follows:

$$\dot{C}_{total} = \frac{43.55\psi^3 + 96.17\psi^2 - 93.68\psi + 26.57}{\psi^2 - 0.78\psi + 0.2} \tag{8.51}$$

This equation is allowable when the efficiency varies between 30% and 70%. The Pareto optimal curve (best rank) is clearly visible in the lower part of the figure (red line), which is separately shown in Figure 8.15. As shown in that figure, the maximum exergy efficiency exists at design point C (49%), while the total cost rate of products is the greatest at this point (33.2 $/hr). But, the minimum value for the total cost rate of product occurs at design point A, which is about 21.6 $/hr. Design point A is the optimal situation when the total cost rate of the product is the sole objective function, while design point C is the optimum point when exergy efficiency is the sole objective function. In multi-objective optimization, a process of decision-making for selection of the final optimal solution from the available solutions is required. The process is usually performed with the aid of a hypothetical point in Figure 8.13 (the ideal point), at which both objectives have their optimal values independent of the other objectives. It can be clearly seen that it is not feasible to have both objectives at their optimum point simultaneously and, as shown in Figure 8.13, the ideal point is not a solution located on the Pareto frontier. The closest point of the Pareto frontier to the ideal point might be considered as a desirable final solution. Nevertheless, in this case, the Pareto optimum frontier exhibits a weak equilibrium, that is, a small change in exergy efficiency from varying the operating parameters causes a large variation in the total cost rate of the product. Therefore, the ideal point cannot be utilized for decision-making in this problem. In selection of the final optimum point, it is desired to achieve a better magnitude for each objective than its initial value for the base case problem.

Note that in multi-objective optimization and the Pareto solution, each point can be utilized as the optimized point. Therefore, the selection of the optimum solution depends on the preferences and criteria of the decision maker, suggesting that each may select a different point as for the optimum solution depending on their needs. Table 8.5 shows all the design parameters for points $A-C$. To study the variation of thermodynamic characteristics, three different points (A to C) on the Pareto frontier are considered. Table 8.6 shows the total cost rate of the system, the total exergy destruction rate, the system exergy efficiency, and hydrogen production rates of the system.

Table 8.5 Optimized values for design parameters of the system based on multi-objective optimization.

Design parameter	A	B	C
A_{FP} (m^2)	402	403	800
$\dot{m}_{WS}$ (kg/s)	70.26	70.22	74.00
PP (°C)	4.80	2.66	2.00
η_T (%)	0.73	0.83	0.84
η_p (%)	0.74	0.78	0.80
T_{cond} (°C)	8.80	6.46	5.22

Table 8.6 Thermodynamic characteristics of three points on the Pareto frontier.

Point	$\dot{W}_{net}$ (kW)	Ψ	$\dot{Ex}_{D,tot}$ (kW)	$\dot{C}_{tot}$ ($/h)	$\dot{m}_{H_2}$ (kg/h)
A	19	17.91	666.25	21.62	0.33
B	36.34	35.23	635.3	28.2	0.63
C	54	49	601	33.2	0.91

From point A to point C in this table, both the total cost rate of the system and exergy efficiencies increase. As previously stated, point A is preferred when the total cost rate is the sole objective function, and design point C when exergy efficiency is the sole objective function. Design point B exhibits better results for both objective functions. Other thermodynamic properties correctly confirm this trend. For instance, from point B to C, the total exergy destruction rate decreases when the exergy efficiency increases.

To better understand the variations of all design parameters, the scattered distribution of the design parameters are shown in Figures 8.16 and 8.17. The results show that warm surface mass flow rate (Figure 8.15a) tends to become as low as possible within its range. This observation means that a decrease in these parameters leads to better optimization results. For example, a decrease in these design parameters leads to improvement for both objective functions in multi-objective optimization. In Figures 8.16 to 8.17, we see that the pinch point temperature (Figure 8.16b), the turbine isentropic efficiency (Figure 8.16c), the pump isentropic efficiency (Figure 8.16d), the solar collector area (Figure 8.17a), and the condenser temperature (Figure 8.17b), have scattered distributions in their allowable domains, suggesting that these parameters have important effects on the trade-off between exergy efficiency and total cost rate. Design parameters selected with their maximum values shown that they do not exhibit a conflict between two objective functions, indicating that increasing those design parameters leads to an improvement of both objective functions.

8.3 Solar Based Energy System

One potential renewable energy source is electromagnetic radiation coming from the sun. In this section, we model a solar based integrated system for electricity generation, heating in winter and cooling in summer. Figure 8.18 shows a schematic of a solar based combined cooling, heating, and power (CCHP) system, which is investigated in summer and winter modes. For stable operation of the CCHP system, a thermal storage tank is installed to balance the mismatch between the supply of solar energy and the thermal demand from the system. An auxiliary boiler is also applied when the desired temperature drops at nights. In summer, superheated vapor produced by the solar thermal collectors and auxiliary boiler generates electrical power through an expansion process in the turbine connected to the electric generator. The extracted vapor stream of the turbine enters an ejector supersonic nozzle and is mixed with the outlet stream of evaporator1 after causing a cooling effect. The turbine and mixer1 outlet streams discharge to the condenser, which rejects heat to cooling water. The saturated liquid is then pumped

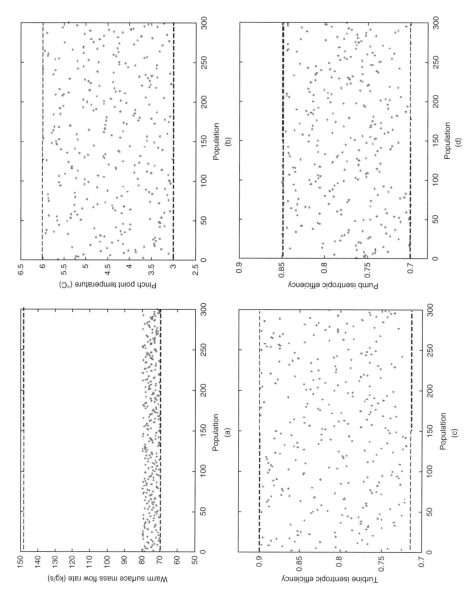

Figure 8.16 Scattered distribution of decision variables with population in Pareto frontier: (a) warm surface mass flow rate, (b) pinch point temperature, (c) turbine isentropic efficiency, (d) isentropic efficiency.

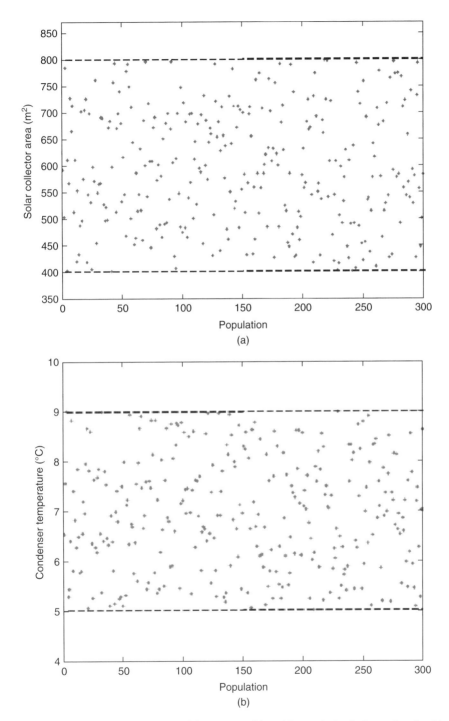

Figure 8.17 Scattered distribution of decision variables with population in Pareto frontier: (a) solar collector area, (b) condenser temperature.

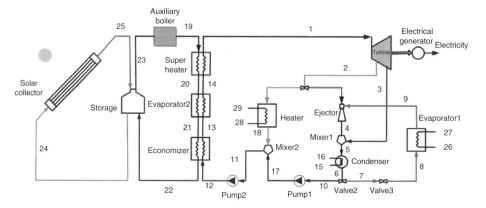

Figure 8.18 A schematic diagram of a solar based CCHP system.

Table 8.7 Input data for the system.

Parameter	Summer	Winter
Turbine inlet pressure (kPa)	1000	1000
Turbine inlet temperature (°C)	130	130
Turbine back pressure (kPa)	300	300
Cooling load (kW)	4.5	–
Heating load (kW)	–	11
Electrical power generated (kW)	2.7	2.7
Turbine isentropic efficiency	0.85	0.85
Evaporator temperature (°C)	−5	–
Pump isentropic efficiency	0.7	0.7
Cooling water inlet pressure (kPa)	300	300
Heater temperature difference (°C)	–	20
Heater outlet temperature (°C)	–	80
Monthly average insolation, H (MJ/m² day)	28.5 (July)	7.99 (December)
Tilt angle of the solar collector (°)	37.4	37.4

into the economizer, evaporator2, and superheater, sequentially, to complete the cycle. In winter, the extracted flow of the turbine enters the heater to supply the heat. The heater and condenser outlet streams are mixed in mixer2 and then pumped into the economizer to absorb the heat from solar collector. The auxiliary boiler utilizes natural gas and the working fluid of the organic Rankine cycle is R123.

For thermodynamic modeling of the system, some data are input. They are listed in Table 8.7 for summer and winter modes.

8.3.1 Thermodynamic Analysis

In the thermodynamic analysis, each component in the system can be treated as a control volume. Energy rate balances and related expressions for each component of the system follow:

Turbine:

$$\dot{W}_{turb} = \dot{m}_1 h_1 - (\dot{m}_2 h_2 + \dot{m}_3 h_3) \tag{8.52}$$

$$\eta_{is,turb} = \frac{\dot{W}_{turb}}{\dot{W}_{is,turb}}$$

Evaporator1:

$$\dot{Q}_{CL} = \dot{m}_8(h_9 - h_8) \tag{8.53}$$

Evaporator2:

$$\dot{m}_{13}(h_{14} - h_{13}) = \dot{m}_{20}(h_{20} - h_{21}) \tag{8.54}$$

Pump1:

$$\dot{W}_{p1} = \dot{m}_{10} v_{10} (P_{17} - P_{10})/\eta_{is,p1} \tag{8.55}$$

Pump2:

$$\dot{W}_{p2} = \dot{m}_{11} v_{11} (P_{12} - P_{11})/\eta_{is,p2} \tag{8.56}$$

Heater:

$$\dot{Q}_{HL} = \dot{m}_2(h_2 - h_{18}) \tag{8.57}$$

Economizer:

$$\dot{m}_{12}(h_{13} - h_{12}) = \dot{m}_{21}(h_{21} - h_{22}) \tag{8.58}$$

Superheater:

$$\dot{m}_{14}(h_1 - h_{14}) = \dot{m}_{19}(h_{19} - h_{20}) \tag{8.59}$$

Condenser:

$$\dot{m}_5(h_5 - h_6) = \dot{m}_{15}(h_{16} - h_{15}) \tag{8.60}$$

Storage Tank:

$$\dot{m}_{25} c_p (T_{25} - T_{24}) = \dot{m}_{22} c_p (T_{23} - T_{22}) + \dot{Q}_{L,ST} \tag{8.61}$$

Auxiliary Boiler:

$$\dot{m}_f LHV_f \eta_{AB} = \dot{m}_{23} c_p (T_{19} - T_{23}) \tag{8.62}$$

The ejector modeling was already explained in previous chapters. The entrainment ratio μ can be expressed as

$$\mu = \frac{\dot{m}_9}{\dot{m}_2} \tag{8.63}$$

The useful heat gained by solar collector, $\dot{Q}_u$ is calculated from the heat balance in the solar collector [24]:

$$\dot{Q}_u = \eta_{coll} \times A_{coll} \times G_t \tag{8.64}$$

where η_{coll} is defined as the ratio of the useful heat gain to the incident solar radiation and G_t is the instantaneous radiation. Details on the calculation of the instantaneous radiation are elsewhere [24].

Table 8.8 Fuel and products for exergy analysis of the system.

Component	Fuel	Product
Turbine	$\dot{E}x_1 - \dot{E}x_2 - \dot{E}x_3$	$\dot{W}_{turb}$
Ejector	$\dot{E}x_2 + \dot{E}x_9$	$\dot{E}x_4$
Evaporator1	$\dot{E}x_8 - \dot{E}x_9$	$\dot{E}x_{27} - \dot{E}x_{26}$
Pump1	$\dot{W}_{pump1}$	$\dot{E}x_{17} - \dot{E}x_{10}$
Economizer	$\dot{E}x_{21} - \dot{E}x_{22}$	$\dot{E}x_{13} - \dot{E}x_{12}$
Heater	$\dot{E}x_2 - \dot{E}x_{18}$	$\dot{E}x_{29} - \dot{E}x_{28}$
Superheater	$\dot{E}x_{19} - \dot{E}x_{20}$	$\dot{E}x_1 - \dot{E}x_{14}$
Condenser	$\dot{E}x_5 - \dot{E}x_6$	$\dot{E}x_{16} - \dot{E}x_{15}$
Auxiliary boiler	$\dot{E}x_{NG}$	$\dot{E}x_{19} - \dot{E}x_{23}$
Storage tank	$\dot{E}x_{25} - \dot{E}x_{24}$	$\dot{E}x_{23} - \dot{E}x_{22}$
Solar collector	$\dot{E}x_S$	$\dot{E}x_{25} - \dot{E}x_{24}$

Overall system:

$$\eta_{CCHP,summer} = \frac{\dot{W}_{elec} + \dot{Q}_{CL}}{\dot{Q}_u + \dot{Q}_{AB}} \tag{8.65}$$

$$\eta_{CCHP,winter} = \frac{\dot{W}_{elec} + \dot{Q}_{HL}}{\dot{Q}_u + \dot{Q}_{AB}} \tag{8.66}$$

The exergy destruction rates for each component of the system are obtained by defining the fuel and product of the components through the second law of thermodynamics, and are shown in Table 8.8.

8.3.2 Exergoeconomic Analysis

Exergoeconomics is an exergy-aided cost analysis method. In this method, a cost rate balance is applied in which the sum of cost rates associated with all exiting exergy streams is equal the sum of cost rates of all entering exergy streams plus the capital investment (Z^{CI}) and operating and maintenance expenses ($\dot{Z}^{OM}$). Cost rate balances for each component of the system are listed in Table 8.9 along with auxiliary equations, and expressions for the capital cost (Z^{CI}) of components are provided in Table 8.10.

The investment cost rate for the components are calculated as below [25, 26]:

$$\dot{Z}_k = Z_k^{CI} \times CRF \times \phi/t \tag{8.67}$$

where t is the number of hours per year, ϕ is the maintenance factor, i is the interest rate and N is the component lifetime. Values for these are assumed to be 7446 hr, 1.06, 10% and 20 year, respectively. Also,

$$CRF = \frac{i(1+i)^N}{(1+i)^N - 1} \tag{8.68}$$

8.3.3 Results and Discussion

In this section, the results of energy and exergy assessments of the solar based CCHP system are presented. The energy and exergy efficiencies respectively of the system are

Table 8.9 Cost rate balances and auxiliary equations for components.

Component	Cost rate balance	Auxiliary equation
Turbine	$\dot{C}_1 + \dot{Z}_{\text{turb}} = \dot{C}_2 + \dot{C}_3 + \dot{C}_{w,\text{turb}}$	$c_1 = c_2 = c_3$
Ejector	$\dot{C}_2 + \dot{C}_9 + \dot{Z}_{\text{ejc}} = \dot{C}_4$	–
Evaporator 1	$\dot{C}_7 + \dot{C}_{26} + \dot{Z}_{\text{eval}} = \dot{C}_9 + \dot{C}_{27}$	$c_8 = c_9$
Pump 1	$\dot{C}_{10} + \dot{C}_{w,\text{pump1}} + \dot{Z}_{\text{pump1}} = \dot{C}_{17}$	–
Heater	$\dot{C}_2 + \dot{C}_{28} + \dot{Z}_H = \dot{C}_{18} + \dot{C}_{29}$	$c_2 = c_{18}$
Economizer	$\dot{C}_{12} + \dot{C}_{21} + \dot{Z}_{\text{eco}} = \dot{C}_{13} + \dot{C}_{22}$	$c_{21} = c_{22}$
Superheater	$\dot{C}_{14} + \dot{C}_{19} + \dot{Z}_{SH} = \dot{C}_1 + \dot{C}_{20}$	$c_{19} = c_{20}$
Condenser	$\dot{C}_5 + \dot{C}_{15} + \dot{Z}_{\text{cond}} = \dot{C}_6 + \dot{C}_{16}$	$c_5 = c_6$
Auxiliary boiler	$\dot{C}_{23} + \dot{C}_{NG} + \dot{Z}_{AB} = \dot{C}_{19}$	–
Storage tank	$\dot{C}_{22} + \dot{C}_{25} + \dot{Z}_{ST} = \dot{C}_{23} + \dot{C}_{24} + \dot{C}_{L,ST}$	$c_{23} = c_{24}$
Solar collector	$\dot{C}_{24} + \dot{C}_s + \dot{Z}_{\text{coll}} = \dot{C}_{25}$	$c_s = 0$

Table 8.10 Capital cost expressions for system components [25, 27–29].

Component	Capital cost
Turbine	$\log_{10}(Z_{\text{Turb}}^{\text{CI}}) = 2.6259 + 1.4398\log_{10}(\dot{W}_{\text{Turb}}) - 0.1776[\log_{10}(\dot{W}_{\text{Turb}})]^2$
Electric generator	$Z_{\text{Elec}}^{\text{CI}} = 60\dot{W}_{\text{Elec}}^{0.95}$
Pump	$Z_{\text{pump}}^{\text{CI}} = 3540\dot{W}_{\text{pump}}^{0.71}$
Heat exchanger	$Z_{\text{HE}}^{\text{CI}} = 130\left(\frac{A_{\text{HE}}}{0.093}\right)^{0.78}$
Condenser	$Z_{\text{Cond}}^{\text{CI}} = 1773\dot{m}_5$
Storage tank	$Z_{\text{ST}}^{\text{CI}} = 4042V_{\text{ST}}^{0.506}$

24.4% and 9.8% in summer and 48.9% and 11.7% in winter. Component exergy destruction rates are shown in Figures 8.19 and 8.20, on seasonal bases:

- Summer: Figure 8.19 illustrates that in summer the solar collector and auxiliary boiler are the major sources of exergy destruction, with the solar collector accounting for 13.09 kW of the exergy destruction rate (44.1% of total exergy input rate), and the auxiliary boiler accounting for 9.87 kW of the exergy destruction rate (33.3% of total exergy input rate). The exergy destruction rates of all remaining components are significantly lower.
- Winter: Figure 8.20 illustrates that in winter the auxiliary boiler and solar collector are again the main sources of exergy destruction. The auxiliary boiler is responsible for 15.04 kW of the exergy destruction rate (49.7% of total exergy input rate) and solar collector 8.25 kW of the exergy destruction rate (27.2% of total exergy input rate). Again, all other components exhibit much lower exergy destruction rates.

In the solar collector, the irreversibility is due to the large temperature difference between solar heat and the fluid in the tubes. In the auxiliary boiler, the irreversibility

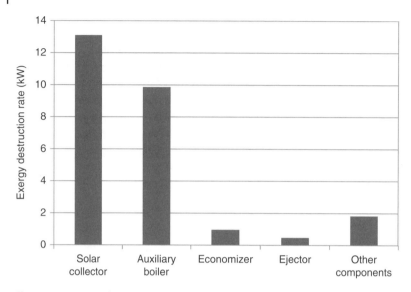

Figure 8.19 Exergy destruction rates of components of solar based CCHP system in summer.

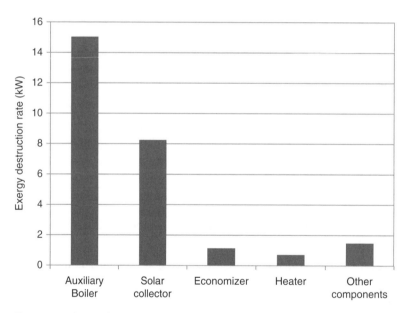

Figure 8.20 Exergy destruction rates of components of solar based CCHP system in winter.

is caused by the combustion process within it, which is typically a major source of irreversibility in a process [30]. Since high destruction rates are observed in both summer and winter for the solar collector and auxiliary boiler, careful design and selection of these components is important in designing a solar CCHP system.

8.3.3.1 Exergoeconomic Results

The exergoeconomic analysis of the solar CCHP cycle is conducted based on the first and second laws of thermodynamics, using the SPECO method [31]. With the equations in

Table 8.11 Exergoeconomic parameters of solar based CCHP system in summer.

Component	$\dot{C}_D$ ($/year)	$\dot{C}_L$ ($/year)	$\dot{Z}$ ($/year)	$\dot{Z} + \dot{C}_D + \dot{C}_L$ ($/year)
Auxiliary boiler	1676.69	0	42.11	1718.80
Solar collector	0	0	1110.56	1110.56
Condenser	478.48	0	50.74	529.22
Economizer	585.97	0	87.84	673.81
Ejector	402.35	0	0	402.35
Evaporator1	148.93	0	272.88	421.81
Evaporator2	458.91	0	214.50	673.41
Pump1	28.12	0	139.93	168.05
Superheater	59.37	0	55.68	115.05
Electric generator	151.61	0	19.19	170.80
Storage tank	158.42	0.20	528.34	686.96
Turbine	372.06	0	226.61	598.67

Table 8.12 Exergoeconomic parameters of solar based CCHP system in winter.

Component	$\dot{C}_D$ ($/year)	$\dot{C}_L$ ($/year)	$\dot{Z}$ ($/year)	$\dot{Z} + \dot{C}_D + \dot{C}_L$ ($/year)
Auxiliary boiler	2244.70	0	55.78	2300.48
Solar collector	0	0	1110.56	1110.56
Condenser	345.52	0	19.72	365.24
Heater	335.07	0	89.34	424.41
Pump1	3.75	0	30.05	33.80
Pump2	14.21	0	121.99	136.20
Economizer	648.96	0	84.20	733.16
Evaporator2	556.22	0	214.18	770.40
Electric generator	176.46	0	19.19	195.65
Superheater	71.92	0	55.59	127.51
Storage tank	107.97	0.32	528.34	636.31
Turbine	431.03	0	226.05	657.08

Tables 8.9 and 8.10, exergoeconomic parameters of the CCHP system are calculated for summer and winter (see Tables 8.11 and 8.12). The results demonstrate that the CCHP product cost rate is 5114.5 $/year in summer and 5688.1 $/year in winter. According to exergoeconomic evaluation guidelines, in designing a new system, more attention must be paid to the components for which the sum $\dot{Z} + \dot{C}_D + \dot{C}_L$ is highest. In summer, Table 8.11 shows that the auxiliary boiler, solar collector, and storage tank have the highest values of $\dot{Z} + \dot{C}_D + \dot{C}_L$ and are, therefore, the most important components from an exergoeconomic point of view. In winter, Table 8.12 demonstrates that the auxiliary boiler, solar collector, and evaporator 2 have the highest values of $\dot{Z} + \dot{C}_D + \dot{C}_L$ and are, therefore, the most important components exergoeconomically.

8.3.4 Sensitivity Analysis

For better understanding of the impact of various design parameters on system performance, a sensitivity analysis is conducted. Figure 8.21a shows the effect of turbine inlet pressure on exergy efficiency and total cost rate in summer. When turbine inlet pressure increases from 700 kPa to 1400 kPa, the exergy efficiency increases by about 40% due to the decrement in system exergy destruction, and the cost rate decreases by 23%. Figure 8.21b shows that, in winter, when the turbine inlet pressure increases from 850 kPa to 1000 kPa, the exergy efficiency decreases by about 5% because of the increment of system exergy destruction, and the total cost rate decreases by 3%.

The influences of turbine inlet temperature on exergy efficiency and total cost rate are illustrated in Figures 8.22a and 8.22b for summer and winter, respectively. In Figure 8.22a, it can be observed that, when the turbine inlet temperature increases to 30°C, the exergy efficiency rises only 2% and the total cost rate decreases. Figure 8.22b

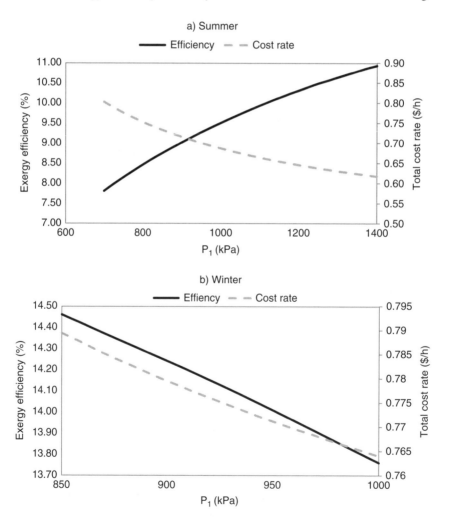

Figure 8.21 Effect of turbine inlet pressure on exergy efficiency and total cost rate for summer (a) and winter (b).

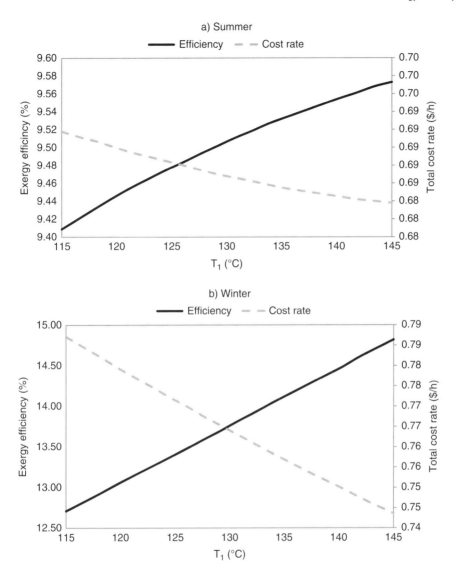

Figure 8.22 Effect of turbine inlet temperature on exergy efficiency and total cost rate for summer (a) and winter (b).

shows that, by increasing the turbine inlet temperature, the exergy efficiency rises by 6% and total cost rate decreases by 5%.

Figures 8.23a and 8.23b illustrate the effects of turbine back pressure on exergy efficiency and total cost rate in summer and winter. In Figure 8.23a, the variation of turbine back pressure is seen in summer to result in a decrease in exergy efficiency of about 18% due to an increase in system exergy destruction. However, total cost rate increases by about 18%. Figure 8.23b indicates that, for a rise of about 150 kPa in turbine back pressure, the exergy efficiency and total cost rate increase by about 9% and 6%, respectively.

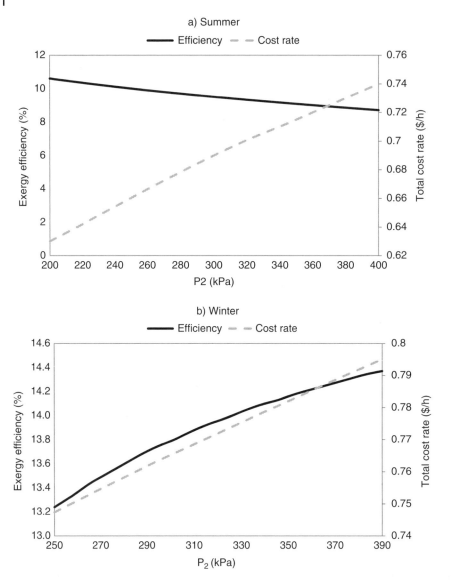

Figure 8.23 Effect of turbine back pressure on exergy efficiency and total cost rate for summer (a) and winter (b).

Figures 8.24 shows the effects of evaporator temperature on exergy efficiency and total cost rate of the system in summer. When temperature increases from −5°C to 5°C, the system exergy efficiency decreases by 6% because of a slight change in the system exergy destruction, while the total cost rate is seen to be independent of evaporator temperature.

Figure 8.25 shows the effect of heater outlet temperature on system exergy efficiency and total cost rate in winter. As the heater outlet temperature rises from 50 to 100°C, the system exergy efficiency increases by 21% due to the decrease in the system exergy destruction; the total cost rate decreases by 0.8%.

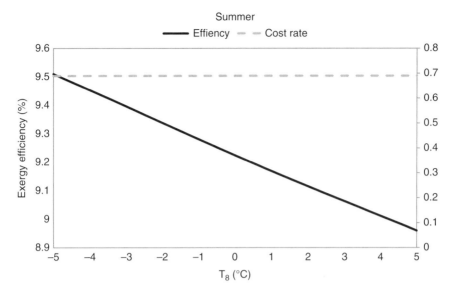

Figure 8.24 Effect of evaporator temperature on exergy efficiency and total cost rate in summer.

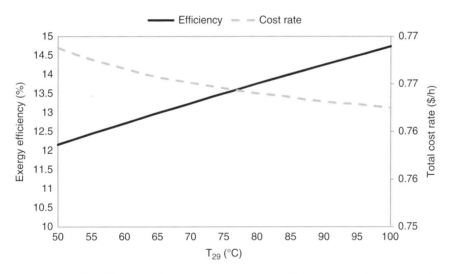

Figure 8.25 Effect of heater outlet temperature on exergy efficiency and total cost rate in winter.

8.3.5 Optimization

Efficiency and cost are two conflicting objectives and the optimum status needs to be obtained by a trade-off process. Exergy efficiency and total cost rate are considered as two objective functions here to determine the optimum condition of the solar based CCHP system. In this multi-objective optimization, exergy efficiency is to be maximized while the total cost rate of the system is to be minimized, while observing constraints related to decision variables. Expressions for these objective functions have been shown in previous sections. The decision variables are selected to be turbine inlet temperature (T_1), turbine inlet pressure (P_1), turbine back pressure (P_2) and evaporator temperature

Table 8.13 Decision variables and their feasible ranges.

Decision variable	Summer	Winter
Turbine inlet pressure	$700 \leq P_1 (\text{kPa}) \leq 1400$	$850 \leq P_1 (\text{kPa}) \leq 1000$
Turbine inlet temperature	$115 \leq T_1 (°\text{C}) \leq 145$	$115 \leq T_1 (°\text{C}) \leq 145$
Turbine back pressure	$200 \leq P_2 (\text{kPa}) \leq 400$	$250 \leq P_2 (\text{kPa}) \leq 400$
Evaporator temperature	$-5 \leq T_8 (°\text{C}) \leq 5$	–
Heater outlet temperature	–	$50 \leq T_{28} (°\text{C}) \leq 100$

(T_8) for summer mode, and turbine inlet temperature (T_1), turbine inlet pressure (P_1), turbine back pressure (P_2) and heater outlet temperature (T_{28}) for winter mode. Feasible ranges for the decision variables are listed in Table 8.13.

8.3.6 Optimization Results

The genetic algorithm (GA) is selected to perform the multi-objective optimization. The Pareto frontier solutions for the solar based CCHP system are shown in Figure 8.26a for summer and Figure 8.26b for winter.

As shown in Figure 8.25a, while the total exergy efficiency of the cycle increases to about 10.5%, the total cost rate increases slightly. In addition, an increase in the exergy efficiency beyond 10.5% leads to a sharp increase in the total cost rate of the system. The maximum exergy efficiency exists at design point C (12%), while the total cost rate is the greatest at this point (0.5 $/hr). But, the minimum value for the total product cost rate occurs at design point A which is about 0.39 $/hr. Design point A is the optimal situation when total cost rate of product is the sole objective function, while design point C is the optimum point when exergy efficiency is the sole objective function. In multi-objective optimization, a process of decision-making for selection of the final optimal solution from the available solutions is required. The process of decision-making is usually performed with the aid of a hypothetical point in Figure 8.26a (the ideal point), at which both objectives have their optimal values independent of the other objectives. It is clear that it is impossible to have both objectives at their optimum point simultaneously and, as shown in Figure 8.26a, the ideal point is not a solution located on the Pareto frontier. The closest point of the Pareto frontier to the ideal point might be considered as a desirable final solution. A similar discussion and explanation applies for the winter mode, shown in Figure 8.26b.

Since each point on the Pareto frontier can be utilized as the optimum point, the importance of each of the objectives and engineering selection plays an important role in choosing a point as an optimum solution. Table 8.14 shows all the design parameters for points A, B, and C.

To better understand the variations of all design parameters, scattered distributions of the design parameters are shown in Figures 8.27 to 8.31, for both summer and winter modes. The results demonstrate that in summer the turbine inlet temperature tends toward its lower values as this results in a decrease in total cost rate (Figure 8.27a), while in winter the turbine inlet temperature tends toward its upper values as this results in an increase in exergy efficiency (Figure 8.27b).

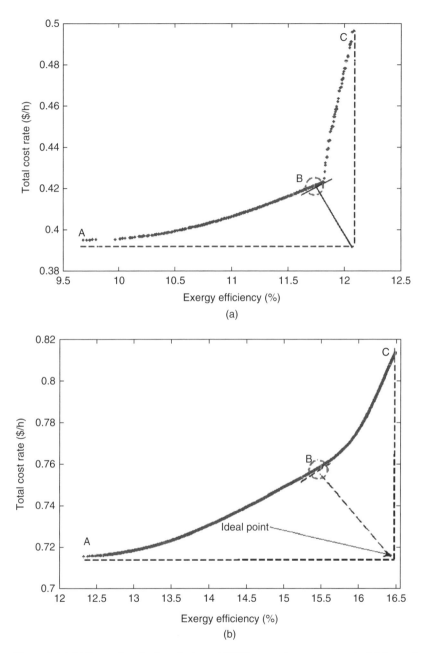

Figure 8.26 (a) Pareto frontier for solar based CCHP system in summer mode. (b) Pareto frontier for solar based CCHP system in winter mode.

Table 8.14 Optimized values for design parameters of the solar based CCHP system for three points on the Pareto frontier from multi-objective optimization.

Point	$\varepsilon_{CCHP}(\%)$		$\dot{C}_p \left(\frac{\$}{h}\right)$		T_1 (°C)		P_1 (kPa)		P_2 (kPa)		T_8 (°C)	T_{29} (°C)
	Summer	Winter	Summer	Winter	Summer	Winter	Summer	Winter	Summer	Winter	Summer	Winter
A	9.7	12.3	0.5	0.8	144	136	1400	1250	201	200	−5	50
B	11.7	15.4	0.6	0.85	115	140	1371	920	207	350	−5	65
C	12.1	16.5	0.7	0.9	145	145	851	850	399	398	−5	100

Figure 8.28a shows that in summer the turbine inlet pressures have a scattered distribution between their allowable ranges. This is due to the fact that an increase in turbine inlet pressure results in a rise in exergy efficiency of the system, while in winter (Figure 8.28b) the turbine inlet pressure tends to reach its lower values as this results in a decrease in total cost rate.

In both Figures 8.29a and 8.29b, the turbine back pressure exhibits a scattered distribution between its allowable ranges because an increase in turbine back pressure results in a decrease in exergy efficiency of the system in both seasons.

Figure 8.30 shows that evaporation temperature exhibits a scattered distribution between its allowable ranges. This is due to the fact that an increase in evaporation temperatures decreases the exergy efficiency of the system. Similarly, Figure 8.31 shows that heater outlet temperature exhibits a scattered distribution between its allowable ranges, again due to the fact that an increase in heater outlet temperatures increases the exergy efficiency of the system.

8.4 Hybrid Wind–Photovoltaic–Battery System

8.4.1 Modeling

In this section, we consider a hybrid solar–wind system with battery storage, as shown in Figure 8.32. It is mainly comprised of solar PV arrays, wind fields, and a battery bank, as well as an inverter to convert the direct electrical current to alternating current. Modeling of the components is described in the following subsections.

8.4.1.1 Photovoltaic (PV) Panel

The total solar radiation incident on a tilted panel is a function of the direct beam and diffuse radiation. The latter comes from all areas of the sky except the solar position, and includes circumsolar diffuse radiation, diffuse radiation from the horizon, and reflected radiation from the surroundings. The total solar radiation incident on a tilted surface can be written as [32]:

$$\dot{I}_t = \left(\dot{I}_b + \dot{I}_d \frac{\dot{I}_b}{\dot{I}_h}\right) R_b + \dot{I}_d \left(1 - \frac{\dot{I}_b}{\dot{I}_h}\right)\left(\frac{1 + \cos \beta}{2}\right)\left(1 + \sqrt{\dot{I}_b/\dot{I}_h}\sin^3\left(\frac{\beta}{2}\right)\right)$$
$$+ \dot{I}_h \rho_g \left(\frac{1 - \cos \beta}{2}\right) \tag{8.69}$$

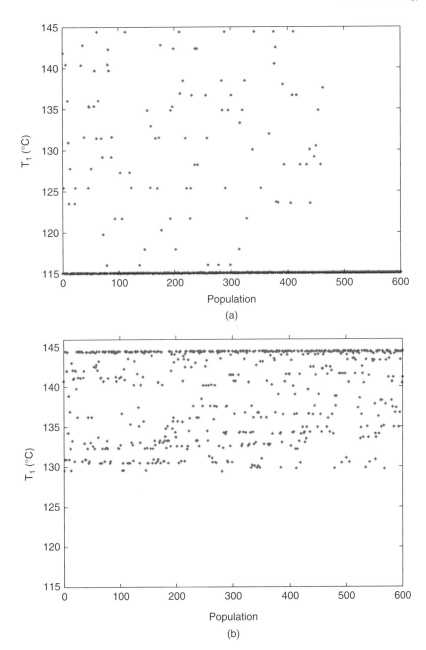

Figure 8.27 (a) Scattered distribution of turbine inlet temperature with population in Pareto frontier in summer. (b) Scattered distribution of turbine inlet temperature with population in Pareto frontier in winter.

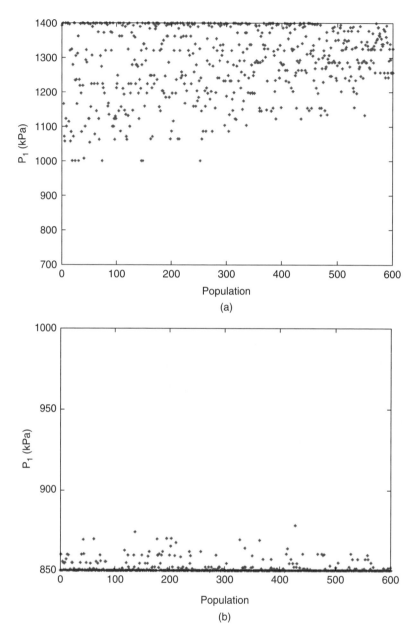

Figure 8.28 (a) Scattered distribution of turbine inlet pressure with population in Pareto frontier in summer. (b) Scattered distribution of turbine inlet pressure with population in Pareto frontier in winter.

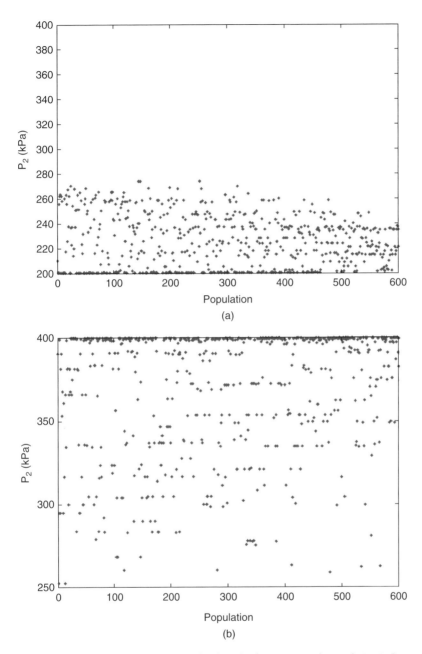

Figure 8.29 (a) Scattered distribution of turbine back pressure with population in Pareto frontier in summer. (b) Scattered distribution of turbine back pressure with population in Pareto frontier in winter.

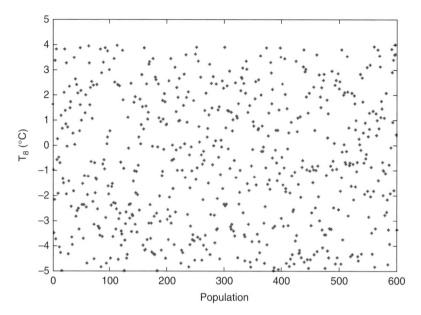

Figure 8.30 Scattered distribution of evaporation temperature with population in Pareto frontier in summer.

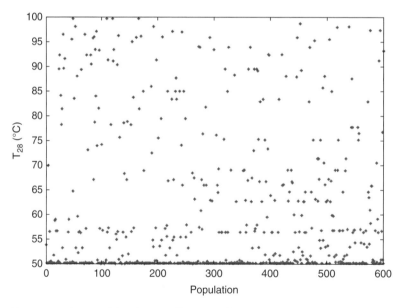

Figure 8.31 Scattered distribution of heater outlet temperature with population in Pareto frontier in winter.

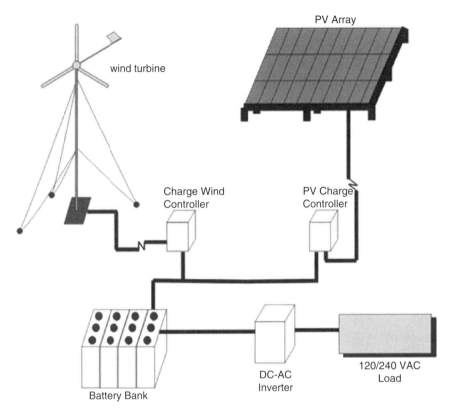

Figure 8.32 Schematic diagram of hybrid solar–wind system with a battery storage system.

Here, $\dot{I}_b, \dot{I}_d, \dot{I}_h, R_b, \beta$, and ρ_g denote direct beam radiation, diffuse radiation, sum of direct and diffuse beam, the ratio of total radiation incident on the titled surface to that incident on a horizontal surface, surface tilt angle, and reflectance from the surroundings. Details of computing direct beam and diffuse radiation are described elsewhere [33].

The net output power of a PV panel (P_{pv}) with an open circuit voltage V_{oc_real} and short circuit current I_{sc_real} under real operating conditions can be written as follows [34]:

$$P_{pv} = f\!f(V_{oc-real} \times I_{sc-real}) \tag{8.70}$$

where $f\!f$ is the fill factor, given by:

$$f\!f = P_{max}/(V_{oc}I_{sc}) \tag{8.71}$$

Here, P_{max} denotes the maximum output power of a PV collector, V_{oc} is the open circuit voltage of the PV collector in laboratory conditions, and I_{sc} is the PV short circuit current under laboratory conditions. Note that values of P_{max}, V_{oc}, and I_{sc} are typically obtained from PV module manufacturers. The quantities V_{oc_real} and I_{sc_real} can be expressed as follows [34]:

$$V_{oc-real} = V_{oc} + f_{Voc-T} \times T_c \tag{8.72}$$

$$I_{sc-real} = [I_{SC} + f_{Isc-T}(T_c - T_{std})](\dot{I}_t/\dot{I}_{std})\dot{I}_t \tag{8.73}$$

where f_{Voc_T} and f_{Isc_T} are temperature correction coefficients of current and voltage, respectively. Also, T_{std} (in $^\circ$C) and $\dot{I}_{std}$ (in W/m^2) are the temperature and solar radiation at standard conditions, respectively, and T_c ($^\circ$C) is the surface temperature of the PV panel. The parameters $f_{Voc\_T}, f_{Isc\_T}, T_{std}$ and G_{std} depend on the type of module used and are usually obtained from PV module manufacturers.

8.4.1.2 Wind Turbine (WT)

Wind turbine manufacturers usually provide turbine power curves at different wind speeds. If the power curve of the turbine is not available, the following can be used to estimate the power output of a wind turbine [35]:

$$P_{tur} = \begin{vmatrix} 0 & if & V < V_c \\ P_{er}\left(\dfrac{V^n - V_c^n}{V_r^m - V_c^n}\right) & if & V_c < V < V_r \\ P_{er} & if & V_r < V < V_f \end{vmatrix} \tag{8.74}$$

where I_{er} is the rated electrical power, V (in m/s) is the wind speed, V_c is the cut-in wind speed; V_r is the rated wind speed, and V_f is the cut-out wind speed. Setting the exponents m and n to 2 and 3 is often sufficiently accurate for analysis of wind power systems [35].

8.4.1.3 Battery

Due to the intermittency of solar collectors and wind turbines, the battery capacity constantly changes in PV/WT/battery-based hybrid systems. In such a system, the state of charge (SOC) of the battery is evaluated under two possible states. When the total power output of the PV panels and wind turbines exceeds the demand load, the battery bank is in a charging situation. Quantities involved in battery charging can be determined with the following expression [36]:

$$P_{bat}(t) = P_{bat}(t-1) \times (1 - \sigma) + (P_{pv}(t)\eta_{inv} + P_{WT}(t) \times \eta_{inv}^2 - P_{dmn}(t)/\eta_{inv}) \times \eta_{bat} \tag{8.75}$$

where $P_{bat}(t)$ and $P_{bat}(t-1)$ are the charge quantities of the battery bank at times t and $t - 1$, respectively. In addition, σ denotes the hourly self-discharge rate, η_{inv} the inverter efficiency, P_{dmn} the demand, and η_{bat} the battery charge efficiency.

When the total output of the PV panels and wind turbines is lower than the demand load, the battery bank is in a discharging situation. For simplicity, the discharge efficiency of battery bank is assumed here to be 1. As a result, the discharge quantity of the battery bank at time t can be expressed as [36]:

$$P_{bat}(t) = P_{bat}(t-1) \times (1 - \sigma) - (-P_{pv}(t)\eta_{inv} - P_{WT}(t) \times \eta_{inv}^2 + P_{dmn}(t)/\eta_{inv})/\eta_{inv} \tag{8.76}$$

8.4.2 Objective Function, Design Parameters, and Constraints

In this section, the total annual cost (TAC) is considered as the objective function. TAC includes the capital cost of equipment including wind turbines, PV panels, batteries and the inverter as well as the maintenance cost, and can be written as:

$$TAC = aC_{in} + C_m \tag{8.77}$$

where

$$C_{in} = n_1 C_{PV} + n_2 C_{WT} + n_3 C_{bat} + n_4 C_{inv} \tag{8.78}$$

$$C_m = n_1 C_{PV,m} + n_2 C_{WT,m} \tag{8.79}$$

Here, n_1, n_2, n_3 and n_4 are the numbers of PV panels, wind turbines, batteries, and inverters, respectively, while C_{PV}, C_{WT}, C_{bat} and C_{inv} are unit cost of the PV panels, the wind turbines, the batteries, and the inverters, respectively. Also, $C_{PV,m}$ and $C_{WT,m}$ are the unit costs of maintenance for the PV panels and wind turbines, respectively, and a denotes the annual cost coefficient, defined as [37]:

$$a = \frac{i}{1 - (1+i)^{-y}} \tag{8.80}$$

where i and y are the interest rate and the depreciation time, respectively.

Some equipment in the PV/WT/battery system needs to be replaced several times during the project lifetime. Here, the battery lifetime is considered to be five years. Using the single payment present value factor, the present value of battery C_{bat} can be expressed as follows:

$$C_{bat} = P_{bat} \left(1 + \frac{1}{(1+i)^5} + \frac{1}{(1+i)^{10}} + \frac{1}{(1+i)^{15}} \right) \tag{8.81}$$

where P_{bat} is the price of the battery. Also, the lifetime of the inverter is considered here to be 10 years, so the present worth of inverter C_{inv} can be expressed using the single payment present value factor as follows:

$$C_{inv} = P_{inv} \left(1 + \frac{1}{(1+i)^{10}} \right) \tag{8.82}$$

where P_{inv} is the inverter price.

The numbers of PV panels, wind turbines, and inverters are considered as design parameters. For the PV/WT/battery system, at any time, the charge quantity of the battery bank should be selected in the range of $P_{bat,min}$ to $P_{bat,max}$. The maximum charge quantity of the battery bank takes on the value of the nominal capacity of the battery bank, and the minimum charge quantity of the battery bank is obtained by maximum depth of discharge (DOD) which can be calculated as:

$$P_{bat} = (1 - DOD)S_{bat} \tag{8.83}$$

here S_{bat} is the nominal capacity of the battery bank (in Wh).

8.4.3 Real Parameter Genetic Algorithm

Binary coded genetic algorithms (GAs) are used in many engineering optimization problems. But, application of the binary GA in a continuous search space has two difficulties. The first is the presence of "Hamming cliffs" related to certain strings in which a transition to a neighboring solution needs a variation of many bits [38]. The second problem is the inability to have an arbitrary precision in the optimal solution. To overcome these difficulties, the Real Parameter Genetic Algorithm (RPGA) is used. The main difference between binary and real parameter genetic algorithms is in the crossover and mutation operators. In fact, the decoding operator in binary coding is eliminated in RPGA and the optimization problem is a step easier compared with the binary coded GAs. Since the selection operator works with the fitness value, any selection operator used with binary coded GAs can also be used in real parameter GAs.

The steps in the RPGA, which is used for optimization in this analysis of the system, are as follows:

1) The initial population with M chromosomes is randomly generated using lower and upper bounds of the design parameters, x^{min} and x^{max}, as follows:

$$x_0^t = x^{min} + rand\,(x^{max} - x^{min}) \tag{8.84}$$

where *rand* is a uniformly distributed random function.

2) Each chromosome is exported to the thermoeconomic modeling section and returned back with a value of objective function (total annual cost).

3) The selection operator is performed to choose the better chromosomes in the genetic algorithm method [38].

4) The crossover operation is performed using the following relations [38]:

$$x_i^{(1,t+1)} = 0.5\{(1 + \beta_i)x_i^{(1,t)} + (1 - \beta_i)x_i^{(2,t)}\} \tag{8.85}$$

$$x_i^{(2,t+1)} = 0.5\{(1 - \beta_i)x_i^{(1,t)} + (1 + \beta_i)x_i^{(2,t)}\} \tag{8.86}$$

Here, β_i is:

$$\beta_i = \begin{cases} (2rand)^{1/(1+a_c)} & \text{if } rand \leq 0.5 \\ \left(\dfrac{1}{2 - 2rand}\right)^{1/(1+a_c)} & \text{otherwise} \end{cases} \tag{8.87}$$

where α_c is a crossover constant parameter.

5) Then, the mutation operation is performed on the population as follows [38]:

$$x_i^{(1,t+1)} = x_i^{(1,t+1)} + (x_i^{max} - x_i^{min})\delta \tag{8.88}$$

where δ is:

$$\delta = \begin{cases} (2rand)^{1/(1+a_m)} - 1 & \text{if } rand < 0.5 \\ 1 - [2(1-rand)]^{1/(1+a_m)} & \text{if } rand \geq 0.5 \end{cases} \tag{8.89}$$

in which α_m is a mutation constant parameter.

6) This procedure is repeated from step 2 until the convergence criterion is met.

8.4.4 Case Study

The PV/WT/battery system optimization procedure is applied for a residential area located in three provinces in Iran. These include Tabriz, Tehran, and Zahedan, representing respectively cold, moderate, and hot climates. The values of equipment lifetime (k) and interest rate (i) are considered to be 15 years and 15%, respectively. In addition, a wind turbine with a nominal capacity of 9.8 kW, a battery with nominal capacity of 200 Ah, and PV panels with a nominal capacity of 240 W are used. The constants of investment cost in Equation 8.78 are considered to be $n = (3200, 614, 130, 40)$ based on equipment available in the marketplace. Hourly and daily variations in electrical demand load for the studied case are shown in Figure 8.33.

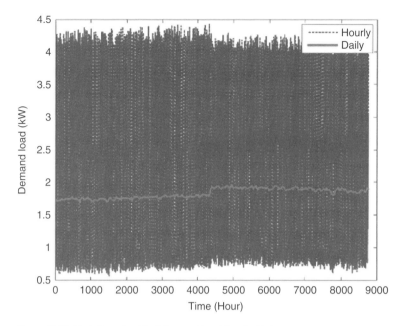

Figure 8.33 Variation electrical demand load for studied case.

8.4.5 Results and Discussion

The hourly variation of solar radiation, wind velocity, and ambient temperature for the three studied climates—including cold (Tabriz), moderate (Tehran), and hot (Zahedan)—are depicted in Figure 8.34.

Then, the RPGA is applied for 200 iterations, using 100 chromosomes, a mutation factor $\alpha_m = 2$, and a crossover factor $\alpha_c = 2$. Note that the RPGA is run three times using a core i7-3200GHz processor for each case and the best results are presented here. To ensure a reasonable result, hourly analysis over a year is performed. As a result of there being 8760 hours during a year, each optimization process needs about 5 hours to complete. To accelerate the optimization process, six separate optimization programs are run simultaneously on the mentioned processor.

The optimum results for the TAC and corresponding values of the design parameters for each studied case are listed in Table 8.15. Zahedan with its hot climate has a significantly lower TAC compared with the other studied cases. Tabriz has the next lowest TAC, followed by Tehran.

The annual average ambient temperature, incident solar radiation, and wind speed are listed in Table 8.16. Zahedan is observed to have the highest average of incident solar radiation and wind velocity. Although the higher temperature decreases the PV efficiency for the case of Zahedan compared with other studied cases, the higher potential solar and wind power more than compensate for this reduction. Due to the low wind velocity and low incident solar radiation in Tehran, the highest TAC is obtained in this case.

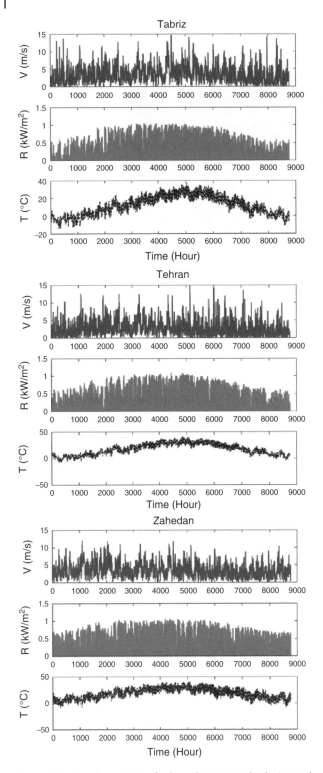

Figure 8.34 Hourly variation of solar radiation, wind velocity, and ambient temperature for the three studied cities.

Table 8.15 Values of optimum design parameters and objective function for three studied cases.

Region	Number of PV units	Number of WT units	Number of battery units	TAC ($/year)
Tabriz	52	10	30	10 145
Tehran	122	1	23	10 516
Zahedan	45	2	27	6186

Table 8.16 Annual average of selected parameters for the studied cases.

Region	Ambient temperature (°C)	Incident solar radiation (kW/m²)	Wind speed (m/s)
Tabriz	12	0.20	3.27
Tehran	17	0.20	2.45
Zahedan	18	0.22	3.33

To obtain a good insight into the hourly variation of power supply by PV and wind, power demand, and battery charging and discharging, variations of these parameters at the optimum condition for the three studied cases during six selected months are shown in Figures 8.35–8.37. Note that the variation is shown for the middle week of each month. The following points can be seen in Figures 8.35–8.37:

- Discharging is high in the cold months of a year such as January and November. The insolation angle on PV panels during these months is high and as a result lower radiation levels are received by the panels. As a result, stored electricity is released. The minimum number of batteries required is determined by the months in which lower levels of solar radiation are available.
- Because of the delay between periods of PV power generation and peak demand, excess electricity is generated and stored during periods with the maximum incident solar radiation.
- The best matching between PV collectors and wind turbines occurs when the maximum solar radiation is received at different times than when maximum wind speeds occur. For this situation, the lowest amount of energy must be stored and, as a result, a lower capacity storage system is needed which reduces the TAC. For the three studied regions, Zahedan exhibits the best matching between solar radiation and wind speed and thus has a much lower TAC compared with Tehran and Tabriz.
- Regular fluctuations in battery charging and discharging are observed for months with high levels of solar radiation such as May and September due to the dominant share of PV power output and roughly regular variations of solar radiation during these months.
- Due to the low level of incident solar radiation in Tabriz in January, more wind turbines should be employed to compensate for the reduction of solar based electrical power. Alternatively, other means could be employed to provide the lacking electrical power for these months, to decrease the number of wind turbines.

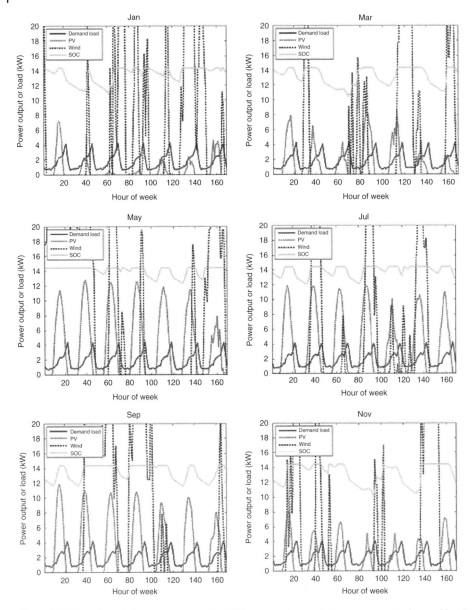

Figure 8.35 Hourly variation over one week of PV power output, wind power output, demand load, and battery state of charge for several months in Tabriz.

- A significant portion of the electricity generated by this plant is in general wasted due to the large variations between electrical supply and load demand during the year.

8.5 Concluding Remarks

This chapter discusses renewable energy systems and examines numerous novel renewable energy systems for various locations. Optimization is applied to determine optimal

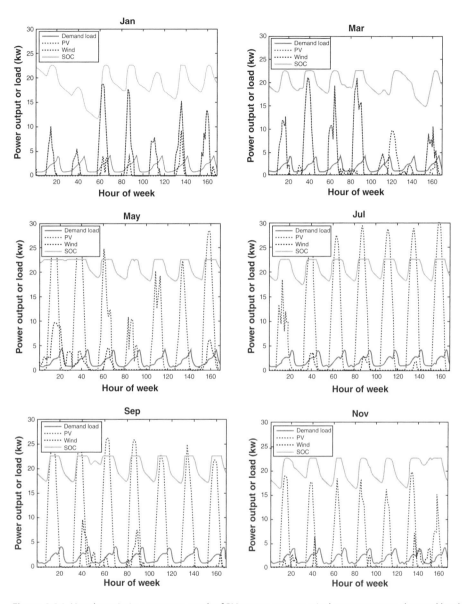

Figure 8.36 Hourly variation over one week of PV power output, wind power output, demand load, and battery state of charge for several months in Tehran.

design parameters. Ocean thermal energy conversion (OTEC), solar based combined cooling, heating and power, and solar/wind/ battery energy systems are introduced, analyzed and optimized in this chapter. Exergy and exergoeconomic analyses are also conducted and the results described. Emphasis is placed on the benefits of exergy methods and the importance of defining objective functions and constraints reasonably to ensure reliable results. An evolutionary algorithm based optimization is applied to each system and the optimal design parameters are carefully determined. To enhance understanding

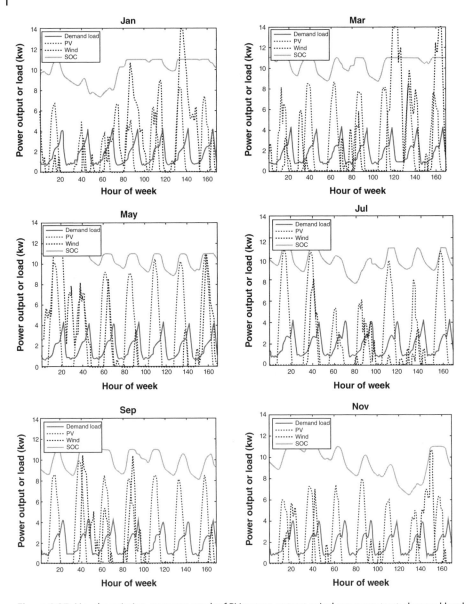

Figure 8.37 Hourly variation over one week of PV power output, wind power output, demand load, and battery state of charge for several months in Zahedan.

of the design criteria, sensitivity analyses are conducted to investigate how each objective function varies when small changes are made to selected design parameters.

References

1 Ahmadi, P. (2013) Modeling, analysis and optimization of integrated energy systems for multigeneration purposes. PhD thesis. Faculty of Engineering and Applied Science, University of Ontario Institute of Technology.

2 Ahmadi, P., Rosen, M. A. and Dincer, I. (2011) Greenhouse gas emission and exergo-environmental analyses of a trigeneration energy system. *International Journal of Greenhouse Gas Control* **5(6)**:1540–1549.

3 Tchanche, B. F., Lambrinos, G., Frangoudakis, A. and Papadakis, G. (2011) Low-grade heat conversion into power using organic Rankine cycles – A review of various applications. *Renewable and Sustainable Energy Reviews* **15(8)**:3963–3979.

4 Faizal, M. and Rafiuddin Ahmed, M. (2011) On the ocean heat budget and ocean thermal energy conversion. *International Journal of Energy Research* **35(13)**:1119–1144.

5 Meegahapola, L., Udawatta, L. and Witharana, S. (2007) *The Ocean Thermal Energy Conversion Strategies and Analysis of Current Challenges*. International conference on industrial and information systems, 2007 *(ICIIS 2007), Aug 9–11, 2007, University of Peradeniya, Peradeniya, Sri Lanka*. IEEE.

6 Esteban, M. and Leary, D. (2012) Current developments and future prospects of offshore wind and ocean energy. *Applied Energy* **90(1)**:128–136.

7 Farahat, S., Sarhaddi, F. and Ajam, H. (2009) Exergetic optimization of flat plate solar collectors. *Renewable Energy* **34(4)**:1169–1174.

8 Sukhatme, K. and Sukhatme, S. P. (1996) *Solar Energy: Principles of Thermal Collection and Storage*, Tata McGraw-Hill Education.

9 Nihous, G. and Vega, L. (1993) Design of a 100 MW OTEC-hydrogen plantship. *Marine structures* **6(2)**: p. 207–221.

10 Cengel, A.Y. and Boles, M.A. (2008) *Thermodynamics: An Engineering Approach*, McGraw-Hill, New York.

11 Esmaili, P., Dincer, I. and Naterer, G. (2012) Energy and exergy analyses of electrolytic hydrogen production with molybdenum-oxo catalysts. *International Journal of Hydrogen Energy* **37(9)**:7365–7372.

12 Ni, M., Leung M. K. and Leung, D. Y. (2008) Energy and exergy analysis of hydrogen production by a proton exchange membrane (PEM) electrolyzer plant. *Energy conversion and management* **49(10)**:2748–2756.

13 Nieminen, J., Dincer, I. and Naterer, G. (2010) Comparative performance analysis of PEM and solid oxide steam electrolysers. *International Journal of Hydrogen Energy* **35(20)**:10842–10850.

14 Ameri, M., Ahmadi, P. and Khanmohammadi, S. (2008) Exergy analysis of a 420 MW combined cycle power plant. *International Journal of Energy Research* **32(2)**:175–183.

15 Ahmadi, P. and Dincer, I. (2010) Exergoenvironmental analysis and optimization of a cogeneration plant system using Multimodal Genetic Algorithm (MGA). *Energy* **35(12)**:5161–5172.

16 Rosen, M.A., Dincer, I. and Kanoglu, M. (2008) Role of exergy in increasing efficiency and sustainability and reducing environmental impact. *Energy policy* **36(1)**:128–137.

17 Bejan, A., Tsatsaronis, G. and Moran, M. J. (1996) *Thermal design and optimization*, John Wiley & Sons, Inc., New York.

18 Ahmadi, P., Rosen, M. A. and Dincer, I. (2012) Multi-objective exergy-based optimization of a polygeneration energy system using an evolutionary algorithm. *Energy* **46(1)**:21–31.

19 Wang, J., Dai, Y. and Gao, L. (2009) Exergy analyses and parametric optimizations for different cogeneration power plants in cement industry. *Applied Energy* **86(6)**:941–948.

20 Peters, M. S., Timmerhaus, K. D., West, R. E., Timmerhaus, K. and West, R. (1968) *Plant Design and Economics for Chemical Engineers*, Vol. **4**, McGraw-Hill, New York.

21 Uehara, H. and Nakaoka, T. (1984) OTEC using plate-type heat exchanger (using ammonia as working fluid). *Transactions of JSME* **50(453)**:1325–1333.

22 Uehara, H., Dilao C. O. and Nakaoka, T. (1988) Conceptual design of ocean thermal energy conversion (OTEC) power plants in the *Philippines*. **41(5)**:431–441.

23 Nag, P. and Gupta, A. (1998) Exergy analysis of the Kalin Solar energy cycle. *Applied Thermal Engineering* **18(6)**:427–439.

24 Kalogirou, S. (2009) *Solar Energy Engineering: Processes and Systems*, Elsevier, UK.

25 Mohammadkhani, F., Shokati, N., Mahmoudi, S. M. S., Yari, M. and Rosen, M. A. (2014) Exergoeconomic assessment and parametric study of a Gas Turbine-Modular Helium Reactor combined with two Organic Rankine Cycles. *Energy* **65(0)**:533–543.

26 Farshi, L.G., Mahmoudi, S. and Rosen, M. (2013) Exergoeconomic comparison of double effect and combined ejector-double effect absorption refrigeration systems. *Applied Energy* **103**:700–711.

27 Martínez-Lera, S., Ballester, J. and Martínez-Lera, J. (2013) Analysis and sizing of thermal energy storage in combined heating, cooling and power plants for buildings. *Applied Energy* **106**:127–142.

28 El-Emam, R.S. and Dincer, I. (2013) Exergy and exergoeconomic analyses and optimization of geothermal organic Rankine cycle. *Applied Thermal Engineering* **59(1)**:435–444.

29 Pierobon, L., Nguyen, T.-V., Larsen, U., Haglind, F. and Elmegaard, B. (2013) Multi-objective optimization of organic Rankine cycles for waste heat recovery: Application in an offshore platform. *Energy* **58**:538–549.

30 Boyaghchi, F.A. and Heidarnejad, P. (2015) Thermodynamic analysis and optimisation of a solar combined cooling, heating and power system for a domestic application. *International Journal of Exergy* **16(2)**:139–168.

31 Lazzaretto, A. and Tsatsaronis, G. (2006) SPECO: A systematic and general methodology for calculating efficiencies and costs in thermal systems. *Energy* **31(8–9)**: 1257–1289.

32 Hajabdollahi, H., Ganjehkaviri, A. and Jaafar, M. N. M. (2015) Thermo-economic optimization of RSORC (regenerative solar organic Rankine cycle) considering hourly analysis. *Energy* **87**:369–380.

33 Duffy, J. and Beckman, W. (1991) *Solar Engineering of Thermal Processes*, John Wiley & Sons, Inc., Chapters I–II.

34 Häberlin, H. (2012) *Photovoltaics System Design and Practice*, John Wiley & Sons Ltd, UK.

35 Manwell, J. F., McGowan, J. G. and Rogers, A. L. (2010) *Wind Energy Explained: Theory, Design and Application*, John Wiley & Sons, Ltd., UK.

36 Ai, B., Yang, H., Shen, H. and Liao, X. (2003) Computer-aided design of PV/wind hybrid system. *Renewable Energy* **28(10)**:1491–1512.

37 Hajabdollahi, H., Ganjehkaviri, A. and Jaafar, M. N. M. (2015) Assessment of new operational strategy in optimization of CCHP plant for different climates using evolutionary algorithms. *Applied Thermal Engineering* **75**:468–480.

38 Deb, K. (2001) *Multi-Objective Optimization Using Evolutionary Algorithms*, John Wiley & Sons, Ltd., UK.

Study Questions/Problems

1   A biomass based renewable energy system is shown in Figure 8.38. Apply mass, energy, entropy, and exergy rate balance equations to determine the exergy destruction rate and exergy efficiency of the system, and determine the optimal design parameters when exergy efficiency is an objective function. Make any reasonable assumptions you deem necessary.

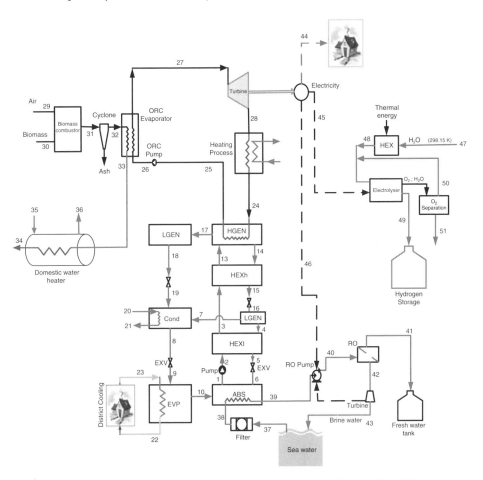

Figure 8.38 Schematic of biomass based renewable energy system for the provision of heating, cooling, electricity, hydrogen, fresh water, and hot water.

2   An extension of the OTEC system discussed in this chapter for multigeneration is shown in Figure 8.39. Apply mass, energy, entropy, and exergy rate balance equations to determine the exergy destruction rate and exergy efficiency of the system, and find the optimal exergy efficiency of the system. Make any reasonable assumptions you deem necessary.

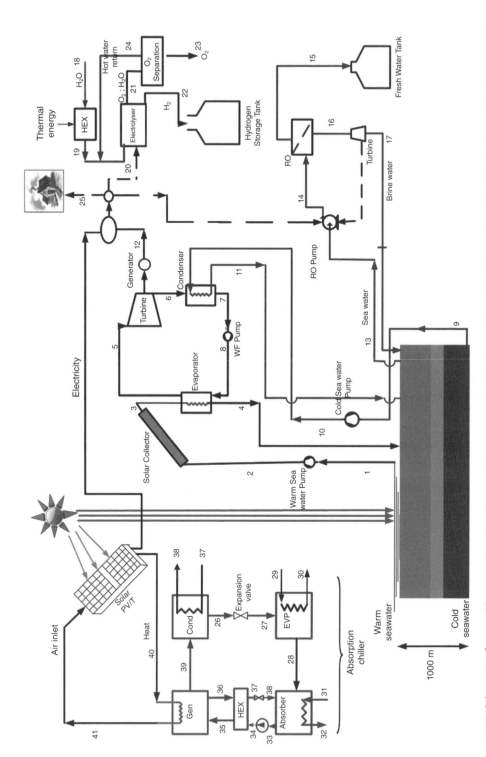

Figure 8.39 Schematic of renewable energy-based multigeneration energy system for the provision of heating, cooling, electricity, hydrogen, fresh water, and hot water.

9

Modeling and Optimization of Power Plants

9.1   Introduction

Electrical power plants are one the most important technologies in modern society as electricity plays a significant role in most aspects of our lives, from charging cell phones to powering our homes and businesses. Although new technologies for power generation are being introduced, conventional thermal power plants, such as steam, gas, and combined cycle plants, remain at the core of global electricity generation. Power generation cycles are typically complex and composed of a wide range of devices, such as heat exchangers, boilers, heaters, pumps, cooling systems, and valves. As populatios grow, the need for energy, and in particular electricity, will likely become increasingly important [1].

Problems with energy supply and use are related not only to global warming, but also to such other environmental concerns as air pollution (e.g., emissions of such pollutants as SO_2, and NO_x), acid precipitation, ozone depletion, forest destruction, and emission of radioactive substances. These issues must be accounted for if humanity is to achieve a sustainable energy future with little environmental impact. Electricity generation accounts for approximately 25% of total worldwide greenhouse gas (GHG) emissions [2]. Therefore, more efficient and cost effective power generation systems with lower GHG emissions are being considered or installed worldwide to support efforts to reduce GHG emissions. For such complex systems, optimization methods can assist through attaining more beneficial system designs, in terms of efficiency and sustainability.

The long-term trends of rising energy prices and decreasing fossil fuel resources make the optimum application and management of energy and energy resources of great importance. In most countries, numerous conventional power plants driven by fossil fuels such as oil, coal, and natural gas or by other energy resources like uranium are in service today. During the past decade, many power generation companies have developed and introduced a range of process improvements to steam power plants, especially measures that improve plant efficiencies and/or reduce environmental impacts [3].

Among conventional power plants, steam power plants, gas turbine power plants, and combined cycle power plants are the most common. They exhibit relatively low costs and good flexibility, although their efficiencies and emissions vary by plant. As a consequence, it is important to model the systems and determine the optimal design parameters for each, and this constitutes the focus of this chapter.

Optimization of Energy Systems, First Edition. Ibrahim Dincer, Marc A. Rosen, and Pouria Ahmadi.
© 2017 John Wiley & Sons Ltd. Published 2017 by John Wiley & Sons Ltd.

A simple steam power plant uses water as the working fluid and exploits fossil fuels or nuclear energy in a boiler to produce high temperature and high pressure steam. The steam expands in a steam turbine, causing this to rotate and generate electrical power. Saturated vapor exits the steam turbine and enters the condenser. The liquid condensate enters a pump and preheaters, after which it is at the desired boiler inlet conditions.

A simple gas turbine power plant consists of an air compressor, a combustion chamber, and a gas turbine. In gas turbine power plants, the working fluid is air and the fuel is usually natural gas. Since much energy is wasted in simple gas turbine power plants, a heat exchanger (which is called a heat recovery steam generator) is often added to the power plant to facilitate utilization of the waste energy and generation of additional electrical power in a bottoming cycle. Such an integrated system is called a combined cycle power plant (CCPP), and usually exhibits a higher thermal efficiency than simple steam and gas turbine power plants.

In this chapter, we describe the thermodynamic modeling, analysis, and optimization of steam, gas, and combined cycle power plants. We apply exergy analysis to each power plant to pinpoint the locations of the most significant loses in each system. To better understand the role of key design parameters, a performance assessment of each system is conducted. Then, the optimization of such plants is explained, including defining appropriate objective functions and establishing reasonable constraints. An evolutionary algorithm based optimization is applied to each system and the optimal design parameters are determined. To enhance understanding of the design criteria, a sensitivity analysis is performed to determine how each objective function varies when changes are made to selected design parameters. Finally, closing remarks are provided on the optimal design of thermal power plants.

9.2 Steam Power Plants

Steam power plants are one of the common systems for electrical power generation. Real plants are quite complex and can generate up to 1000 MW of electricity in units with large steam turbines [2]. One of the main technologies for electricity generation, especially in countries where fossil fuels like coal or natural gas or oil are abundant, steam power plants can use various fuels. Since steam power plants are responsible for most of the electricity generation in the world, a small increase in thermal efficiency can lead to large fuel savings and GHG emission reductions.

A simple, idealized Rankine cycle for a steam power plant is shown in Figure 9.1. The idealized Rankine cycle consists of the following main processes:

- Isentropic compression in a pump
- Combustion of fuel in a boiler and heat addition at constant pressure
- Isentropic expansion in a turbine
- Heat rejection in a condenser at constant pressure

Water enters the pump of the Rankine cycle in Figure 9.1 at point 1 as a saturated liquid and is compressed to the boiler pressure. The temperature increases during isentropic compression, and there is a slight decrease in specific volume. High pressure water enters the boiler at point 2 where heat is added, usually by combusting fuel in air. The boiler is essentially a large heat exchanger composed of several heating elements, for example, economizers, evaporators, and superheaters. The combustion reaction takes

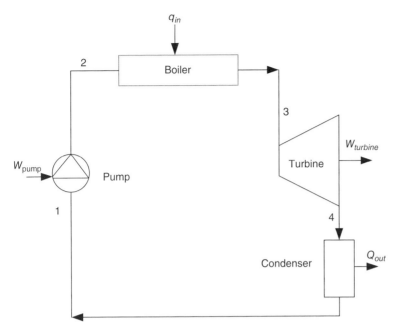

Figure 9.1 A simple idealized Rankine cycle.

place in the boiler if it is fossil fuel-driven, and heat is transferred to the water to produce steam at high temperature. In the idealized system, no pressure losses are assumed across the boiler. The high temperature and high pressure steam from the boiler enters the steam turbine where it expands isentropically and produces work by rotating a shaft connected to a generator. Both temperature and pressure decline to the values at point 4 during this process. The steam is then changed back to a liquid in the condenser, where heat q_{out} is rejected to a cooling medium, using a lake, a river, or a cooling tower. Steam exits the condenser as a saturated liquid and then enters the pump where its pressure increases to the desired point 2.

The efficiency of this simple steam power plant is defined as:

$$\eta = \frac{w_{turbine} - w_{pump}}{q_{in}} \tag{9.1}$$

In advanced real steam power plants, additional components are added to the simple cycle in Figure 9.1 to increase the thermal efficiency and provide economic and other benefits. One method involves increasing the temperature entering the boiler using feedwater heaters. Another effective method is to decrease the average temperature at which heat is rejected from the working fluid in the condenser. A process flow diagram of a real steam power plant is shown in Figure 9.2.

The steam power plant represented in Figure 9.2 is located 25 km from the city of Qazvin in Iran, and is a typical plant. The steam power plant uses natural gas as the main fuel and diesel oil as the secondary fuel. The power plant can produce up to 250 MW of electrical power at full load conditions. The main components of the steam power plant are a three stage steam turbine (i.e., a turbine with high pressure, intermediate pressure, and low pressure stages), a steam generator, a drum boiler, feedwater heaters, and a condenser. The natural gas fuel is a mixture of 98.57% methane (CH_4), 0.63% ethane (C_2H_6),

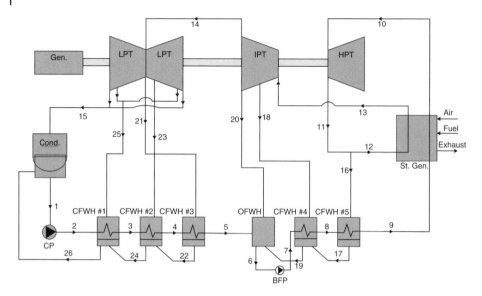

Figure 9.2 Schematic of real steam power plant.

Table 9.1 Operating conditions of a steam power plant.

Parameter	Value
Fuel mass flow rate (kg/s)	13.89
Plant gross electrical power (MW)	263
Stack flue gas temperature (°C)	115
Lower heating value of fuel (kJ/kg)	49 433.96

0.1% propane (C_3H_8), 0.05% butane (C_4H_{10}), 0.04% pentane (C_5H_{12}), 0.6% nitrogen (N_2) and 0.01% carbon dioxide (CO_2), where the values represent volume fractions [4]. Data on the operation of this power plant are listed in Table 9.1.

9.2.1 Modeling and Analysis

For each component, rate balances of mass, energy, exergy, and cost can be applied in order to determine such quantities as work, heat, exergy flow, thermodynamic properties, and energy and exergy efficiencies. Several rate balances based on energy and exergy for the system component are given below. For a steady state process, mass, energy, and exergy rate balance equations can be written as follows:

$$\sum \dot{m}_{in} = \sum \dot{m}_{out} \tag{9.2}$$

$$\dot{Q} - \dot{W} = \sum \dot{E}_{out} - \sum \dot{E}_{in} \tag{9.3}$$

$$\dot{Ex}_Q - \dot{E}_W = \sum \dot{Ex}_e - \sum \dot{Ex}_i + \dot{Ex}_D \tag{9.4}$$

Since the differences in elevation and velocity in a steam power plant are usually not significant, potential and kinetic energy can be assumed negligible. With this assumption,

Equation 9.3 becomes:

$$\dot{Q} - \dot{W} = \sum \dot{m}_e h_e - \sum \dot{m}_i h_i \tag{9.5}$$

In Equation 9.4, $\dot{Ex}_Q$ and $\dot{E}_W$ are the exergy rates associated with heat transfer across the system boundary and work produced, respectively. These terms can be expressed as follows:

$$\dot{Ex}_Q = \sum \left(1 - \frac{T_o}{T_b}\right) \dot{Q} \tag{9.6}$$

$$\dot{E}_W = \dot{W} \tag{9.7}$$

Neglecting potential and kinetic exergy, the total exergy rate $\dot{Ex}$ can be expressed as

$$\dot{Ex} = \dot{m}(ex_{ph} + ex_{ch}) \tag{9.8}$$

where the specific physical exergy is:

$$ex_{ph} = (xh_o) - T_o(s - s_o) \tag{9.9}$$

The specific chemical exergy of the fuel for the steam power plant can be written as:

$$ex_{fuel} = \xi_{fuel} \times LHV \tag{9.10}$$

Here, LHV represents the lower heating value of the fuel, which is natural gas for this power plant (see Table 9.1). For gaseous fuels with a general chemical formula of $C_x H_y$, a chemical to heating value ratio (ξ_{fuel}) can be determined from the following empirical relation [5]:

$$\xi_{fuel} = 1.033 + 0.0169\frac{y}{x} - \frac{0.0698}{x} \tag{9.11}$$

The exergy rate balance equations and exergy efficiency expressions for the main components of the steam power plant in Figure 9.2 are listed in Table 9.2. The energy rate balance equations, which are based on the first law of thermodynamics, can be used to determine energy related properties. Thermodynamic properties and quantities for the steam power plant in Figure 9.2 are listed in Table 9.3.

Simulation code is developed and used in the analyses. In order to verify the simulation code, the results are compared with actual data provided for the power plant in Figure 9.2. The comparison results are listed in Table 9.4. The simulation results are found to be in good agreement with actual data, with most differences due to the assumptions considered in the modeling and the steady state approximation.

Table 9.2 Equations for exergy destruction rate and exergy efficiency for the power plant components.

Component	Exergy destruction rate	Exergy efficiency
Boiler	$\dot{Ex}_{D,boiler} = \dot{Ex}_{fuel} + \dot{Ex}_i - \dot{Ex}_e$	$\eta_{boiler} = (\dot{Ex}_i - \dot{Ex}_e)/\dot{Ex}_{fuel}$
Turbine	$\dot{Ex}_{D,turbine} = \dot{Ex}_i - \dot{Ex}_e - \dot{W}_{turbine}$	$\eta_{turbine} = 1 - (\dot{Ex}_{turbine}/\dot{Ex}_i - \dot{Ex}_e)$
Pump	$\dot{Ex}_{D,pump} = \dot{Ex}_i - \dot{Ex}_e + \dot{W}_{pump}$	$\eta_{pump} = 1 - (\dot{Ex}_{pump}/\dot{W}_{pump})$
Feed water heater	$\dot{Ex}_{D,fwh} = \dot{Ex}_i - \dot{Ex}_e$	$\eta_{fwh} = 1 - (\dot{Ex}_{fwh}/\dot{Ex}_i)$
Condenser	$\dot{Ex}_{D,condenser} = \sum_k \dot{Ex}_i - \dot{Ex}_e$	$\eta_{condenser} = \frac{\sum_i \dot{Ex}_e}{\sum_i \dot{Ex}_i}$

Table 9.3 Thermodynamic quantities for the steam power plant in Figure 9.2.

Point	P (bar)	T (°C)	ṁ (kg/s)	h (kJ/kg)	s (kJ/kg.K)	ex (kJ/kg)
1	0.19	59.2	173.8	247.8	0.8211	7.543
2	0.22	61.2	173.8	256.2	0.8462	8.43
3	0.21	82.6	173.8	345.9	1.105	20.99
4	0.21	105.7	173.8	443.3	1.363	41.51
5	0.21	125.3	173.8	526.2	1.581	59.26
6	0.21	164.6	215	695.8	1.985	108.7
7	168.90	167.7	215	718.4	2.000	126.7
8	168.90	203.6	215	874.9	2.341	181.5
9	168.90	243.1	215	1054	2.701	252.9
10	137.20	538	215	3430	6.535	1487
11	36.60	349.1	211.4	3098	6.628	1126
12	36.60	349.1	192.7	3098	6.628	1126
13	33.60	538	192.7	3539	7.286	1371
14	7.37	318.3	171.1	3096	7.338	912.8
15	0.29	68.87	154.4	2484	7.361	293.9
16	36.60	349.1	17.46	3098	6.628	1126
17	18.53	208.6	17.45	894.2	2.417	177.9
18	2.69	318.3	11.81	3097	7.341	913.2
19	8.57	173.3	29.36	733.8	2.074	119.8
20	2.69	318.3	11.81	3097	7.341	913.2
21	1.43	210.1	5.948	2888	7.413	681.8
22	1.43	111.3	5.948	466.8	1.433	44.18
23	0.62	149.1	6.802	2772	7.468	550.4
24	0.56	88.2	12.75	369.4	1.172	24.59
25	61.27	86.7	6.127	2637	7.478	411.8
26	0.27	66.8	18.88	279.6	0.9157	11.14

Table 9.4 Comparison of simulated and actual data for the steam power plant.

Component	Simulated value	Actual value	Difference (%)
High pressure turbine produced power (MW)	80.6	79 983	0.87
Intermediate pressure turbine produced power (MW)	83.4	83 114	0.37
Low pressure turbine produced power (MW)	99.9	100 878	0.88
Condensate pump consumed power (MW)	0.19	208.5	8.9
Boiler feed pump consumed power (MW)	4.85	4860	0.04
Mass flow rate of fuel (kg/s)	13.86	13.89	0.21

The exergy analysis results are presented and the results show that the highest exergy destruction rate is exhibited by the boiler. The boiler irreversibility is associated with two important processes: the irreversibility that arises from the combustion reaction and the irreversibility that arises during heat transfer to the working fluid. In the process where combustion occurs, there is a large entropy generation rate, which corresponds to a high exergy destruction rate. In addition, there are two separate streams passing through the boiler, combustion gases at high temperature and cold water; the large temperature difference between these streams also contributes to a high entropy generation rate and a high exergy destruction rate. The results help identify possible improvements for this device, like increasing the water temperature entering the boiler, preheating the combustion air, and replacing the water tubes in the boiler with ones having better heat transfer characteristics.

9.2.2 Objective Functions, Design Parameters, and Constraints

To apply a multi-objective optimization, we first define objective functions and establish appropriate constraints. We also need to consider the major design parameters for the system. In a steam power plant, several design parameters are possible, but we consider only the major ones here. For the given steam power plant, exergy efficiency and total cost rate are taken to be the objective functions. They are defined, respectively, as follows:

$$\psi = \frac{\dot{W}_{\text{net}}}{\xi \dot{m}_{\text{fuel}} LHV} \tag{9.12}$$

$$\dot{Z} = \frac{\alpha \Phi Z_{\text{in}}}{\tau \times 3600} + \dot{C}_f \tag{9.13}$$

Here, the maintenance factor is $\phi = 1.06$, and Z_{in} is the total purchase equipment cost (in US dollars), listed in Table 9.5. Also, α is the annual cost coefficient, evaluated as

$$\alpha = \frac{i}{1 - (1 + i)^{-n}} \tag{9.14}$$

Here, i and n are the interest rate and depreciation time, respectively [6]. In Equation 9.13, $\dot{C}_f$ is the fuel cost rate which is calculated as

$$\dot{C}_f = c_f \dot{m}_f LHV \tag{9.15}$$

The design parameters are selected for consideration when maximizing the exergy efficiency and minimizing the total cost rate are steam turbine inlet pressure, steam turbine inlet temperature, extraction pressures for the turbines, isentropic efficiencies for the turbines and pumps, reheat pressure, and condenser pressure. The constraints considered for the optimization study are listed in Table 9.6.

The steam cycle optimum design parameters are obtained for the operating power plant described in the previous section. As already explained, the exergy efficiency and total cost rate are the objective functions. The number of iterations for finding the global extremum in the overall searching domain is about 3×10^{33}. The system is optimized for a depreciation time $n = 20$ years, and an interest rate $i = 0.11$. The developed genetic algorithm code is applied for optimization for 600 generations using a search population size of $M = 100$ individuals, a crossover probability of $p_c = 0.9$, a gene mutation probability of $p_m = 0.03$ and a controlled elitism value $c = 0.55$.

Table 9.5 Cost functions in terms of thermodynamic parameters for the steam power plant components [7].

System component	Capital or investment cost functions
Boiler	$Z_{Boiler} = a_1 (\dot{m}_{boiler})^{a_2} \Phi_p \Phi_T \Phi_\eta \Phi_{SH/RSH}$ $\Phi_p = \exp\left(\dfrac{P_e - \overline{P}_e}{a_3}\right), \Phi_T = 1 + a_5 \exp\left(\dfrac{T_e - \overline{T}_e}{a_6}\right), \Phi_\eta = 1 + \left(\dfrac{1 - \overline{\eta}_1}{1 - \eta_1}\right)^{a_4}$ $\Phi_{SH/RSH} = 1 + \dfrac{T_e - T_{iSH}}{T_e} + \dfrac{\dot{m}_{RSH}}{\dot{m}_{boiler}} \cdot \dfrac{T_{eRSH} - T_{iRSH}}{T_{eRSH}}$ $\overline{T}_e = 593°c, \overline{P}_e = 28\ bar, \overline{\eta}_1 = 0.9, a_1 = 208582\ \$/kgs$ $a_2 = 0.8, a_3 = 150\ bar, a_4 = 7, a_5 = 5, a_6 = 10.42°c$
Deaerator	$Z_{deaerator} = a_1 (\dot{m}_{water})^{a_2}$ $a_1 = 145315\ \$/kW^{0.7} \quad a_2 = 0.7$
Steam Turbine (ST)	$Z_{ST} = a_{51}.P_{ST}^{0.7}\left(1 + \left(\dfrac{0.05}{1 - \eta_{ST}}\right)^3\right) \times \left(1 + 5.\exp\left(\dfrac{T_a - 866\,K}{10.42\,K}\right)\right),$ $a_{51} = 3880.5\$.kW^{-0.7}$
Condenser	$Z_{COND} = a_{61}.\dfrac{\dot{Q}_{cond}}{k.\Delta T_{in}} + a_{62}.\dot{m}_{CW} + 70.5.\dot{Q}_{cond} \times (-0.6936.Ln(\overline{T}_{CW} - T_b) + 2.1898)$ $a_{61} = 280.74\$.m^{-2}, a_{62} = 746\$.(kg.s)^{-1}, k = 2200\,W.(m^2.K)^{-1}$
Pump	$Z_{pump} = a_{71}.P_{pump}^{0.71}\left(1 + \dfrac{0.2}{1 - \eta_{pump}}\right), a_{71} = 705.48\$.(kg.s)^{-1}$

Table 9.6 Design parameters for the steam power plant and their ranges.

Design parameter	Lower bound	Upper bound
T_{10} (°C)	500	550
P_{10} (MPa)	8	17
P_{16} (MPa)	2	4
P_{18} (MPa)	1	2
P_{20} (MPa)	0.4	1
P_{21} (MPa)	0.3	0.4
P_{23} (MPa)	0.1	0.3
P_{25} (MPa)	0.05	0.1
P_1 (MPa)	0.005	0.5
η_t	0.7	0.9
η_p	0.5	0.8

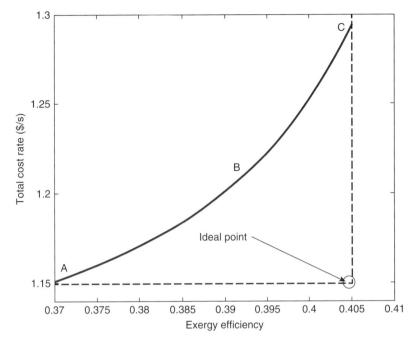

Figure 9.3 Pareto Frontier, identifying best trade-off values for the objective functions.

The results for the multi-objective optimization of the steam power plant are shown in Figure 9.3. The Pareto frontier solution is shown for this system with objective functions as shown in Equations 9.12 and 9.13 in multi-objective optimization. It can be seen in this figure that the total cost rate of products increases moderately as the total exergy efficiency of the cycle increases to about 40%.

As shown in Figure 9.3, the maximum exergy efficiency exists at design point C (40.05%), while the total cost rate of products is the greatest at this point (1.28 $/hr). But, the minimum value for the total cost rate of product occurs at design point A which is about 1.15 $/hr. Design point A is the optimal situation when total cost rate of product is the sole objective function, while design point C is the optimum point when exergy efficiency is the sole objective function. In multi-objective optimization, a process of decision-making for selection of the final optimal solution from the available solutions is required. The process of decision-making is usually performed with the aid of a hypothetical point (marked as the "ideal point" in Figure 9.3), at which both objectives have their optimal values independent of the other objectives. The optimal design parameters for points on the Pareto curve are listed in Table 9.7. The optimization results for this power plant demonstrate that if the design parameter point C in Figure 9.3 is selected, the exergy efficiency of the plant increases by approximately 4% which is a significant improvement.

9.3 Gas Turbine Power Plants

Gas turbine (GT) plants are widely utilized throughout the world for electricity generation, and natural gas is often used as the fuel in such plants. Today, many electrical

Table 9.7 Values of design parameters for optimum selected points A–C in Pareto optimum front.

Design parameter	A	B	C
T_{10} (°C)	545	540	550
P_{10} (MPa)	9	9	16.71
P_{16} (MPa)	2.07	2.08	2.2
P_{18} (MPa)	1	1.08	1.05
P_{20} (MPa)	0.5	0.52	0.55
P_{21} (MPa)	0.33	0.34	0.35
P_{23} (MPa)	0.28	0.29	0.3
P_{25} (MPa)	0.05	0.05	0.0521
P_1 (MPa)	0.007	0.007	0.007
η_t	0.88	0.87	0.9
η_P	0.69	0.0.66	0.65

generating utilities are striving to improve the efficiency (or heat rate) at their existing thermal electric generating stations, many of which are over 25 years old. Often, a heat rate improvement of only a few percent appears desirable, since the costs and complexity of such measures often are more manageable than more expensive options. In this regard, exergy based optimization is considered a useful tool for performance assessment of power plants [8]. Energy utilization is very much governed by thermodynamic principles (particularly by exergy), and therefore, an understanding of exergetic aspects can improve understanding of pathways to sustainable development [9]. GT plants are known for their low capital cost to power ratios, high flexibility, high reliability without complexity, short delivery time, rapid commissioning to commercial operation, and short start up time [9].

The gas turbine power plant considered here is shown in Figure 9.4. It includes an air preheater that is part of a heat exchanger that recovers heat from the exhaust gas. During operation of the gas turbine power plant, air at ambient conditions enters an air compressor, which is treated as adiabatic, at point 1. After compression to point 2, the air enters the air preheater (AP) where its temperature is increased to point 3. The hot air exiting the air preheater is supplied to the combustion chamber (CC). The fuel at point 9 is injected into the combustion chamber and hot combustion gases exit (point 4) and pass through a gas turbine to produce power. The hot flue gases exit the gas turbine at point 5 and pass through the heat recovery heat exchanger, which increases the temperature of the air before it enters the combustion chamber, and then exit the plant as flue gases at point 6 and as emissions to the ambient surroundings.

9.3.1 Thermodynamic Modeling

Energy balances and governing equations for the components of the gas turbine power plant are utilized to model the device. An energy analysis for the air compressor is provided in the previous section. Here, the thermodynamic modeling of the other

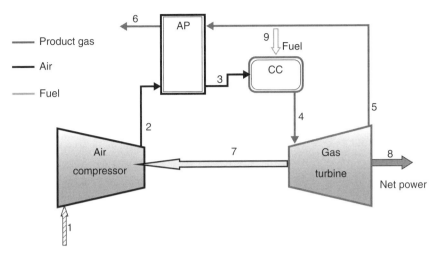

Figure 9.4 Gas turbine power plant.

components is explained. Several simplifying assumptions are made to render the modeling and analysis more tractable, while retaining adequate accuracy to ensure meaningful results [10]:

- All processes take place at steady state.
- Air and combustion products are taken to be ideal gas mixtures.
- The fuel injected into the combustion chamber is taken to be natural gas.
- Heat loss from the combustion chamber is taken to be 2% of the fuel energy entering the combustion chamber based on lower heating value, and all other components are considered adiabatic.
- The dead state is $P_0 = 1.01$ bar and $T_0 = 293.15$ K.

An energy rate balance equation is described below for each plant device:

9.3.1.1 Air Compressor

Air at ambient pressure and a temperature T_1 enters the compressor. The compressor outlet temperature is a function of compressor isentropic efficiency (η_{AC}), compressor pressure ratio (r_{AC}) and the specific heat ratio. That is,

$$T_2 = T_1 \times \left(1 + \frac{1}{\eta_{AC}}\left(r_{AC}^{\frac{\gamma_a - 1}{\gamma_a}} - 1\right)\right) \tag{9.16}$$

The compressor work rate is a function of air mass flow rate ($\dot{m}_a$), the air specific heat, and the temperature difference, and can be expressed as follows:

$$\dot{W}_{AC} = \dot{m}_a C_{pa}(T_2 - T_1) \tag{9.17}$$

where γ_a is the air specific heat ratio, and C_{pa} is treated as a function of temperature as follows [11]:

$$C_{pa}(T) = 1.048 - \left(\frac{3.83T}{10^4}\right) + \left(\frac{9.45T^2}{10^7}\right) - \left(\frac{5.49T^3}{10^{10}}\right) + \left(\frac{7.92T^4}{10^{14}}\right) \tag{9.18}$$

9.3.1.2 Air Preheater (AP)

We can write the following energy rate balance for a control volume around the air preheater:

$$\dot{m}_a(h_3 - h_2) = \dot{m}_g(h_5 - h_6)\eta_{AP} \tag{9.19}$$

where η_{AP} is air preheater efficiency. A pressure drop through the preheater ΔP_{AP} is considered, where

$$\frac{P_3}{P_2} = (1 - \Delta P_{AP}) \tag{9.20}$$

9.3.1.3 Combustion Chamber (CC)

The outlet properties of the combustion chamber are a function of air mass flow rate, fuel lower heating value (LHV), and combustion efficiency, and are related as follows:

$$\dot{m}_a h_2 + \dot{m}_f LHV = \dot{m}_g h_3 + (1 - \eta_{cc})\,\dot{m}_f LHV \tag{9.21}$$

The combustion chamber outlet pressure is defined by considering a pressure drop across the combustion chamber as follows:

$$\frac{P_3}{P_2} = 1 - \Delta P_{CC} \tag{9.22}$$

where ΔP_{cc} is the pressure loss across the combustion chamber and η_{cc} is the combustion efficiency.

The combustion reaction occurring and its species coefficients can be expressed as follows:

$$\lambda C_{x1} H_{y1} + (x_{O_2} O_2 + x_{N_2} N_2 + x_{H_2O} H_2O + x_{CO_2} CO_2 + x_{Ar} Ar) \rightarrow y_{CO_2} CO_2$$
$$+ y_{N_2} N_2 + y_{O_2} O_2 + y_{H_2O} H_2O + y_{NO} NO + y_{CO} CO + y_{Ar} Ar), \tag{9.23}$$

where

$$y_{CO_2} = (\lambda x_1 + x_{CO_2} - y_{CO_2}),$$

$$y_{N_2} = (x_{N_2} - y_{NO}),$$

$$y_{H_2O} = \left(x_{H_2O} + \frac{\lambda x y_1}{2}\right),$$

$$y_{O_2} = \left(x_{O2} - \lambda \times x_1 - \frac{\lambda x y_1}{4} - \frac{y_{CO}}{2} - \frac{y_{NO}}{2}\right),$$

$$y_{Ar} = x_{Ar}, \text{ and}$$

$$\lambda = \frac{n_f}{n_{air}}.$$

9.3.1.4 Gas Turbine

The gas turbine outlet temperature can be written as a function of gas turbine isentropic efficiency (η_{GT}), the gas turbine inlet temperature (T_3), and gas turbine pressure ratio (P_3/P_4) as follows:

$$T_5 = T_4(1 - \eta_{GT}\left[1 - \left(\frac{P_4}{P_5}\right)^{\frac{1-\gamma_g}{\gamma_g}}\right] \tag{9.24}$$

The gas turbine output power is also found as:

$$\dot{W}_{GT} = \dot{m}_g C_{pg}(T_4 - T_5) \tag{9.25}$$

Here, $\dot{m}_g$ is the gas turbine mass flow rate, which is calculated as

$$\dot{m}_g = \dot{m}_f + \dot{m}_a \tag{9.26}$$

The net output power can be expressed as

$$\dot{W}_{Net} = \dot{W}_{GT} - \dot{W}_{AC} \tag{9.27}$$

where C_{pg} is taken to be a function of temperature as follows [5]:

$$C_{pg}(T) = 0.991 + \left(\frac{6.997T}{10^5}\right) + \left(\frac{2.712T^2}{10^7}\right) - \left(\frac{1.2244T^3}{10^{10}}\right) \tag{9.28}$$

9.3.2 Exergy and Exergoeconomic Analyses

We now focus on the exergy destruction and efficiency for the components of the gas turbine system. Expressions for these exergy parameters, which are used in the optimization procedures discussed subsequently, are given in Table 9.8. A complete exergy analysis is not provided, as extensive details regarding exergy analyses of gas turbine based power generation are given elsewhere [5, 9] and in previous chapters.

Exergoeconomics is the branch of engineering that appropriately combines thermodynamic evaluations based on an exergy with economic principles, at the level of system components, in order to provide information that is useful to the design and operation of a cost effective system, but not obtainable by conventional energy and exergy analyses and economic analysis [6, 12]. Some suggest that, when exergy costing is not applied, the general term thermoeconomics is more appropriate as it characterizes any combination of thermodynamic and economic analyses [38, 39]. In order to define a cost function that is dependent on the optimization parameters of interest, the component cost is expressed a function of thermodynamic design parameters [39].

Table 9.8 Expressions for exergy destruction rate and exergy efficiency for components of the gas turbine power plant.

Component	Exergy destruction rate	Exergy efficiency
Compressor	$\dot{Ex}_1 + \dot{W}_{AC} = \dot{Ex}_2 + \dot{Ex}_D$	$\psi_{AC} = \dfrac{\dot{Ex}_2 - \dot{Ex}_1}{\dot{W}_{ac}}$
Air preheater	$\dot{Ex}_2 + \dot{Ex}_5 = \dot{Ex}_3 + \dot{Ex}_6 + \dot{Ex}_D$	$\psi_{AP} = \dfrac{\dot{Ex}_3 - \dot{Ex}_2}{\dot{Ex}_5 - \dot{Ex}_6}$
Combustion chamber	$\dot{Ex}_3 + \dot{Ex}_9 = \dot{Ex}_4 + \dot{Ex}_D$	$\psi_{CC} = \dfrac{\dot{Ex}_4}{\dot{Ex}_3 + \dot{Ex}_9}$
Gas turbine	$\dot{Ex}_4 = \dot{Ex}_5 + \dot{W}_{GT} + \dot{Ex}_D$	$\psi_{GT} = \dfrac{\dot{W}_{GT}}{\dot{Ex}_4 - \dot{Ex}_5}$

For each flow in a system, a parameter called flow cost rate $\dot{C}$ ($/h) is defined, and a cost rate balance is written for each component as

$$\dot{C}_{q,k} + \sum_i \dot{C}_{i,k} + \dot{Z}_k = \sum_e \dot{C}_{e,k} + \dot{C}_{w,k} \tag{9.29}$$

The cost rate balances are generally written so that all terms are positive. Using Equation 9.29, one can write [12]:

$$\sum (c_e \dot{Ex}_e)_k + c_{w,k} \dot{W}_k = c_{q,k} \dot{Ex}_{q,k} + \sum (c_i \dot{Ex}_i)_k + \dot{Z}_k \tag{9.30}$$

$$\dot{C}_j = c_j \dot{Ex}_j \tag{9.31}$$

In this analysis, the fuel and product exergy need to be defined. The product exergy is defined according to the component under consideration, while the fuel represents the source that is consumed in generating the product. Both product and fuel are expressed in terms of exergy. The cost rates associated with the fuel ($\dot{C}_F$) and product ($\dot{C}_P$) of a component are obtained by replacing the exergy rates ($\dot{Ex}$).

In the cost rate balance formulation (Equation 9.31), there is no cost term directly associated with the exergy destruction of each component. Accordingly, the cost associated with the exergy destruction in a component or process is a hidden cost. If one combines both exergy and cost rate balances, one can obtain the following:

$$\dot{Ex}_{F,k} = \dot{Ex}_{P,k} + \dot{Ex}_{D,k} \tag{9.32}$$

Accordingly, the cost rate of exergy destruction is can be expressed as follows:

$$\dot{C}_{D,k} = c_{F,k} \dot{Ex}_{D,k} \tag{9.33}$$

In addition, several methods have been proposed to express the purchase cost of equipment in terms of the design parameters in Equation 9.30 [3, 12, 13]. Here, we use the cost functions suggested by Ahmadi *et al.* [20, 26, 29] and Rosen *et al.* [13], modified to fit the regional conditions of Iran and the inflation rate. We convert the capital investment into cost per time unit as follows:

$$\dot{Z}_k = \frac{Z_k . CRF . \phi}{N \times 3600} \tag{9.34}$$

where Z_k is the purchase cost of the kth component, and CRF is the capital recovery factor. CRF depends on the interest rate and equipment life time, and is determined here as follows [12, 14]:

$$CRF = \frac{i \times (1 + i)^n}{(1 + i)^n - 1} \tag{9.35}$$

where i denotes the interest rate and n the total operating period of the system in years. Also, N is the annual number of operation hours for the unit, and φ is the maintenance factor, which is often 1.06 [6]. To determine the component exergy destruction cost rate, the exergy destruction rate, $\dot{Ex}_{D,k}$, is evaluated using exergy rate balances from the previous section.

To estimate the cost rate of exergy destruction in each component of the plant, we first solve the cost rate balances for each component. In the cost rate balance (Equation 9.30), there is more than one inlet and outlet flow for some components. In such cases, the number of unknown cost parameters is higher than the number of cost balances for that component. Auxiliary exergoeconomic equations are developed to allow such problems to be solved. Writing Equation 9.30 for each component together with the auxiliary equations forms a system of linear equations as follows:

$$[\dot{E}x_k] \times [c_k] = [\dot{Z}_k] \tag{9.36}$$

where $[\dot{E}x_k]$, $[c_k]$ and $\dot{Z}_k$ are the exergy rate matrix (obtained via exergy analysis), the exergetic cost vector (to be evaluated), and the vector of $\dot{Z}_k$ factors (obtained via economic analysis), respectively. That is,

$$
\begin{bmatrix}
\dot{E}x_1 & 0 & 0 & 0 & 0 & 0 & 0 & 0 & 0 \\
\dot{E}x_1 & -\dot{E}x_2 & 0 & 0 & 0 & 0 & \dot{E}x_7 & 0 & 0 \\
0 & \dot{E}x_2 & -\dot{E}x_3 & 0 & -\dot{E}x_5 & -\dot{E}x_6 & 0 & 0 & 0 \\
0 & 0 & \dot{E}x_3 & -\dot{E}x_4 & 0 & 0 & 0 & 0 & \dot{E}x_9 \\
0 & 0 & 0 & \dot{E}x_4 & -\dot{E}x_5 & 0 & -\dot{E}x_7 & -\dot{E}x_8 & 0 \\
0 & 0 & 0 & 1 & -1 & 0 & 0 & 0 & 0 \\
0 & 0 & 0 & 0 & 0 & 0 & 1 & -1 & 0 \\
0 & 0 & 0 & 0 & 1 & -1 & 0 & 0 & 0 \\
0 & 0 & 0 & 0 & 0 & 0 & 0 & 0 & 1
\end{bmatrix}
\times
\begin{bmatrix}
c_1 \\ c_2 \\ c_3 \\ c_4 \\ c_5 \\ c_6 \\ c_7 \\ c_8 \\ c_9
\end{bmatrix}
=
\begin{bmatrix}
0 \\ -\dot{Z}_{AC} \\ -\dot{Z}_{AP} \\ -\dot{Z}_{CC} \\ -\dot{Z}_{GT} \\ 0 \\ 0 \\ 0 \\ F_c
\end{bmatrix}
\tag{9.37}
$$

9.3.3 Environmental Impact Assessment

A primary objective for minimizing environmental impact is to increase the efficiency of energy conversion processes and thereby decrease fuel use. In recent years, a special focus has been placed on releases of carbon dioxide, since it is the main greenhouse gas, and optimization of thermal systems based on this parameter has received much attention. A focus of the present analysis is to consider emissions of pollutants (e.g., CO and NO_x) since most approaches reported for optimizing power plants pay little attention to environmental impacts.

The adiabatic flame temperature in the primary zone of the combustion chamber can be expressed as follows [15]:

$$T_{pz} = A\sigma^{\alpha} \exp(\beta(\sigma + \lambda)^2)\pi^{x^*} \theta^{y^*} \psi^{z^*} \tag{9.38}$$

where π is dimensionless pressure (P/P_{ref}), θ is dimensionless temperature (T/T_{ref}), ψ is the H/C atomic ratio, $\sigma = \phi$ for $\phi \le 1$ (where ϕ is mass or molar ratio) and $\sigma = \phi - 0.7$ for $\phi \ge 1$. Moreover, x^*, y^* and z^* are quadratic functions of σ based on the following equations:

$$x^* = a_1 + b_1\sigma + c_1\sigma^2 \tag{9.39}$$
$$y^* = a_2 + b_2\sigma + c_2\sigma^2 \tag{9.40}$$
$$z^* = a_3 + b_3\sigma + c_3\sigma^2 \tag{9.41}$$

where A, α, β, λ, a_i, b_i, and c_i are constants. Values for these parameters are listed in Table 9.9, and more details on the methodology are presented elsewhere [15, 16].

The amount of CO and NO_x produced in a combustion chamber and combustion reaction is dependent on the adiabatic flame temperature [16]. The pollutant emissions

Table 9.9 Values for parameters in Equations 9.38 to 9.41.

Parameter	$0.3 \leq \varphi \leq 1.0$		$1.0 \leq \varphi \leq 1.6$	
	$0.92 \leq \theta \leq 2$	$2 \leq \theta \leq 3.2$	$0.92 \leq \theta \leq 2$	$2 \leq \theta \leq 3.2$
A	2361.76	2315.75	916.82	1246.17
α	0.1157	−0.049	0.288	0.381
β	−0.948	−1.114	0.145	0.347
λ	−1.097	−1.180	−3.277	−2.0365
a_1	0.014	0.010	0.031	0.036
b_1	−0.055	−0.045	−0.078	−0.085
c_1	0.052	0.0482	0.0497	0.0517
a_2	0.395	0.568	0.025	0.009
b_2	−0.441	−0.550	0.260	0.502
c_2	0.141	0.132	−0.131	−0.247
a_3	0.005	0.011	0.004	0.017
b_3	−0.129	−0.129	−0.178	−0.189
c_3	0.082	0.084	0.098	0.1037

(in grams per kilogram of fuel) can be determined as follows [16]:

$$m_{\text{NO}_x} = \frac{0.15E16 \; \tau^{0.5} \exp\left(-\frac{71\,100}{T_{\text{pz}}}\right)}{P_3^{0.05}\left(\frac{\Delta P_3}{P_3}\right)^{0.5}} \tag{9.42}$$

$$m_{\text{CO}} = \frac{0.179E9 \; \tau^{0.5} \exp\left(\frac{7800}{T_{\text{pz}}}\right)}{P_3^2 \tau \left(\frac{\Delta P_3}{P_3}\right)^{0.5}} \tag{9.43}$$

where τ is the residence time in the combustion zone (assumed constant and equal to 0.002 s) [16], T_{pz} is the primary zone combustion temperature, P_3 is the combustor inlet pressure, and $\Delta P_3/P_3$ is the non-dimensional pressure drop in the combustion chamber.

9.3.4 Optimization

In this section we consider objective functions, design parameters, and constraints, as well as the overall optimization procedure.

9.3.4.1 Definition of Objective Functions

The objective functions considered here for multi-objective optimization are exergy efficiency (to be maximized), the total cost rate of the product and environmental impact (to be minimized), and the CO_2 emission (to be minimized). The second objective function expresses the environmental impact as the total pollution damage ($/h) due to CO and NO_x emissions by multiplying their respective flow rates by their corresponding unit damage costs, which have been reported elsewhere as $C_{\text{CO}} = 0.02086$ $/kg CO and $C_{\text{NOx}} = 6.853$ $/kg NO_x [6]. In the present analysis, the cost of pollution damage is assumed to be added directly to the expenditures that must be paid, making the second objective function the sum of thermodynamic and environmental objectives. Consequently, the objective functions in this analysis can be expressed as follows:

Gas Turbine Power Plant Exergy Efficiency

$$\psi_{\text{total}} = \frac{\dot{W}_{\text{net}}}{\dot{m}_{f,CC} \times LHV \times \xi} \tag{9.44}$$

Here, $\dot{W}_{\text{net}}$, $\dot{m}_{f,CC}$ and ξ denote respectively the gas turbine net output electrical power, the mass flow rate of fuel injected into the combustion chamber, and a fuel exergy/energy correlation parameter for gaseous fuels of composition (C_xH_y).

Total Cost Rate

$$\dot{C}_{\text{tot}} = \dot{C}_f + \sum_k \dot{Z}_k + \dot{C}_D + \dot{C}_{\text{env}} \tag{9.45}$$

where the cost rates of environmental impact and fuel are expressed as

$$\dot{C}_{\text{env}} = C_{CO}\dot{m}_{CO} + C_{NO_x}\dot{m}_{NO_x}, \ \dot{C}_f = c_f \dot{m}_f \times LHV \tag{9.46}$$

Here, $\dot{Z}_K$ and $\dot{C}_D$ are the purchase cost rate of each component and the cost rate of exergy destruction, respectively. In addition, $\dot{m}_{CO}$ and $\dot{m}_{NO_x}$ are calculated with Equations 9.42 and 9.43, respectively.

9.3.4.2 Decision Variables

The decision variables (design parameters) in this case study are compressor pressure ratio R_{AC}, compressor isentropic efficiency η_{AC}, gas turbine isentropic efficiency η_{GT}, combustion chamber inlet temperature T_3, and gas turbine inlet temperature GTIT. Even though the decision variables may be varied in the optimization procedure, each decision variable is normally required to be within a reasonable range so constraints are applied. These constraints and the reasons of their applications are listed in Table 9.10, based on previous analyses [6, 17].

9.3.4.3 Model Validation

To simulate the performance of the gas turbine power plant, simulation code developed with Matlab software is used. To validate the simulation code, the results of this study are compared with an actual operating gas turbine power plant located in the Shahid Salimi Power Plant, Neka, Iran. This power plant is located close to Mazandaran city

Table 9.10 Constraints for optimization of gas turbine power plant.

Constraint	Reason
GTIT < 1550 K	Material temperature limit
$P_2/P_1 < 20$	Commercial availability
$\eta_{GT} < 0.92$	Commercial availability
$\eta_{AC} < 0.92$	Commercial availability
$T_7 > 400$ K	Avoid formation of sulfuric acid in exhaust gases

Table 9.11 Comparison of power plant data and simulation code results.

Quantity	Unit	Measured data	Simulation code result	Difference (%)
T_2	°C	321.4	332.0	3.2
T_6	°C	500	529.0	5.5
T_7	°C	448	487.0	8.0
$\dot{m}_f$	kg/s	8.44	8.18	3.2
ψ	%	32.0	30.4	5.0

near the Caspian Sea, in one of the northern provinces in Iran. A schematic diagram of this power plant was shown earlier (Figure 9.4).

Based on power plant data obtained in 2005, the incoming air has a temperature of 20°C and a pressure of 1 bar. The pressure increases to 10.1 bar in the compressor, which has an isentropic efficiency of 82%. The gas turbine inlet temperature is 971°C. The turbine has an isentropic efficiency of 86%, and the regenerative heat exchanger has an effectiveness of 87%. The pressure drop through the air preheater is considered 3% of the inlet pressure for both flow streams and through the combustion chamber is 3% of the inlet pressure. The fuel (natural gas) is injected at 20°C and 30 bar. Thermodynamic properties of the power plant based on modeling and simulation are listed in Table 9.11 and contrasted with given power plant data. The data from the simulation code are observed to vary from those for the actual power plant by about 3% to 8%, with the maximum difference applicable to the air preheater outlet temperature. This good agreement suggests that data generated by the simulation code for the gas turbine power plant is reasonably valid.

9.3.5 Results and Discussion

Figure 9.5 shows the Pareto frontier solution obtained by multi-objective optimization for the gas turbine power plant, based on the objective functions described in Equations 9.44 and 9.45. It is observed that, while the total exergy efficiency of the cycle is increased to about 41%, the total cost rate of products increases very slightly. An increase in the total exergy efficiency from 41% to 43.5% corresponds to a moderate increase in the cost rate of the product. In addition, an increase in the exergy efficiency from 43.5% to the higher value leads to a significant increasing of the total cost rate. It is seen in Figure 9.5 that the maximum exergy efficiency exists at design point (C) (43.0%), while the total cost rate of products is the greatest at this point. But, the minimum value for total cost rate of product occurs at design point (A). Design point C is the optimal situation if efficiency is the sole objective function, while design point A is the optimum condition if total cost rate of product is the sole objective function. The specifications of these three possible design points A–C in the Pareto optimal fronts are listed in Table 9.12.

In multi-objective optimization, a process of decision-making for selection of the final optimal solution from the available solutions is required. The process of decision-making is usually performed with the aid of a hypothetical point, called the equilibrium point in Figure 9.5, at which both objective functions have their optimal values, independent of the other objective functions. It is clear that it is impossible to

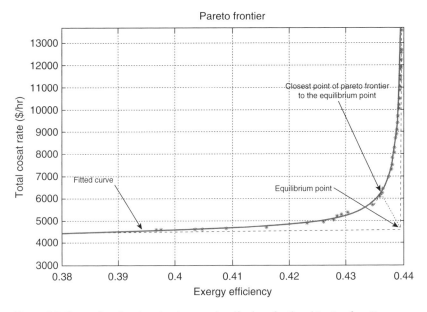

Figure 9.5 Pareto frontier, showing best trade-off values for the objective functions.

Table 9.12 Optimum design values for A to C Pareto optimal fronts for input value.

Property	Unit	A	B	C
η_{ex}	%	39.59	43.5	43.89
$\dot{C}_D$	$/hr	1227.1	1354	2309
Unit CO_2 emission	Kg/MWhr	201.5	183.4	181.8
$\dot{C}_{env}$	$/hr	16.92	11.88	11.78

have both objectives at their optima simultaneously, and as shown in Figure 9.5 the equilibrium point is not a solution located on the Pareto frontier. The closest point of the Pareto frontier to the equilibrium point might be considered as a desirable final solution. Nevertheless, in this case, the optimum Pareto frontier exhibits weak equilibrium, that is, a small change in the exergetic function due to variations in operating parameters causes a large variation in the cost rate of the product. Therefore, the equilibrium point cannot be utilized for decision-making in this problem. In selection of the final optimum point, it is desired to achieve a better value for each objective than its initial value for the base case problem. Hence, because the optimized points in the *B-C* region exhibit a maximum exergy efficiency increment of about 1% and a minimum total cost rate increment of 82.53% relative to design *C*, this region is eliminated from the Pareto curve, leaving just the region *A-B* as shown in Figure 9.6

Since each point can be the optimal point in the Pareto solution obtained via multi-objective optimization, selection of the optimum solution is dependent on the preferences and criteria of each decision maker. Hence, each decision maker may select a different point as the optimum solution so as to best suit his/her objectives.

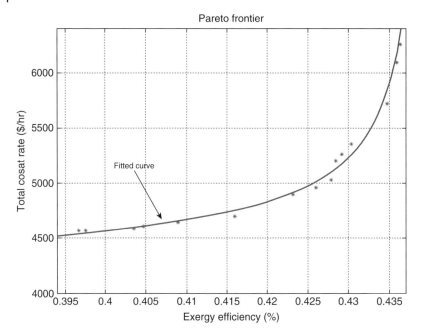

Figure 9.6 Selecting the optimal solution from Pareto Frontier.

To provide a helpful tool for the optimal design of the gas turbine cycle, the following equation is derived for the Pareto optimal points curve shown in Figure 9.5.

$$\dot{C}_{Tot} = \frac{7.42189\eta^3 + 16.3579\eta^2 - 18.7497\eta + 4.45071}{\eta^4 + 21.3513\eta^3 - 7.18236\eta^2 - 6.40907\eta + 2.35422} \tag{9.47}$$

This equation is valid in the range of $0.38 < \eta < 0.44$.

9.3.6 Sensitivity Analysis

A sensitivity analysis can provide enhanced insights into the results. This analysis considers changes in a relevant parameter as well as some other modeling parameters, and helps predict the effects of the modifications on modeling and optimization. Here, a sensitivity analysis of the Pareto optimum solution is performed considering the fuel specific cost and the interest rate. Figure 9.7 shows the sensitivity of the optimal Pareto frontier to variations of specific fuel cost. This figure shows that the Pareto frontier shifts upward as the specific fuel cost increases. At lower exergy efficiency, the sensitivity of the optimal solutions to fuel cost is much higher than the location of the Pareto frontier when the weight of the thermoenvironomic objective is high than when it is low. In fact the exergetic objective does not have a significant effect on the sensitivity of the objective functions to economic parameters such as the fuel cost and interest rate. Moreover, at a higher exergy efficiency, the purchase cost of equipment in the plant increases, causing the cost rate of the plant to increase too. At a constant exergy efficiency, increasing the fuel cost causes the total cost rate of product to increase due to the fact that the fuel cost plays a significant role in this objective function.

Figure 9.8 displays the sensitivity of the Pareto optimum solution for the total cost rate to the specific CO_2 emissions and the fuel cost rate. It can be seen in this figure that

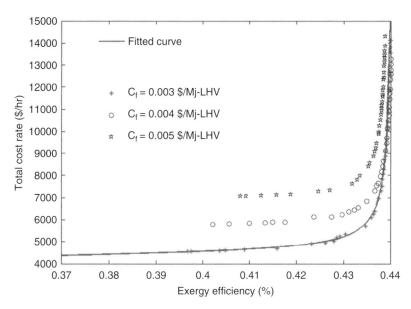

Figure 9.7 Sensitivity of Pareto optimum solution to exergy efficiency and total cost rate in multi-objective optimization, for a fixed interest rate ($i = 13\%$).

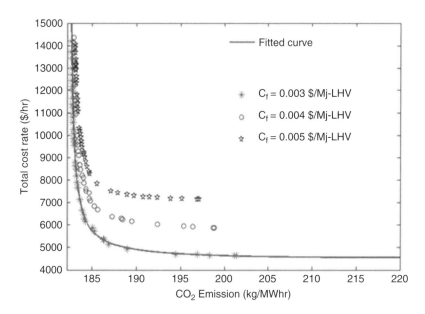

Figure 9.8 Sensitivity of Pareto optimum solution for total cost rate from multi-objective optimization to specific CO_2 emission and fuel cost, for a fixed interest rate ($i = 13\%$).

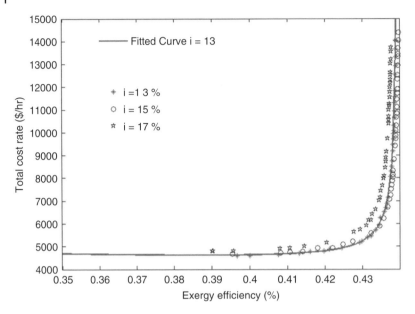

Figure 9.9 Sensitivity of Pareto optimum solution for total cost rate from multi-objective optimization to exergy efficiency and interest rate, for a fixed fuel cost ($C_f = 0.003$ $/MJ).

obtaining a cycle that produces less CO_2 involves selecting components that have higher thermodynamic merit measures (such as isentropic efficiency). Such selections typically lead to increases in the purchase costs of the equipment. But, by increasing the fuel cost, the total cost rate of the product is increased because of the importance of the fuel cost in this objective function. Similar behavior is observed in the sensitivity of the Pareto optimal solution to interest rate in Figures 9.9 and 9.10. The final optimal solution that is selected here belongs to the region of the Pareto frontier with significant sensitivity to the costing parameters. The region with lower sensitivity to the costing parameters is not reasonable for the final optimum solution due to the weak equilibrium of the Pareto frontier, in which a small change in the plant exergy efficiency due to a variation of operating parameters risks increasing the product cost rate considerably.

9.3.7 Summary

The modeling, analysis, and optimization results are summarized here. The modeling and analysis results exhibit average differences between simulated and measured parameters values of about 5.1%, with the maximum of 10.3% occurring for the cycle exergy efficiency. An evolutionary algorithm (specifically a genetic algorithm), which is an alternative to previously presented calculus-based optimization approaches, is utilized for multi-objective optimization of typical gas turbine power plant. The evolutionary algorithm is shown to be a powerful and effective tool for finding the set of the optimal solutions and the corresponding choices of optimum design variables in the power plant, relative to conventional mathematical optimization algorithms. The importance of quantifying environmental impacts is accommodated through the introduction of pollution related costs in the economic objective function. Hence, the environmental objective is transformed into a cost function that incorporates the costs of environmental impacts. The new environmental cost function is merged into

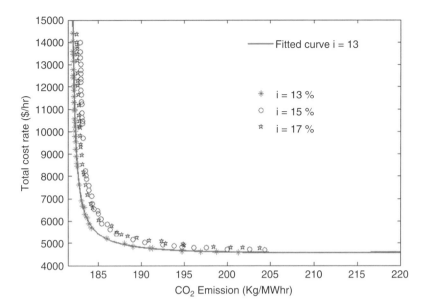

Figure 9.10 Sensitivity of Pareto optimum solution for total cost rate from multi-objective optimization to CO_2 emission and interest rate, for a fixed fuel cost ($C_f = 0.003$ $/MJ).

the thermoeconomic objective and a new thermoenvironomic function is obtained. To provide good insights into CO_2 emission in the plant, the emission of this gas is considered as a distinct objective function, that is, the CO_2 emission per unit of electricity generation of the plant is minimized. Hence the four-objective optimization problem is transformed to a three objective problem, facilitating the decision-making process. Further, results for the optimized plant and the actual operating power plant are compared. Relative to the actual power plant, the optimization results attain increases in the overall exergoeconomic factor of the system, raising it from 32.79 to 62.24%, implying that the optimization process mostly improves the associated cost of thermodynamic inefficiencies. The sensitivity of the obtained Pareto solutions to interest rate and fuel cost are examined. The selection of the final optimum solution from the Pareto frontier requires a decision-making process that is dependent on the preferences and criteria considered by the decision maker.

9.4 Combined Cycle Power Plants

Combined cycle power plants (CCPPs) are attractive for electrical power generation mainly due to their higher thermal efficiencies than independent steam or gas turbine cycles. The optimal design of such cobined cycles is of increasing importance due to long-term increases in fuel prices and decreases in fossil fuel resources [18]. The main challenge in designing a combined cycle is proper utilization of the gas turbine exhaust heat in the steam cycle so as to achieve optimum steam turbine output from the perspective of the overall plant. In addition to higher thermal efficiencies than gas turbine cycle and steam cycles, combined cycles typically have higher net output power outputs and lower unit emissions (per unit of electricity generation). The benefits of CCPPs have led to increases in the number and output power of such cycles recently, around the world.

In this section, we develop a thermodynamic model and apply thermoeconomic optimization to a combined cycle power plant, based on actual data from the Neka combined cycle power plant in the northern part of Iran close the Caspian Sea. This power plant, shown in Figure 9.11, uses a dual pressure combined cycle and has a net output power of 420 MW. Exergy and exergoeconomic analyses and optimization of the CCPP are conducted using energy and exergy approaches, and parametric studies are undertaken to investigate how system inputs and outputs are affected by operating conditions. The model results are compared with actual data for validation purposes. An optimization program is developed to determine the best design parameters for the cycle.

9.4.1 Thermodynamic Modeling

As with the steam and gas turbine power plants considered in the previous two sections, energy and mass rate balance equations are used for the modeling of the CCPP. Since a gas turbine power plant and the topping cycle of a CCPP are both based on the gas turbine cycle, which was already explained in detail in section 9.3.1, we focus here on the bottoming cycle of the CCPP.

In this analysis, several assumptions are made [18]:

- All processes are steady state and steady flow.
- Air and combustion products are treated as ideal gas mixtures.
- The fuel injected into the combustion chamber is natural gas.
- Heat loss from the combustion chamber is considered to be 3% of the fuel lower heating value [6], and all other components are considered adiabatic.
- The dead state conditions are $P_0 = 1.01$ bar and $T_0 = 293.15$ K.

Details on the thermodynamic modeling of the relevant components follow:

9.4.1.1 Duct Burner

Additional fuel is burned in the supplementary firing duct burner to increase the temperature of the exhaust gas that passes through the HRSG. In the duct burner:

$$\dot{m}_g h_D + \dot{m}_{DB} LHV = (\dot{m}_g + \dot{m}_{DB}) h_5 + (1 - \eta_{DB}) \dot{m}_{DB} LHV$$

where LHV is the lower heating value of the natural gas and η_{DB} is the duct burner combustion efficiency, which is considered to be 93% [18].

9.4.1.2 Heat Recovery Steam Generator (HRSG)

A dual pressure HRSG, which is common for CCPPs, is considered here. Applying the energy rate balance for the gas and water flows in each part of the HRSG, the gas and water properties can be calculated. Doing so involves solving the following equations:

High pressure superheater:

$$\dot{m}_g c_p (T_{11} - T_{12}) = \dot{m}_{s,HP} (h_{10} - h_9)$$

High pressure evaporator:

$$\dot{m}_g c_p (T_{12} - T_{13}) = \dot{m}_{s,HP} (h_9 - h_8)$$

High pressure economizer:

$$\dot{m}_g c_p (T_{13} - T_{14}) = \dot{m}_{s,HP} (h_8 - h_7)$$

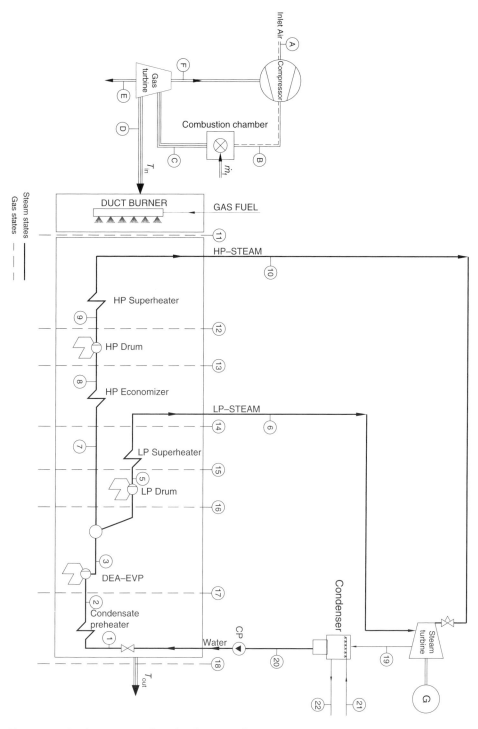

Figure 9.11 Dual pressure combined cycle power plant.

Low pressure superheater:

$$\dot{m}_g c_p (T_{14} - T_{15}) = \dot{m}_{s,LP} (h_6 - h_5)$$

Low pressure evaporator:

$$\dot{m}_g c_p (T_{15} - T_{16}) = \dot{m}_{s,LP} (h_5 - h_4)$$

Deaerator evaporator:

$$\dot{m}_g c_p (T_{16} - T_{17}) = \dot{m}_{s,LP} (h_3 - h_2)$$

Condensate preheater:

$$\dot{m}_g c_p (T_{17} - T_{18}) = \dot{m}_{s,LP} (h_2 - h_1)$$

These combinations of energy and mass rate balance equations are numerically solved, allowing the temperature profiles of the gas and water/steam sides of the HRSG to be determined.

9.4.1.3 Steam Turbine (ST)

An energy rate balance for the steam turbine shown in Figure 9.11, and an expression for its isentropic efficiency, follow:

$$\dot{m}_w h_5 = \dot{W}_{ST} - \dot{m}_w h_6$$

$$\eta_{ST} = \frac{\dot{W}_{ST,act}}{\dot{W}_{ST,is}}$$

9.4.1.4 Condenser

An energy rate balance for the condenser follows:

$$\dot{m}_6 h_6 = \dot{Q}_{cond} - \dot{m}_7 h_7$$

9.4.1.5 Pump

An energy rate balance for the pump, and an expression for its isentropic efficiency, follow:

$$\dot{m}_w h_7 + \dot{W}_{pump} = \dot{m}_w h_8$$

$$\eta_{pump} = (w_{is})/(w_{act})$$

These energy and mass rate balance equations are numerically solved and the temperature and enthalpy of each flow in the plant are determined.

9.4.2 Exergy Analysis

Exergy analysis can help improve understanding of efficiencies and losses, and thereby assist in developing strategies and guidelines for more efficient and effective use of energy. Exergy methods have been utilized on various thermal processes, including power generation, CHP, trigeneration, and multigeneration. The exergy of a substance is often divided into four components: physical and chemical exergy (which are the most common ones), as well as kinetic and potential exergy (which, as is done here, are often assumed to be negligible because elevation changes are small and speeds are relatively low [6, 12, 19]). Physical exergy is defined as the maximum useful work obtainable as a system interacts with a reference environment at an equilibrium state. Chemical exergy is associated with the departure of the chemical composition of a system from chemical equilibrium with the reference environment and is considered important in processes involving combustion and other chemical changes [3]. A general exergy rate balance can be written as:

$$\dot{Ex}_Q + \sum_i \dot{m}_i ex_i = \sum_e \dot{m}_e ex_e + \dot{Ex}_W + \dot{Ex}_D$$

where subscripts i and e denote the control volume inlet and outlet flows, respectively, $\dot{Ex}_D$ is the exergy destruction rate, and other terms are as follows:

$$\dot{Ex}_Q = \left(1 - \frac{T_0}{T_i}\right) \dot{Q}_i$$

$$\dot{Ex}_w = \dot{W}$$

$$ex = ex_{ph} + ex_{ch}$$

Here, $\dot{Ex}_Q$ is the exergy rate of heat transfer crossing the boundary of the control volume at absolute temperature T, the subscript 0 refers to the reference environment conditions and $\dot{Ex}_W$ is the exergy rate associated with shaft work. Also, ex_{ph} is defined as follows:

$$ex_{ph} = (h - h_0) - T_0(s - s_0)$$

The specific chemical exergy of a mixture can be expressed as follows [3]:

$$ex_{mix}^{ch} = \sum_{i=1}^n x_i ex_i^{ch} + RT_0 \sum_{i=1}^n x_i \ln x_i$$

In the present analysis, the exergy of the flows are calculated at all states and the destruction in exergy is determined for the major components. The cause of exergy destruction (or irreversibility) in the combustion chamber is mainly combustion (chemical reaction) and thermal losses in the flow path [6]. However, the exergy destruction in the system heat exchangers (i.e., condenser and HRSG) is due to the large temperature differences between the hot and cold fluids.

9.4.3 Optimization

A genetic algorithm (GA) optimization method based on an evolutionary algorithm is applied to the combined cycle power plant shown in Figure 9.11 to determine the best design parameters for the system. The objective functions, design parameters and constraints, and overall optimization study are described in this section.

9.4.3.1 Definition of Objectives

As noted in the previous section, the objective function considered here is sum of the fuel cost and the cost of exergy destruction, as well as purchase cost. This objective function is to be minimized using the genetic algorithm. The objective function can be written as follows:

$$\dot{C}_{tot} = \sum_k \dot{Z}_k + \dot{C}_f + \dot{C}_{env} \tag{9.48}$$

where the cost rates of environmental impact and fuel are expressed as:

$$\dot{C}_{env} = C_{CO}\dot{m}_{CO} + C_{NOx}\dot{m}_{NOx} + C_{CO_2}\dot{m}_{CO_2} \text{ and } \dot{C}_f = c_f(\dot{m}_f + \dot{m}_{fDB})\,LHV$$

Here, $\dot{Z}_K$ is the purchase cost rate of component K. Further details about equipment purchase costs can be found elsewhere [12, 13]. Also c_f is the fuel unit cost which is taken to be 0.003 \$/MJ in this analysis. Also, we express the environmental impact as the total cost rate of pollution damage (\$/s) due to CO, NO_x and CO_2 emissions by multiplying their respective flow rates by their corresponding unit damage costs (C_{CO}, C_{NOx}, and C_{CO2}, which are taken to be equal to 0.02086 \$/kg, 6.853 \$/kg and 0.024 \$/kg, respectively) [6]. The cost of pollution damage is assumed here to be added directly to other system costs.

9.4.3.2 Decision Variables
The design parameters selected for this analysis are compressor pressure ratio (r_{AC}), compressor isentropic efficiency (η_{AC}), gas turbine isentropic efficiency (η_{GT}), gas turbine inlet temperature (GTIT), duct burner mass flow rate ($\dot{m}_{DB}$), high pressure steam pressure (P_{HP}), low pressure steam pressure (P_{LP}), HP main steam temperature (T_{HP}), LP main steam temperature (T_{LP}) HP pinch point temperature difference (PP_{HP}), LP pinch point temperature difference (PP_{LP}), condenser pressure (P_{Cond}), steam turbine isentropic efficiency (η_{ST}), and pump isentropic efficiency (η_{pump}). Values for these decision parameters are sought such that the objective function is minimized.

9.4.3.3 Constraints
To render the optimization realistic, constraints are applied during the optimization procedure. These constraints are listed in Table 9.13 and are applied in the Matlab software program. Note that, without constraints, a better result for the objective function may be attainable, but the constraints exclude such a point from consideration during the optimization procedure.

Table 9.13 Constraints for optimization of combined cycle power plant.

Constraint	Rationale
$GTIT < 1550$ K	Material temperature limitation
$r_{comp} < 22$	Commercial availability
$\eta_{CompC} < 0.9$	Commercial availability
$\eta_{GT} < 0.9$	Commercial availability
$P_{main} < 110$ bar	Commercial availability
η_{ST}	Commercial Availability
η_p	Commercial availability
$\dot{m}_{DB} < 2$ kg/s	Super heater temperature limitation
5 bar $< P_{cond} < 15$ bar	Thermal efficiency limitation
T_{main}	Material temperature limitation
$T_{18} \geq 120°C$	Avoid formation of sulfuric acid in exhaust gases
$5°C < PP < 30°C$	Second law of thermodynamics limitation

Source: Ahmadi 2011. Reproduced with permission of Elsevier.

9.4.4 Results and Discussion

To verify the modeling results, the HRSG, a significant component of the CCPP, is examined and the results of HRSG are compared with the corresponding measured values obtained from the actual operating HRSG in the Neka combined cycle power plant in Iran. Values of thermal modeling parameters for the HRSG at the Neka CCPP are listed in Table 9.14.

The gas temperature variation of the Neka dual pressure HRSG with a duct burner obtained with the simulation program, and the corresponding measured values, are shown and compared in Figure 9.12. The average of the difference between the numerical and measured parameter values at various sections of the HRSG is about 1.14% and the maximum difference is 1.36% for the LP superheater. This good agreement suggests that the accuracy of the results from the simulation code developed for the HRSG as well as the whole plant is adequate.

The genetic algorithm optimization routine is applied to obtain the CCPP optimum design parameters. Figure 9.13 shows the convergence of the objective function with

Table 9.14 Input parameters of the HRSG of the Neka CCPP used for model verification.

Input	Value
Inlet gas temperature, (K)	773.15
Inlet gas flow rate, (kg/s)	500
Inlet water temperature, (K)	320
Total water flow rate, (kg/s)	76.11
Inlet water enthalpy, (kJ/kg)	185
Ambient temperature, (K)	293.15

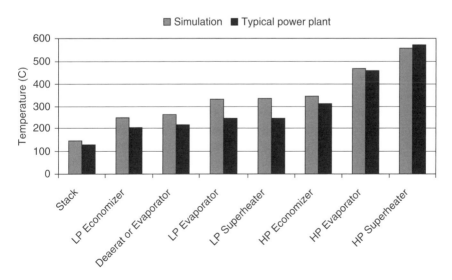

Figure 9.12 Variation of hot gas temperature in various heat transfer elements of HRSG at Neka CCPP, for both modeled and measured values.

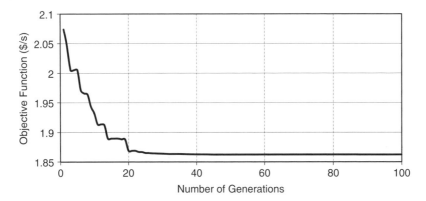

Figure 9.13 Variation of objective function of CCPP system with number of generations (for $C_f = 0.003$ $/MJ).

the number of generations (40 in our case study, beyond which there is no noticeable change in the value of the objective function). From this figure, it can be observed that the developed genetic algorithm has a good convergence rate, leading to two important benefits: a lower computer running time and better optimization results.

By applying the developed genetic algorithm code for this problem and considering both objective functions and constraints, the optimal design parameters of the combined cycle are found. The optimal decision variables of the plant are given in Table 9.14. It can be observed that by selecting these design parameters the objective function defined in Equation 9.48 takes on the minimum value. Note that fuel cost has a significant effect on the objective function. Therefore, from one perspective the design parameters should be selected such that both the combustion chamber and duct burner mass flow rates have minimum values. But from other perspectives it is noted that by decreasing these mass flow rates the environmental emissions are reduced and the total efficiency of the cycle is decreased.

To provide further insight into this real optimization problem, a sensitivity analysis is performed. Here, two important factors are considered: unit cost of fuel and net output power of the combined cycle power plant. The sensitivity analysis is performed by making modifications to these two parameters and applying the genetic algorithm. Note that, to determine the optimal design parameters for each unit cost of fuel and output power, a new optimization procedure is applied. Hence, each optimal value is the best for each cost and output power. An important benefit of this method is the ability to predict the trend of design parameters when any changes in the unit cost of fuel and output power occur. As discussed earlier, the unit cost of fuel has a significant effect on the objective function, so any change in this parameter affects the value of the objective function as well as the design parameters. This implies that, when the unit cost of fuel increases, the optimal design parameters of the plant should be selected in such a way that other terms in the objective function (Equation 9.48) decrease.

Consequently, to investigate the effects of fuel price on the optimum design parameters, the simulation and optimization procedures are repeated with different values of input parameters. For instance, when the fuel price increases, the cost objective function increases. Thus, design parameters should be selected in such a way as to minimize this objective function. Since fuel price is multiplied by the mass flow rate in the power

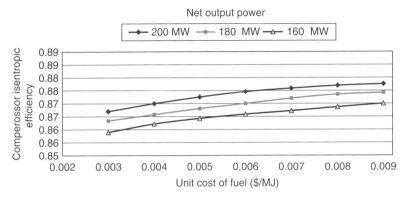

Figure 9.14 Effect of unit cost of fuel and net power output on optimal value of air compressor isentropic efficiency η_{comp}.

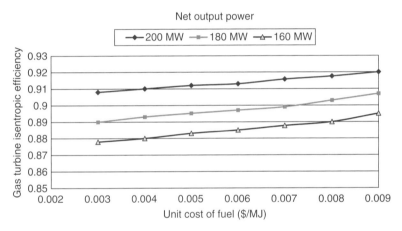

Figure 9.15 Effect of unit cost of fuel and net power output on optimal value of gas turbine isentropic efficiency η_{GT}.

plant, the design parameters tend to be selected in order to decrease mass flow rates (e.g., by utilizing a compressor with higher isentropic efficiency).

Figures 9.14 and 9.15 show that, at constant output power, both the gas turbine isentropic efficiency and compressor isentropic efficiency increase with any increase in the unit cost of fuel. The reason is that, when the unit cost of fuel is increased, the first term in the objective function increases; thus other terms in the objective function need to decrease. According to the literature and other investigations by the authors, an increase in the isentropic efficiency leads to a decrease in the cost of exergy destruction. Thus, the last term in Equation 9.48 decreases. This study reveals that, by describing the variation of the optimal decision variables versus fuel unit cost, increasing the fuel cost causes the optimal decision variables generally to shift to a more thermodynamically efficient design.

As seen in Figures 9.14 to 9.17, the values of decision variables r_{comp}, η_{comp}, η_{GT}, and GTIT increase with an increase in the fuel unit cost for several net power outputs. An increase in the inlet gas turbine temperature causes a decrease in the exergy destruction rate of the combustion chamber. But, according to the cost of exergy destruction rate,

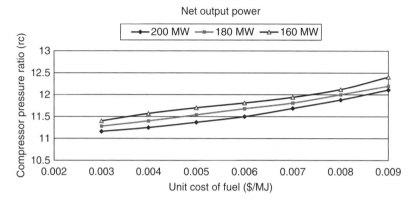

Figure 9.16 Effect of unit cost of fuel and net power output on optimal value of compressor ratio r_{comp}.

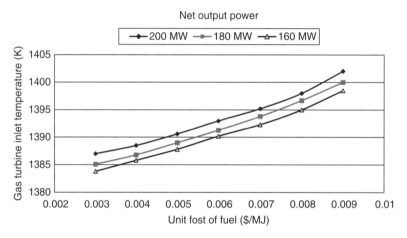

Figure 9.17 Effect of unit cost of fuel and net power output on optimal value of gas turbine temperature GTIT.

which is proportional to the exergy destruction rate, the last term of the objective function decreases. Moreover, an increase in the turbine inlet temperature, GTIT, reduces the exergy destruction rate in the combustion chamber and turbine. Since increasing GTIT results in a higher temperature of the exhaust gases, the constraint $T_{18} > 120°C$ does not cause any limitation as GTIT rises. However, due the fact that any increase in GTIT affects the turbine investment cost, GTIT can only increase within a certain limit. As a result, these changes result in decrease in the objective function.

Figure 9.18 shows the effect on the duct burner mass flow rate versus of varying the unit cost of fuel for several net power outputs. An increase in the unit cost of fuel is observed to cause the mass flow rate through the duct burner to decrease, due to the fact that the objective function should decrease correspondingly. Since raising the mass flow rate has a positive effect on (i.e., increases) the first term of the objective function, the genetic algorithm tends to optimally determine the design parameters that decrease the duct burner mass flow rate. Additionally, this reduction in mass flow rate can decrease the environmental impacts, as discussed elsewhere [5].

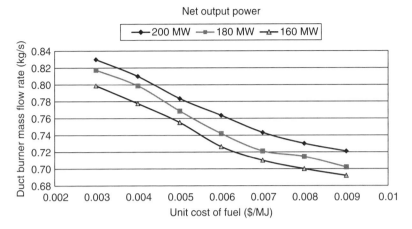

Figure 9.18 Effect of unit cost of fuel and net power output on optimal value of duct burner mass flow rate.

One of the most important parameters in designing heat recovery steam generators is the pinch point temperature difference. The pinch temperature is defined as the temperature difference between the outlet gas from the evaporator and the saturation temperature. A smaller pinch temperature corresponds to a larger heat transfer surface area and a more costly system, as well as a higher exergy efficiency and lower operating cost. A good HRSG typically has a pinch temperature at a minimum value; however, based on the second law of thermodynamics, this temperature cannot be zero. Therefore, a decrease in the pinch temperature results in a decrease in the HRSG cost of exergy destruction. Since the HRSG in this study is dual pressure, it has two pinch point temperatures: high pressure and low pressure. Figures 9.19 and 9.20 show the variations of pinch temperatures with unit cost of fuel. It is seen that, at constant output power, an increase in the unit cost of fuel results in a decrease in the pinch temperature due to an increase in the HRSG exergy efficiency, as well as a decrease in the HRSG cost of exergy destruction.

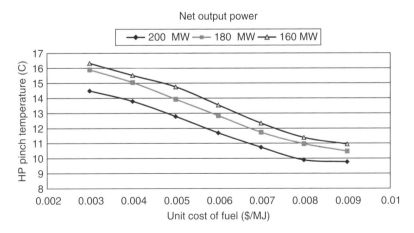

Figure 9.19 Effect of unit cost of fuel and net power output on optimal value of high pressure pinch temperature PP_{HP}.

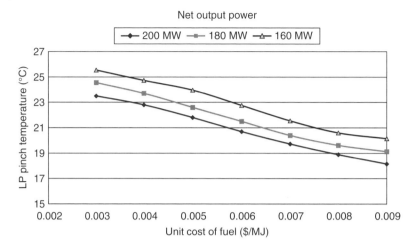

Figure 9.20 Effect of unit cost of fuel and net power output on optimal value of low pressure pinch temperature PP$_{LP}$.

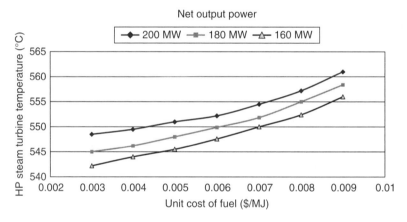

Figure 9.21 Effect of unit cost of fuel and net power output on optimal value of HP steam turbine temperature.

Figures 9.21 and 9.22 show the variation of superheater temperatures with unit cost of fuel. An increase in the unit cost of fuel is observed in these figures to increase both the HP and LP superheater temperatures. Thus, as discussed earlier, an increase in the unit cost of fuel leads to an increase in the first term of the objective function. The genetic algorithm code is as a consequence developed to select the design parameters such that other terms in the objective function decrease. Therefore, any increments in the superheater steam temperature result in a decrease in the last term of the objective function in Equation 9.48, in part because the higher temperature entering the steam turbine increases the output power of the Rankine cycle. Moreover, increasing the main steam temperature from the HRSG results in an increase in the HRSG efficiency as well as a reduction in its exergy destruction rate. The effect of changes in the unit cost of fuel on HP and LP drum pressures are shown in Figures 9.23 and Figure 9.24, respectively. It is observed that, at constant output power, an increase in the unit cost of fuel causes

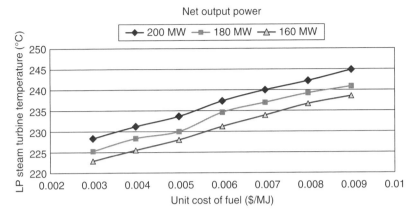

Figure 9.22 Effect of unit cost of fuel and net power output on optimal value of LP steam turbine temperature.

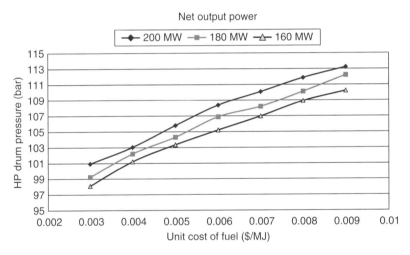

Figure 9.23 Effect of unit cost of fuel and net power output on optimal value of HP drum pressure.

both the HP and LP drum pressures to increase, reducing the last term in the objective function.

Because of increasing drum pressures, the HRSG cost of exergy destruction decreases, reducing the objective function. The main purpose of increasing the drum pressures is to produce steam with a higher enthalpy as well as exergy. In addition, the higher drum pressures cause the steam temperature and mass flow rate to rise, which provides a higher steam turbine power output. This is why a high pressure pump is used in these HRSGs, which are common in Iran. Figure 9.25 shows the effect of variations in condenser pressure with unit cost of fuel. As the unit cost of fuel increases, the condenser pressure decreases in order to lower the objective function. Also, a decrease in the condenser pressure results in an increase in the total exergy efficiency of the cycle. Therefore, it has a positive effect on both the objective function and the combined cycle efficiency.

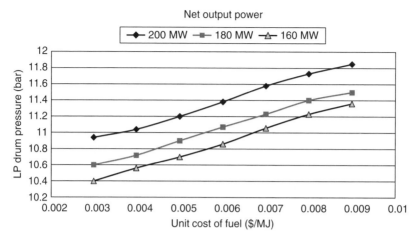

Figure 9.24 Effect of unit cost of fuel and net power output on optimal value of LP drum pressure.

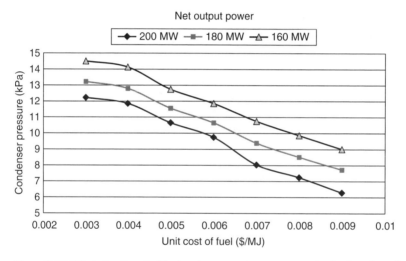

Figure 9.25 Effect of unit cost of fuel and net power output on optimal value of condenser pressure P_{cond}.

Finally, the effects on the steam turbine and pump isentropic efficiencies of varying the unit cost of fuel are shown in Figures 9.26 and 9.27. An increase in the unit cost of fuel results in an increase in both the turbine and pump isentropic efficiencies. In this case, more efficient devices raise the exergy efficiency and reduce the cost of exergy destruction. Therefore, the chief aim of increasing the efficiency is to decrease the objective function, especially the last term. This result reveals that, while the unit cost of fuel increases, more efficient devices are merited to reduce the irreversibilities.

An important parameter in CCPPs is the net output power. To provide enhanced insights, three output power levels are considered throughout; this is accomplished by, for each output power, running the genetic algorithm again to determine the best of the optimal design parameters. The effect of net output power on the design parameters is included in Figures 9.14 to 9.27. Figures 9.14 to 9.18 show that, while keeping the fuel

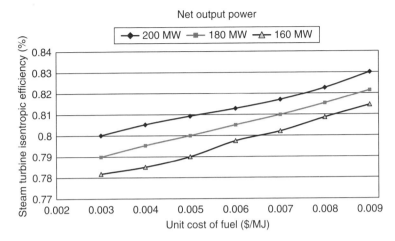

Figure 9.26 Effect of unit cost of fuel and net power output on optimal value of steam turbine isentropic efficiency η_{ST}.

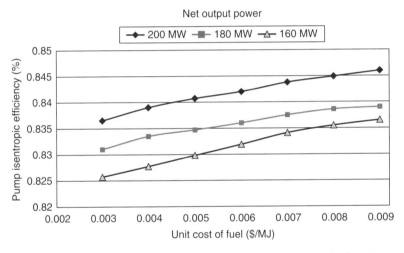

Figure 9.27 Effect of unit cost of fuel and net power output on optimal value of pump isentropic efficiency η_{pump}.

unit cost fixed, the optimal decision variables (except for r_{AC}) generally increase as the net electrical power rises. This is because, when the net output power increases, the devices should be selected thermodynamically such that they can produce the output power. For instance, Figure 9.17 shows that, at constant unit cost of fuel, increasing the net output power raises the gas turbine inlet temperature. Also, increasing the net output power results in increasing mass flow rate in both the combustion chamber and the duct burner. Hence, the first term in the objective function tends to increase. Therefore, the genetic algorithm seeks the optimal design parameter that is able to compensate for this rise in the objective function. Further, as discussed in the previous section, increasing the GTIT results in a decrease in the cost of exergy destruction, which is the last term in the objective function. The same result is seen for the pinch temperature difference. Figures 9.19 and 9.20 show that, at a constant unit cost of fuel, increasing the output power leads to a decrease in the pinch temperature, mainly due to the fact that to increase the steam turbine output power the HRSG efficiency should be increased. This result is achieved by decreasing the pinch temperature differences, as it permits more energy to be extracted from the GT exhaust gases across the HRSG. A similar result is observed for the superheater steam temperature, as can be seen in Figures 9.21 and 9.22.

From these figures it is concluded that increasing the net output power directly affects the objective function, due to the increase in the injected mass flow rate to the combustion chamber and duct burner. As a result, the optimal design parameters are selected so that the other terms in the objective function decrease, an aim that is attained by applying the developed genetic algorithm by considering the mentioned constraints.

9.5 Concluding Remarks

After providing introductory information about power generation systems, modeling and optimization are described comprehensively for three common types of electrical power plants: steam turbine, gas turbine, and combined cycle. The model and simulation computer code are verified with some data from actual power plants to ensure adequate accuracy, and the results are observed to exhibit good agreement. Various objective functions are defined for each of the power plants and appropriate constraints considered. Then the genetic algorithm is applied to determine the best optimal design parameters. A sensitivity analysis is conducted to assess how each design parameter affects the objective functions.

The main concluding remarks drawn from the findings follow:

- The optimization of steam power plants shows that selecting the optimal design parameters with a genetic algorithm can improve the exergy efficiency of a typical power plant by approximately 4% while increasing the total cost rate of the system. In addition, multi-objective optimization can provide useful information for selecting the optimal design parameters based on priorities and policies.
- For the gas turbine power plant, optimization increases the overall exergoeconomic factor of the system from 32.79% to 62.24% compared to an actual power plant, implying that optimization mostly improves the associated cost of thermodynamic inefficiencies. The sensitivity of the obtained Pareto solutions to interest rate and fuel cost is better illustrated, and the selection of the final optimum solution from the Pareto frontier is shown to require decision-making, which is dependent on preferences and criteria of the decision maker.
- For the combined cycle power plant, the determined optimum design parameters for the plant exhibit a trade-off between thermodynamically and economically optimal designs. For example, from the thermodynamic point of view, the decision variable η_T should be selected to be as high as possible, although this leads to an increase in capital cost. It is noted that any change in the numerical values of a decision variable not only affects the performance of the related equipment but also impacts the performance of other equipment. It is concluded that by increasing the fuel price, the numerical values of decision variables in the thermoeconomically optimal design tend to those of the thermodynamically optimal design. The optimized objective functions and their maximum allowable values improve both objective functions simultaneously.

References

1 Ahmadi, P. (2013) Modeling, analysis and optimization of integrated energy systems for multigeneration purposes. PhD thesis. Faculty of Engineering and Applied Science, University of Ontario Institute of Technology.

2 Dinçer, İ. and Zamfirescu, C. (2014) *Advanced Power Generation Systems*, Academic Press.

3 Dincer, I. and Rosen, M. A. (2012) *Exergy: Energy, Environment and Sustainable Development*, Newnes.

4 Ameri, M., Ahmadi, P. and Khanmohammadi, S. (2008) Exergy analysis of a 420 MW combined cycle power plant. *International Journal of Energy Research* **32**:175–183.

5 Ahmadi, P. and Dincer, I. (2010) Exergoenvironmental analysis and optimization of a cogeneration plant system using Multimodal Genetic Algorithm (MGA). *Energy* **35**:5161–5172.

6 Ahmadi, P., Dincer, I. and Rosen, M. A. (2011) Exergy, exergoeconomic and environmental analyses and evolutionary algorithm based multi-objective optimization of combined cycle power plants. *Energy* **36**:5886–5898.

7 Ameri, M., Ahmadi, P. and Hamidi, A. (2009) Energy, exergy and exergoeconomic analysis of a steam power plant: a case study. *International Journal of Energy Research* **33**:499–512.

8 Ahmadi, P. and Dincer, I. (2011) Thermodynamic and exergoenvironmental analyses, and multi-objective optimization of a gas turbine power plant. *Applied Thermal Engineering* **31**:2529–2540.

9 Barzegar Avval, H., Ahmadi, P. and Ghaffarizadeh, A., Saidi, M. (2011) Thermo-economic-environmental multiobjective optimization of a gas turbine power plant with preheater using evolutionary algorithm. *International Journal of Energy Research* **35**:389–403.

10 Ahmadi, P., Dincer, I. and Rosen, M. A. (2013) Thermodynamic modeling and multi-objective evolutionary-based optimization of a new multigeneration energy system. *Energy Conversion and Management* **76**:282–300.

11 Ahmadi, P., Dincer, I. and Rosen, M. A. (2012) Exergo-environmental analysis of an integrated organic Rankine cycle for trigeneration. *Energy Conversion and Management* **64**:447–453.

12 Bejan, A., Tsatsaronis, G. and Moran, M. J. (1996) *Thermal Design and Optimization*, John Wiley & Sons, Inc., New York.

13 Roosen, P., Uhlenbruck, S. and Lucas, K. (2003) Pareto optimization of a combined cycle power system as a decision support tool for trading off investment vs. operating costs. *International Journal of Thermal Sciences* **42**:553–560.

14 Ghaebi, H., Saidi, M. and Ahmadi, P. (2012) Exergoeconomic optimization of a trigeneration system for heating, cooling and power production purpose based on TRR method and using evolutionary algorithm. *Applied Thermal Engineering* **36**:113–125.

15 Rizk, N. K. and Mongia, H. (1992) Semianalytical correlations for NOx, CO and UHC emissions. In *ASME 1992 International Gas Turbine and Aeroengine Congress and Exposition*. American Society of Mechanical Engineers. 92-GT-130.

16 Toffolo, A. and Lazzaretto, A. (2004) Energy, economy and environment as objectives in multi-criteria optimization of thermal system design. *Energy* **29**:1139–1157.

17 Ganjehkaviri, A., Jaafar, M. M., Ahmadi, P. and Barzegaravval, H. (2014) Modelling and optimization of combined cycle power plant based on exergoeconomic and environmental analyses. *Applied Thermal Engineering* **67**:566–578.

18 Ahmadi, P. and Dincer, I. (2011) Thermodynamic analysis and thermoeconomic optimization of a dual pressure combined cycle power plant with a supplementary firing unit. *Energy Conversion and Management* **52**:2296–2308.

19 Kotas, T. J. (1995) *The Exergy Method of Thermal Plant Analysis*, Krieger Melbourne, FL; 1995.

Study Questions/Problems

1 For the steam power plant shown in Figure 9.2, draw a *T-s* diagram and show all state points.

2 In the optimization of the steam power plant, explain how the excess air entering the boiler for combustion affects the efficiency results.

3 Consider the solar powered steam power plant with feedwater heating shown in Figure 9.28. The conditions of steam generation in the boiler are P_{main} and T_{main}. The exhaust pressure of the turbine is 10 kPa. The saturation temperature of the exhaust steam is therefore 45.83°C. Allowing for slight subcooling of the condensate, the temperature of the liquid water from the condenser is fixed at 45°C. The feedwater pump causes a temperature rise of about 1°C, making the temperature of the feedwater entering the series of heaters equal to 46°C. The saturation temperature of steam at the boiler pressure of 8,600 kPa is 300.06°C, and the temperature to which the

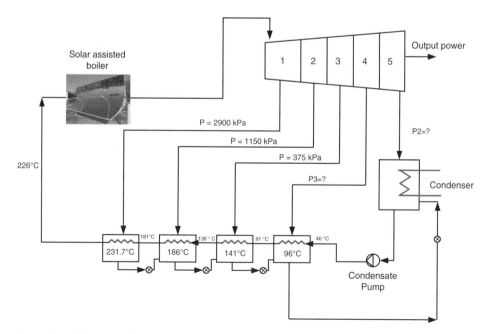

Figure 9.28 Schematic of a solar powered steam power plant with feedwater heating.

feedwater can be raised in the heaters is less. This temperature is a design variable, which is ultimately fixed by economic considerations. However, a value must be chosen before any thermodynamic calculations can be made. We therefore arbitrarily specify a temperature of 226°C for the feedwater stream entering the boiler. We also specify that all four feedwater heaters accomplish the same temperature rise. Thus, the total temperature rise of $226 - 46 = 180$°C is divided into four 45°C increments. This establishes all intermediate feedwater temperatures at the values shown in the figure. The steam supplied to a given feedwater heater must be at a pressure high enough that its saturation temperature is above that of the feedwater stream leaving the heater. We assume a minimum temperature difference for heat transfer of no less than 5°C, and choose extraction steam pressures such that the T_{sat} values shown in the feedwater heaters are at least 5°C greater than the exit temperatures of the feedwater streams. The condensate from each feedwater heater is flashed through a

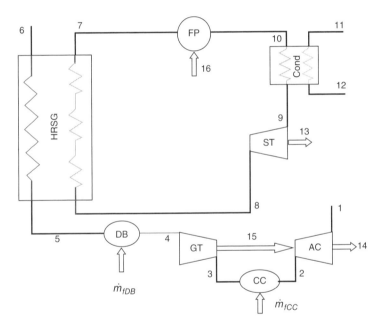

Point	Specification
1	Inlet air to compressor
2	Outlet air from compressor
3	Combustion gases exiting combustion chamber
4	Outlet hot gases exiting gas turbine
5	Combustion outlet gases exiting duct burner
6	Outlet gases exiting HRSG
7	Supply water to HRSG
8	Superheated steam entering steam turbine
9	Outlet steam from steam turbine
10	Saturated liquid entering feedwater pump
11	Inlet cooling water
12	Outlet water exiting condenser
13	Steam turbine net output power
14	Gas turbine net output power

Figure 9.29 Schematic of a combined cycle power plant with supplementary firing.

throttle valve to the heater at the next lower pressure, and the collected condensate in the final heater of the series is flashed into the condenser. Thus, all condensate returns from the condenser to the boiler by way of the feedwater heaters. Calculate the optimized values for the steam turbine main temperature and pressure, P_2 and P_3 when the system exergy efficiency is maximized. Make reasonable assumptions as required.

4 Repeat the multi-objective optimization for a gas turbine power plant when there is no preheater and compare the results with the case including a preheater. What is the main benefit of using the air preheater?

5 For the gas turbine power plant, describe how the compressor air inlet temperature affects the exergy efficiency and net output power. List and explain some options to improve the gas turbine power plant.

6 Repeat the optimization for a gas turbine power plant and consider the total exergy destruction rate and the total cost rate as the two objective functions. Compare the results with exergy efficiency and total cost rate as objective functions.

7 Why are duct burners used in advanced combined cycle power plants?

8 Figure 9.29 shows the schematic of a single pressure combined cycle power plant. Model this plant in a similar way to the CCPP modeling in this chapter, and find the best optimal design parameters when the exergy efficiency of the power plant is maximized.

9 Apply multi-objective optimization to the combined cycle power plant shown in Figure 9.11. Consider exergy efficiency and total cost rate as the two objective functions.

10

Modeling and Optimization of Cogeneration and Trigeneration Systems

10.1 Introduction

With increased demand for energy, society is increasingly considering more efficient and cost effective systems, as well as those that are more sustainable and clean. It is also important to manage energy requirements in a more efficient, cost effective, and environmentally benign manner. Cogeneration, or combined heat and power (CHP), systems produce both electricity and heat simultaneously with a single system, and constitute one type of system that can often be employed to provide energy services efficiently, environmentally benignly, and cost effectively for a wide range of uses. Cogeneration represents a relatively simple case of a multigeneration energy system, and often involves utilizing waste or other heat from thermal electricity generation for heating. The overall energy efficiency of a cogeneration system, defined as the part of the fuel converted to both electricity and useful thermal energy, is typically 40–80% [1].

The benefits of integrating energy systems became increasingly recognized with the application of cogeneration for heat and electricity. The cogenerated heat is usually used to provide heating or, via thermally driven chillers, cooling. Cogeneration often yields considerable reductions in input energy compared to separate processes for the same products. Cogeneration is often associated with the combustion of fossil fuels, but can also be carried out using other thermal power generation processes, for a variety of "fuels," including some renewable energy sources, nuclear energy, and waste thermal energy (obtained directly or by burning waste materials). The recent trend has been to use cleaner fossil fuels for cogeneration, such as natural gas.

The long-term prospects for cogeneration in global energy markets are strong, in large part because of its ability to provide significant operational, environmental, and financial benefits. The product thermal energy from cogeneration can be used for domestic hot water heating, space heating, swimming pool heating, laundry heating processes, and absorption cooling, as well as for a wide range of industrial heating processes. The more the product heat from cogeneration can be used in existing systems, the more financially attractive the system usually is. Cogeneration sometimes helps overcome a drawback of many conventional electrical and thermal systems: significant heat losses, which detract greatly from efficiency [2]. Normally, heat losses are reduced and efficiency is increased when cogeneration is used to supply heat to applications and facilities.

Recently, researchers have extended cogeneration to yield more than two products. One outcome has been that trigeneration systems have become more suitable for energy markets. Trigeneration is the simultaneous production of heating, cooling, and electricity in a single system from a common energy source. Trigeneration is often based on

Optimization of Energy Systems, First Edition. Ibrahim Dincer, Marc A. Rosen, and Pouria Ahmadi.
© 2017 John Wiley & Sons Ltd. Published 2017 by John Wiley & Sons Ltd.

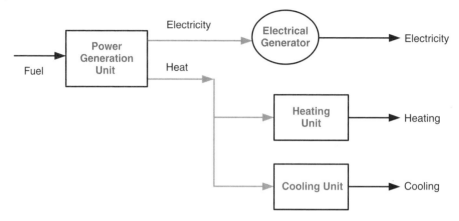

Figure 10.1 A typical trigeneration energy system.

utilizing the waste or other heat of a thermal power plant to improve overall thermal performance. In a trigeneration system, waste heat from the plant's prime mover (e.g., gas turbine, diesel engine, Rankine cycle [3]), sometimes with temperature enhancement, drives heating and cooling devices. Although the heat is typically used for space heating, domestic hot water production, or steam production for process heating, it can also be used for cooling, by driving an absorption chiller. Several investigations and applications of trigeneration have been reported over the last decade, likely due to its benefits. Trigeneration can be applied widely, for example, in chemical and food industries, residential buildings, commercial and institutional facilities like airports, shopping centers, hotels, and hospitals.

Figure 10.1 illustrates a trigeneration energy system and its three main parts:

- A power generation unit, that is, a prime mover, such as a gas turbine
- A cooling unit, such as a single effect absorption chiller
- A heating unit, such as a boiler or heat recovery steam generator

The following processes normally occur in a trigeneration plant:

- Mechanical power is produced via a power generation unit, such as a gas turbine.
- The mechanical power is used to drive an electrical generator.
- Waste heat exits the power generation unit directly or via heated materials like exhaust gases.

As shown in Figure 10.1, with a single prime mover, a trigeneration system can produce heating, cooling, and electricity simultaneously.

Recently, researchers have extended trigeneration to produce more products like hydrogen and potable water using a single prime mover through a process called multigeneration. Although this topic is beyond the scope of this chapter, the material in this chapter can be extended and applied to multigeneration systems.

The overall energy efficiency for a cogeneration system is defined as follows:

$$\eta_{en} = \frac{W_{elec} + Q_{heat}}{E_{fuel}} \tag{10.1}$$

where W_{elec} denotes the electrical product and Q_{heat} the heating product of cogeneration, while E_{fuel} denotes the inlet energy of the fuel.

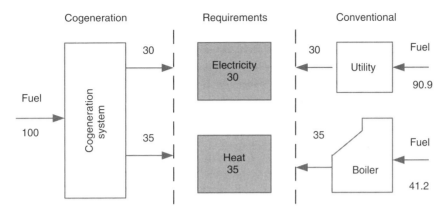

Figure 10.2 Comparison of conventional systems for electricity generation and heat production with cogeneration system. Numerical values denote relative energy units, with the energy entering the cogeneration system with fuel designated as having 100 units of energy.

The main benefit of the combined production of heat and power via cogeneration results from the more efficient use of fuel, and the corresponding reductions in emissions of SO_2, NO_x, CO_2, and other pollutants. Sometimes the energy efficiency of cogeneration (around 80%) is misleadingly compared with the energy efficiency of power generation facilities, for example, condensing power plants (approximately 40%). For a valid comparison, we must consider the efficiency of independent power and heat production processes versus the efficiency of cogeneration.

A comparison between a cogeneration plant and conventional plants to produce electricity and heat separately is shown in Figure 10.2. There, the energy needs to be satisfied are taken to be 30 energy units of electrical energy and 35 energy units of heat. These electricity and thermal needs may be satisfied either by a cogeneration system or by conventional systems. The conventional systems in this example are a standard electricity utility and a common boiler to produce steam. The cogeneration system is shown on the left side of Figure 10.2, and the conventional systems on the right. Assuming an electrical generation efficiency of 30% for the cogeneration system, 30 units of electrical energy are seen to be produced with 100 units of fuel energy. This cogeneration system also supplies 35 energy units of product heat. Hence, for this cogeneration system, the overall efficiency, following Equation 10.1, is

$$\eta_{en} = \frac{30 + 35}{100} = 65\%$$

The conventional systems, shown on the right side of Figure.10.2, satisfy the power and thermal needs by the use of an electric utility and a boiler. In this example, the utility is assumed to be able to deliver the 30 units of electrical energy with a plant efficiency of 33%. This results in a fuel energy input of 90.9 units. The boiler is assumed to supply the 35 units of heat with an 85% efficiency, which requires a fuel energy input of 41.2 units. The overall efficiency is, therefore

$$\eta_{en} = \frac{30 + 35}{(90.9 + 41.2)} = 49\%$$

The practical usage of cogeneration is as old as the generation of electricity. When electrification of broad areas was planned to replace gas and kerosene lighting in residences and commercial facilities, the concept of central station power generation plants was born. The prime movers that drove electric generators emit waste heat that is normally released to the environment. By capturing that heat, for example, by making low pressure steam, that steam could be piped throughout the district; it could be used for heating homes and businesses. Thus, cogeneration on a fairly large scale came about. In the past, most of the cogenerated electricity was produced on site in large industrial plants, and some of the waste heat was relatively easily captured and utilized in industrial processes.

The term "cogeneration" as used in this chapter is defined as the combined production of electrical energy and useful thermal energy by the use of a fuel or fuels. Each term in this definition is important. The word "combined" means the production processes for electricity and thermal energy are linked, and often are accomplished in a series or parallel fashion. The electrical energy is produced by an electrical generator, which is most often powered by a kind of prime mover like a steam turbine, a gas turbine, a reciprocating engine, or a fuel cell. The product thermal energy is used to provide heating or, indirectly, cooling. This thermal energy can be transported in many forms, including hot exhaust gases, hot water, steam, and chilled water. The word "useful" means that the energy is directed at fulfilling an existing need for heating or cooling.

The benefits of cogeneration are often economic as well as thermodynamic. Any facility that uses electricity and needs thermal energy is a candidate for cogeneration. Although many considerations are involved in determining if cogeneration is feasible for a particular facility, the basic consideration often is whether the reduced thermal energy costs are sufficient to justify the capital expenditures for a cogeneration system. Cogeneration technologies are for the most part available in a wide range of sizes: from less than 100 kW to over 100 MW [1]. The main devices comprising a cogeneration system are a prime mover, which is usually the most significant part, an electrical generator, a heat recovery system, and other typical instrumentation. Large cogeneration systems provide hot water or steam and electricity for an industrial site or an entire region (e.g., a town).

Some common types of CHP plants are as follows:

- Gas turbine based CHP plants, which recover and use the waste heat in the flue gas of gas turbines. The fuel used is typically natural gas.
- Gas engine CHP plants, which use a reciprocating gas engine and generally are more competitive than gas turbine based CHP up to about 5 MW electrical power. The fuel is normally natural gas. These plants are generally manufactured as fully packaged units that can be installed within a plant room or external plant compound with simple connections to the site's gas supply and electrical distribution and heating systems.
- Biofuel engine CHP plants use an adapted reciprocating gas engine or diesel engine, depending on which biofuel is used, and are otherwise similar in design to gas engine CHP plants. The primary advantage of using a biofuel is reduced hydrocarbon fuel consumption and thus reduced carbon emissions. These plants are generally manufactured as fully packaged units that can be installed within a plant room or external plant compound with simple connections to the site's electrical distribution and heating systems. Another variant is the wood gasifier CHP plant, in which a wood pellet or wood chip biofuel is gasified in a zero oxygen high temperature environment; the resulting gas is then used to power the gas engine.
- Combined cycle power plants that are adapted for CHP.

- Steam turbine based CHP plants that use the steam condenser for the steam turbine as the heating system.
- Molten carbonate fuel cells and solid oxide fuel cells, which have hot exhausts that are suitable for various heating purposes.
- Coal and nuclear power plants can be fitted with taps after the turbines to provide steam to a heating system. With a heating system temperature of 95°C, it is possible to extract about 10 MW thermal for every 1 MW electrical power lost. With a temperature of 130°C, the gain is slightly smaller, about 7 MW thermal for every 1 MW lost.

As there exist many cogeneration and trigeneration systems, the choice of which to use depends on the application and the setting.

In this chapter we focus on some of the main types. Each system is thermodynamically modeled, and parametric and optimization studies are conducted. Exergy analysis is utilized throughout. We define meaningful objective functions and realistic constraints to ensure the results are useful and reliable. An evolutionary algorithm based optimization is applied to each system and the optimal design parameters are determined. In order to enhance understanding of the design criteria, sensitivity analyses are conducted to determine how each objective function varies when selected design parameters vary. We close the chapter by providing insights for the efficient design of CHP and trigeneration systems.

10.2 Gas Turbine Based CHP System

The gas turbine is a commonly employed prime mover. A gas turbine, also called a combustion turbine, is a type of internal combustion engine that has an upstream rotating compressor coupled to a downstream turbine, with a combustion chamber in between. Energy is added to the gas stream in the combustion chamber, where fuel is mixed with hot air and then ignited. In the high pressure environment of the combustor, combustion of the fuel increases the temperature. The products of the combustion expand in the turbine to produce shaft work. At the entrance of the turbine, the high temperature and pressure gas flow is directed through a nozzle over the turbine blades, rotating the turbine that powers the compressor and, for some turbines, provides a mechanical output. The work developed by the turbine comes from the reduction in the temperature and pressure of the combustion gas.

In most practical gas turbines, air enters and is accelerated in either a centrifugal or radial compressor. The air is then slowed using a diverging nozzle known as a diffuser, increasing the pressure and temperature of the flow. In an ideal system this process is isentropic, but in practice energy is lost in the form of heat, and entropy is produced due to friction and turbulence. The air then passes from the diffuser to a combustion chamber or similar device, where heat is added. In an ideal system this occurs at constant pressure [4]. In practical situations this process is usually accompanied by a slight loss in pressure due to friction. Finally, the combustion gases are expanded and accelerated by nozzle guide vanes before energy is extracted by the turbine.

Micro gas turbines are a kind of gas turbine that has become widespread in distributed power generation units and combined heat and power applications. They range from small units producing less than a kilowatt to commercial sized systems that produce

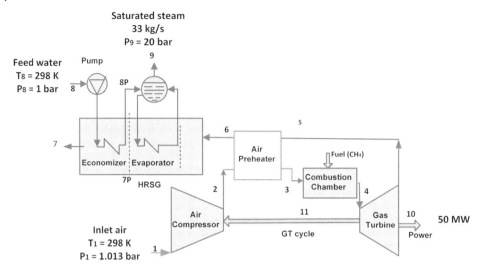

Figure 10.3 Schematic of a CHP system for electricity and steam production.

tens or hundreds of kilowatts. The basic principles of micro-turbines are based on micro combustion [4].

Figure 10.3 shows the schematic of a gas turbine CHP system. This CHP system is used in a paper mill with need for 50 MW of electric power and 33.3 kg/s of saturated steam at 20 bar. Air at ambient conditions enters the air compressor at point 1 and exits after compression (point 2). The hot air enters the air preheater and its temperature increases. The hot air at point 3 enters the combustion chamber (CC) into which fuel is injected, and hot combustion gases exit (point 4) and pass through a gas turbine to produce shaft power. The turbine exhaust gas at point 5 passes through the air preheater, which heats the air exiting the compressor. The combustion gases still have energy when leaving the heater, so they are passed through a heat recovery steam generator (HRSG), which uses the exhaust gas energy to produce saturated steam at 20 bar at point 9. The flue gases leave the HRSG at point 7 and are released to the environment. The system described, although designed to meet the needs of a paper mill, can produce both electricity and saturated steam for other applications.

10.2.1 Thermodynamic Modeling and Analyses

To find the optimum physical and thermal design parameters of the system, a simulation program is developed in the Matlab software. Thus, the temperature profile in the CHP plant and the input and output enthalpy and exergy of each flow are determined in order to develop an understanding of the plant and permit its subsequent optimization. The energy balance equations for the compressor, the combustion chamber, and the gas turbine were provided earlier (see section 9.3). Here the relations for modeling the air preheater and HRSG are provided:

10.2.1.1 Air Preheater

An energy balance for a control volume around the air preheater yields

$$\dot{m}_a(h_3 - h_2) = \dot{m}_g(h_5 - h_6)\eta_{AP} \tag{10.2}$$

where η_{AP} is the effectiveness of the heat exchanger. The pressure drop across the pre-heater ΔP_{AP} can be expressed implicitly as follows:

$$\frac{P_3}{P_2} = (1 - \Delta P_{AP}) \tag{10.3}$$

10.2.1.2 Heat Recovery Steam Generator (HRSG)

Writing energy balances for the economizer and the evaporator in the HRSG yields:

$$\dot{m}_s(h_9 - h_8) = \dot{m}_g(h_6 - h_7) \tag{10.4}$$
$$\dot{m}_s(h_9 - h_{8p}) = \dot{m}_g(h_6 - h_{7p})$$

These combinations of energy and mass balance equations are numerically solved, and the temperature and enthalpy of each flow in the plant are then determined. In addition, several assumptions are made to facilitate the analysis [1, 5]:

- All processes are steady state and steady flow.
- Air and combustion products are treated as ideal gases.
- Heat loss from the combustion chamber is considered to be 3% of the fuel input energy, and all other components are considered adiabatic.
- The dead state properties are $P_0 = 1.01$ bar and $T_0 = 293.15$ K.
- The pressure drop is assumed to be 4% across the preheater and of 3% across the combustion chamber.

Exergy analysis is also used. The exergy destruction rate and the exergy efficiency for each component and for the overall power plant in Figure 10.3 are listed in Table 10.1. The source of exergy destruction (or irreversibility) in the combustion chamber is mainly combustion (chemical reaction) and thermal losses in the flow path [6], while the exergy destruction in the primary system heat exchanger (the HRSG) is due to the large temperature differences between the hot and cold fluids.

Exergoeconomics is applied here. It appropriately combines, at the level of system components, thermodynamic evaluations based on an exergy analysis and economic principles, in order to provide information that is useful to the design and operation of a cost effective system, but not obtainable by conventional energy and exergy analyses and economic analysis. Some suggest that, when exergy costing is not applied, the general term thermoeconomics is more appropriate as it characterizes any combination of thermodynamic and economic analysis. To define a cost function that is dependent on the optimization parameters of interest, the component cost is expressed as a function of thermodynamic design parameters.

Table 10.1 Expressions for exergy destruction rate and exergy efficiency for each component of the gas turbine power plant.

Component	Exergy destruction rate	Exergy efficiency
Compressor	$\dot{Ex}_1 + \dot{W}_{AC} = \dot{Ex}_2 + \dot{Ex}_D$	$\psi_{AC} = \frac{\dot{Ex}_2 - \dot{Ex}_1}{\dot{W}_{ac}}$
Air preheater	$\dot{Ex}_2 + \dot{Ex}_5 = \dot{Ex}_3 + \dot{Ex}_6 + \dot{Ex}_D$	$\psi_{AP} = \frac{\dot{Ex}_3 - \dot{Ex}_2}{\dot{Ex}_5 - \dot{Ex}_6}$
Combustion chamber	$\dot{Ex}_3 + \dot{Ex}_9 = \dot{Ex}_4 + \dot{Ex}_D$	$\psi_{CC} = \frac{\dot{Ex}_4}{\dot{Ex}_3 + \dot{Ex}_9}$
Gas turbine	$\dot{Ex}_4 = \dot{Ex}_5 + \dot{W}_{GT} + \dot{Ex}_D$	$\psi_{GT} = \frac{\dot{W}_{GT}}{\dot{Ex}_4 - \dot{Ex}_5}$
HRSG	$\dot{Ex}_6 + \dot{Ex}_8 = \dot{Ex}_7 + \dot{Ex}_9 + \dot{Ex}_D$	$\psi_{HRSG} = \frac{\dot{Ex}_9 - \dot{Ex}_8}{\dot{Ex}_6 - \dot{Ex}_7}$

For each flow in a system, a parameter called the flow cost rate $\dot{C}$ ($/h) is defined, and a cost balance is written for each component as:

$$\dot{C}_{q,k} + \sum_i \dot{C}_{i,k} + \dot{Z}_k = \sum_i \dot{C}_{e,k} + \dot{C}_{w,k} \qquad (10.5)$$

The cost balances are generally written so that all terms are positive. Using Equation 10.5, one can write [6]

$$\sum (c_e \dot{E}x_e)_k + c_{w,k}\dot{W}_k = c_{q,k}\dot{E}x_{q,k} + \sum (c_i \dot{E}x_i)_k + \dot{Z}_k \qquad (10.6)$$

$$\dot{C}_j = c_j \dot{E}x_j \qquad (10.7)$$

In this analysis, the fuel and product exergy need to be defined. The product exergy is defined for each component, while the fuel represents the source that is consumed in generating the product. Both product and fuel are expressed in terms of exergy. The cost rates associated with the fuel $(\dot{C}_F)$ and the product $(\dot{C}_P)$ of a component are obtained by replacing the exergy rates $(\dot{E}x)$.

In the cost balance formulation (Equation 10.5), there is no cost term directly associated with the exergy destruction of each component. Accordingly, the cost associated with the exergy destruction in a component or process is a hidden cost. If one combines both exergy and cost balances, one can obtain the following:

$$\dot{E}x_{F,k} = \dot{E}x_{P,k} + \dot{E}x_{D,k} \qquad (10.8)$$

Accordingly, the expression for the cost of exergy destruction is defined as follows:

$$\dot{C}_{D,k} = c_{F,k}\dot{E}x_{D,k} \qquad (10.9)$$

In addition, several methods have been proposed to express the purchase cost of equipment in terms of design parameters in Equation 10.5 [6]. Here, we use the cost functions suggested by Ahmadi *et al.* [1, 7, 8] and Rosen *et al.* [9], modified to fit the regional conditions of Iran and the inflation rate. We convert the capital investment into cost per unit time as follows:

$$\dot{Z}_k = \frac{Z_k \cdot CRF \cdot \varphi}{N \times 3600} \qquad (10.10)$$

where Z_k is the purchase cost of the kth component, and *CRF* is the capital recovery factor. *CRF* depends on the interest rate and equipment lifetime, and is determined as follows [6, 10]

$$CRF = \frac{i \times (1+i)^n}{(1+i)^n - 1} \qquad (10.11)$$

where i denotes the interest rate and n the total operating period of the system in years. Also, N is the annual number of operation hours for the unit, and φ is the maintenance factor, which is often 1.06 [10]. To determine the component exergy destruction cost, the exergy destruction, $\dot{E}x_{D,k}$, is evaluated using exergy balances from the previous section.

To estimate the cost rate of exergy destruction in each component of the plant, we first solve the cost balances for each component. In the cost balance (Equation 10.6), there is more than one inlet and outlet flow for some components. In such cases, the number of unknown cost parameters is higher than the number of cost balances for that component. Auxiliary exergoeconomic equations are developed to allow such problems to be solved. Writing Equation 10.5 for each component together with the auxiliary equations forms a system of linear equations as follows:

$$[\dot{Ex}_k] \times [c_k] = [\dot{Z}_k] \tag{10.12}$$

where $[\dot{Ex}_k]$, $[c_k]$ and $\dot{Z}_k$ are the exergy rate matrix (obtained via exergy analysis), the exergetic cost vector (to be evaluated), and the vector of $\dot{Z}_k$ factors (obtained via economic analysis), respectively. That is,

$$
\begin{bmatrix}
1 & 0 & 0 & 0 & 0 & 0 & 0 & 0 & 0 & 0 & 0 & 0 \\
1 & -1 & 0 & 0 & 0 & 0 & 0 & 0 & 0 & 0 & 1 & 0 \\
0 & 1 & -1 & 0 & 1 & -1 & 0 & 0 & 0 & 0 & 0 & 0 \\
0 & 0 & 0 & 0 & \frac{1}{\dot{Ex}_5} & \frac{-1}{\dot{Ex}_6} & 0 & 0 & 0 & 0 & 0 & 0 \\
0 & 0 & 1 & -1 & 0 & 0 & 0 & 0 & 0 & 0 & 0 & 0 \\
0 & 0 & 0 & 0 & 0 & 0 & 0 & 0 & 0 & 1 & 0 & 0 \\
0 & 0 & 0 & 1 & -1 & 0 & 0 & 0 & 0 & 0 & 1 & -1 \\
0 & 0 & 0 & \frac{1}{\dot{Ex}_4} & \frac{-1}{\dot{Ex}_5} & 0 & 0 & 0 & 0 & 0 & 0 & 0 \\
0 & 0 & 0 & 0 & 0 & 0 & 0 & 0 & 0 & \frac{1}{\dot{Ex}_{11}} & \frac{-1}{\dot{Ex}_{12}} \\
0 & 0 & 0 & 0 & 0 & 1 & -1 & 1 & -1 & 0 & 0 & 0 \\
0 & 0 & 0 & 0 & 0 & \frac{1}{\dot{Ex}_6} & \frac{-1}{\dot{Ex}_7} & 0 & 0 & 0 & 0 & 0 \\
0 & 0 & 0 & 0 & 0 & 0 & 0 & 0 & 0 & 0 & 0 & 0
\end{bmatrix}
\times
\begin{bmatrix}
\dot{C}_1 \\ \dot{C}_2 \\ \dot{C}_3 \\ \dot{C}_4 \\ \dot{C}_5 \\ \dot{C}_6 \\ \dot{C}_7 \\ \dot{C}_8 \\ \dot{C}_9 \\ \dot{C}_{10} \\ \dot{C}_{11} \\ \dot{C}_{12}
\end{bmatrix}
=
\begin{bmatrix}
0 \\ -\dot{Z}_{AC} \\ -\dot{Z}_{AP} \\ 0 \\ -\dot{Z}_{CC} \\ C_F \\ -\dot{Z}_{GT} \\ 0 \\ 0 \\ -\dot{Z}_{HRSG} \\ 0 \\ 0
\end{bmatrix}
\tag{10.13}
$$

The environmental impact assessment of this system is similar to that for the gas turbine power plant in section 9.3.3.

10.2.2 Optimization

In this section we consider objective functions, design parameters, and constraints, as well as overall optimization studies for gas turbine cogeneration system. Genetic algorithm code is developed in Matlab software and applied to determine the most advantageous optimal design parameters of this CHP system.

10.2.2.1 Single Objective Optimization

Here, a new objective function is defined as the summation of the operational cost rate, which is related to the fuel expense, the rate of capital cost, which represents the capital investment and maintenance expenses, the corresponding cost for the exergy destruction, and the cost of environmental impacts (NO$_x$ and CO). Therefore, the objective function representing the total cost rate of the plant in terms of monetary unit per unit time is defined as

$$\dot{C}_{tot} = \dot{C}_f + \sum_k \dot{Z}_k + \dot{C}_D + \dot{C}_{env} \tag{10.14}$$

Here, we use $c_f = 0.003$ \$/MJ, which is the regional fuel cost per unit of energy [1, 6], $\dot{m}_f$ is the fuel mass flow rate, and LHV = 50,000 kJ/kg is the lower heating value of natural gas. The last part of the objective function (OF) expresses the environmental impact as the total pollution damage rate (\$/s) due to CO and NOx emissions by multiplying their respective flow rates by their corresponding unit damage costs: $C_{CO} = 0.02086$ \$/kg$_{CO}$ and $C_{NOx} = 6.853$ \$/kg$_{NOx}$ [11]. In the present work the cost of pollution damage is assumed to be added directly to the expenditures that must be paid. Therefore, the objective function is sum of the exergoeconomic and environmental objectives. Since

Table 10.2 Optimization constraints and their rationales for the CHP system.

Constraint	Reason
$T_4 \leq 1600°C$	Material limitation
$\dfrac{P_2}{P_1} \leq 15$	Commercial availability
$\eta_{Ac} \leq 0.9$	Commercial availability
$\eta_{Ac} \leq 0.93$	Commercial availability
$T_4 \geq 400°C$	Avoid formation of sulfuric acid in exhaust gases
$T_{8P} = T_9 - 15.$	Avoid evaporation of water in HRSG economizer

the rates of ultimate products (net power and process steam) are fixed, the objective function is to be minimized in order to obtain the values of the optimal design parameters.

Decision Variables The decision variables (i.e., design parameters) considered in this study are compressor pressure ratio (r_{Ac}), compressor isentropic efficiency (η_{AC}), gas turbine isentropic efficiency (η_{GT}), combustion chamber inlet temperature (T_3), and gas turbine inlet temperature (GTIT). Even though the decision variables may be varied in the optimization procedure, each decision variables is normally required to be within a reasonable range. The list of these constraints and the reasons of their applications are listed in Table 10.2.

Constraints Based on Figure 10.3, the following constraints need to be satisfied in the heat exchangers (air preheater and heat recovery steam generator):

$$T_3 > T_2, \ T_5 > T_3, \ T_4 > T_3, \ T_6 > T_2, \ T_6 > T_9, \ T_{7p} > T_9 + \Delta T_{pinch}$$

Verification of the Optimization Method In order to ensure the validity of the thermodynamic and economic modeling, as well as the optimization procedure (i.e., the developed genetic algorithm), we first consider a CHP unit with the same characteristics as the classical and well known CGAM problem [12–15], which is modeled and optimized by the multimodal genetic algorithm method. As shown in Table 10.3 our model exhibits good agreement with the CGAM results and better results in minimizing the objective function. That is, the results in this table show that our developed code has better optimization results in comparison with other results. Moreover, by applying our developed GA code, a 9.80% improvement in the objective function is achieved, which is a significant improvement in thermal systems optimization and leads to significant reductions in fuel consumption and GHG emissions. We therefore take this comparison to verify the validity of the obtained global optimum as well as other optimization results.

The input parameters of the problem are modified to match the conditions and requirements of the paper mill. The fuel unit cost (C_f) and fuel LHV in this case are 0.003 \$/MJ and 50 000 kJ/kg respectively. In addition, considering the values of i and n to be 14% and 15 years, respectively, CRF is calculated as 16.3%. Also, N, the annual number of the operation hours of the unit, and φ, the maintenance factor, are taken to be 7000 hours and 1.1, respectively.

Table 10.3 Comparison of present optimization results for the CGAM problem with results reported in the literature [12–15].

Decision variable	Optimum design values reported by [12]	Optimum design values reported by [14]	Optimum design values using GA in present study
r_{Ac}	8.597	8.523	6.7
η_{AC}	0.84	0.84	0.83
η_{GT}	0.87	0.87	0.86
$T_3(K)$	913.1	914.2	951.6
$T_4(K)$	1491.9	1492.63	1475.4
Objective function ($/s)	0.36	0.36	0.33

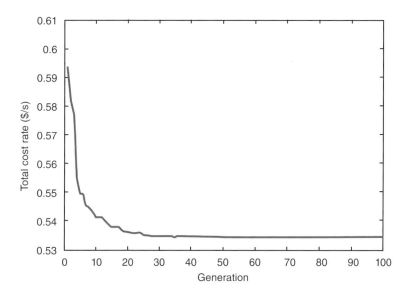

Figure 10.4 Variation of the objective function of the CHP system with the number of generations ($c_f = 0.003$ \$/MJ).

Figure 10.4 shows the variation of the objective function of the system by generation in the genetic algorithm. The genetic algorithm used in this problem is seen in the figure to have a good convergence rate. After 50 generations the final value of objective function is determined. The solution is observed to be obtained with a lower computer run time and better optimization results are attained. The obtained numerical values of the optimum design parameters for the CHP plant are reported in Table 10.4. The temperature and pressure for each state point of the CHP system at their optimal values are listed in Table 10.5.

Since the cost of fuel and the net output of a CHP system can vary, the optimal design parameters vary when determining the optimal values for objective functions. To enhance understanding of how the optimized values vary when the cost of fuel and the output electrical power change, a sensitivity analysis is carried out. Here, sensitivity analyses are performed to assess the changes in design parameters due to variations

Table 10.4 Numerical values of selected dependent variables in the optimal design.

Variable	Optimal design value
$\dot{m}_f$ (kg/s)	2.78
$\dot{m}_g$	191.58
ΔT_{pinch} (K)	12.83
η_{AC}	0.827
η_{GT}	0.862
r_{AC}	6.72
T_3 (K)	938
GTIT (K)	1473
$\dot{W}_{GT}$ (MW)	98.779
$\dot{C}_{env}$ (S/s)	45.95
$\dot{C}_D$ ($/s)	1116
Total cost rate ($/s)	0.535

Table 10.5 Optimal values for temperatures and pressures of CHP Plant.

State point	T (K)	P (bar)
1	298.15	1.013
2	556.7	6.65
3	935.77	6.36
4	1474.17	6.04
5	1034.6	1.089
6	714.4	1.06
7	430.1	1.013
8	298.1	13
9	464.79	13

in unit cost of fuel and net output power. Thus, the simulation and optimization procedures are repeated with the new set of input values. For the same objective function, Figures 10.5 to Figure 10.11 show the effects of changes in net output power on the numerical values of optimal design parameters (decision variables).

Figure 10.5 shows the effect of unit cost of fuel on the optimal value of compressor isentropic efficiency for several values of net output power. An increase in unit cost of fuel is seen to lead to an increase in the first term of the objective function, which is associated with fuel cost, which is why an increase in the unit cost of fuel results in an increase in compressor isentropic efficiency. Therefore, compressor efficiency should

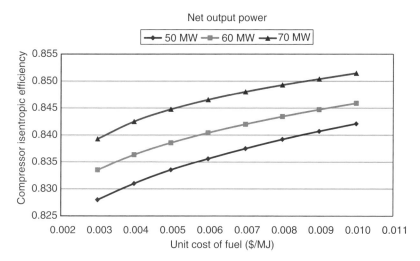

Figure 10.5 Effects of varying unit cost of fuel and net power output on the optimal value of compressor isentropic efficiency.

be selected in a way that decreases the objective function. The higher the compressor efficiency, the lower the compressor consumption power and the lower the compressor cost of exergy destruction which is the last term of Equation 10.10. It is also observed that for fixed unit cost of fuel, an increase in net output power results in an increase in compressor efficiency as well. This is again due to the fact that higher compressor efficiency will reduce the compressor consumption power which in turn increases the net output power.

Similar effects are observed due to varying the unit cost of fuel and net output power on the optimal values for the gas turbine isentropic efficiency as shown in Figure 10.6. The higher the gas turbine efficiency, the higher the exergy efficiency of the plant, which results in a reduction of the total cost of exergy destruction.

Figure 10.7 shows the effects of varying the unit cost of fuel on the optimal value of the compressor pressure ratio. The results show that a rise in the unit cost of fuel increases the optimal compressor pressure ratio. The reason is that when the compressor pressure ratio increases, the compressor outlet temperature rises and air enters the preheater and combustion chamber with a higher temperature, which in turn results in a decrease in the fuel injected into the combustion chamber. Since the first term in the objective function is the product of unit cost of fuel and fuel mass flow rate, a decreases in fuel mass flow rate results in a decrease in this term and consequently the objective function. This is why, in the optimization procedure, higher values for compressor pressure ratio are selected. In addition, since air enters the combustion chamber with a higher temperature, the combustion chamber exergy destruction decreases, which eventually reduces the total plant cost of exergy destruction which is the last term in the objective function. Also, the fuel use reduction decreases pollutant emissions from the combustion reaction in the chamber, which in turn lowers the cost of environmental impact in the objective function.

It can also be observed in Figure 10.7 that, at fixed unit cost of fuel, an increase in system net output power results in a lower compressor pressure ratio. The reason is that when the compressor pressure ratio increases, the outlet temperature rises, which

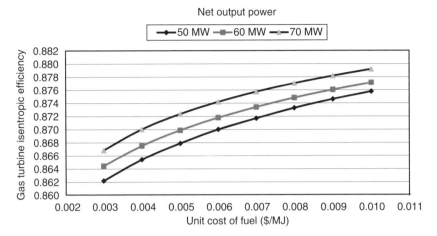

Figure 10.6 Effects of varying unit cost of fuel and net output power on the optimal value of gas turbine isentropic efficiency.

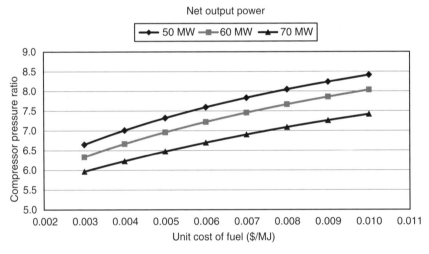

Figure 10.7 Effects of varying unit cost of fuel and net output power on the optimal value of the compressor pressure ratio.

in turn raises the compressor power consumption. Hence, when the output power increases, the optimization routine selects a lower pressure ratio to compensate.

Figure 10.8 shows the effect of varying unit cost of fuel on the gas turbine inlet temperature (GTIT). It is observed that an increase in fuel cost results in an increase in GTIT at constant net output power. When fuel cost increases, the first term in objective function increases, which is why the optimization routine selects higher GTIT values. This outcome leads in turn to a decrease in the exergy destruction for the combustion chamber. As this exergy destruction is far greater compared to that for every other component [10], it leads to a notable decrease in the cost of exergy destruction and at the same time a notable increase in the gas turbine power.

The highest exergy destruction rate in the plant occurs in the combustion chamber. The irreversible chemical reaction and heat transfer across a large temperature

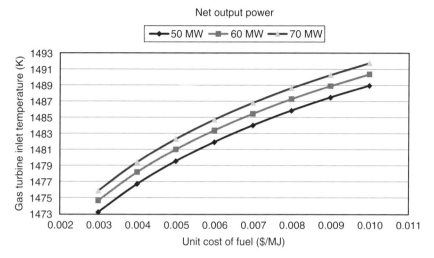

Figure 10.8 Effects of varying unit cost of fuel and net output power on the optimal value of the gas turbine inlet temperature.

difference (between the burners and working fluid) are the main causes of the irreversibility. Hence any rise in gas turbine inlet temperature results in a decrease in the combustion chamber exergy destruction as well as the cost of exergy destruction for this component. Figure 10.9 shows the effects of varying unit cost of the fuel on the fuel mass flow rate to the combustion chamber. An increase in unit cost of fuel is seen to result in a decrease in the fuel mass flow rate, as explained earlier in this paragraph. This decrease has two significant effects. Specifically, a reduced mass flow rate lowers the environmental impacts and decreases the objective function. At a fixed unit cost of fuel, an increase in the net output power results in an increase in the fuel mass flow rate. This is due to the fact that, for a higher gas turbine power, the mass flow rate of the fuel needs to rise.

Finally, Figure 10.10 shows the effect of varying the unit cost of fuel on the objective function which here is total cost rate of the system. It is seen that an increase in the unit cost of fuel results in an increase in the total cost rate of the system. As this is the most important parameter in the objective function, an increase in its value leads to an increase in the objective function. That is why, in the optimization algorithm, other design parameters need to be selected so that this objective function is minimized.

10.2.2.2 Multi-objective Optimization

In the previous section, a single objective optimization of a gas turbine based CHP system was conducted. In CHP systems there are usually several objective functions that need to be optimized simultaneously, and this can be achieved by applying multi-objective optimization. Three objective functions are considered here for multi-objective optimization: exergy efficiency (to be maximized), total cost rate of the system (to be minimized), and CO_2 emission rate (to be minimized). The second objective function includes the environmental impact and expresses this as the total cost rate of pollution damage ($/s) due to CO and NO_x emissions, by multiplying their respective flow rates by their corresponding unit damage costs (C_{CO} and C_{NOx} are taken to be equal to 0.02086 $/kg CO and 6.853 $/kg NOx) [7]. In this part the cost

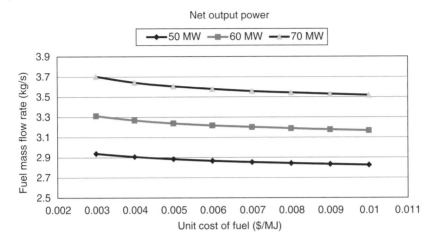

Figure 10.9 Effects of varying unit cost of fuel and net output power on the optimal value of the fuel mass flow rate.

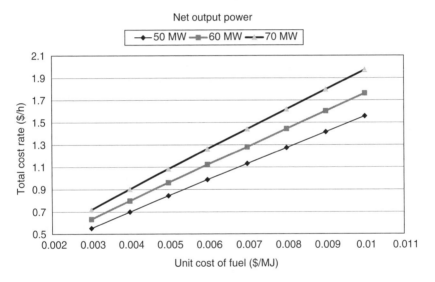

Figure 10.10 Effects of varying unit cost of fuel and net output power on the total cost rate of the system.

of pollution damage is assumed to be added directly to the expenditures that must be paid. Therefore, the second objective function is sum of the thermodynamic and environomic objectives. Due to the importance of environmental effects, especially climate change, the third objective function accounts explicitly for CO_2 emissions that are produced in the combustion chamber.

The objective functions for this analysis are defined as follows:

CHP Plant Exergy Efficiency

$$\psi = \frac{\dot{W}_{net} + \dot{m}_{steam}(ex_9 - ex_8)}{\dot{m}_f LHV \times \xi} \tag{10.15}$$

where $\dot{W}_{net}$, $\dot{m}_f$ and ξ are the gas turbine net output power, the mass flow rate of fuel injected into the combustion chamber, and $\xi = 1.033 + 0.0169\frac{y}{x} - \frac{0.0698}{x}$ for gaseous hydrocarbon fuels having a chemical formula C_xH_y respectively.

Total Cost Rate

$$\dot{C}_{tot} = \dot{C}_f + \sum_k \dot{Z}_k + \dot{C}_D + \dot{C}_{env} \tag{10.16}$$

where $c_f = 0.003$ \$/MJ is the regional cost of fuel per unit of energy [1, 6]. Other terms in this expression are in Equation 10.10 and are already defined.

Unit CO_2 Emission

To have a complete optimization for this CHP system, a unit CO_2 emission rate from the combustion chamber is considered as an objective function. With the combustion equation, one can find the CO_2 emission of the plant. A unit CO_2 emission is thus defined as follows:

$$\varepsilon = \frac{\dot{m}_{CO_2}}{\dot{W}_{net}} \tag{10.17}$$

Decision Variables The decision variables (design parameters) in this study are compressor pressure ratio (r_{AC}), compressor isentropic efficiency (η_{AC}), gas turbine isentropic efficiency (η_{GT}), combustion chamber inlet temperature (T_3), and gas turbine inlet temperature (GTIT). A list of constraints the decision variables and the reasons of their application were listed earlier (see Table 10.2).

10.2.2.3 Optimization Results

Figure 10.11 shows the Pareto frontier solution for the CHP plant used in a paper mill considering the objective functions in Equations 10.11 and 10.12 in multi-objective optimization. It can be observed in this figure that, while the total exergy efficiency of the cycle increases to about 68%, the total cost rate increases very slightly. An increase in the total exergy efficiency from 68% to 69% corresponds to a moderate increase in the product cost rate, while an increase in the exergy efficiency from 69% to higher values leads to a more significant increase of the total cost rate.

It is shown in Figure 10.11 that the maximum exergy efficiency exists at design point C (69.89%), while the total cost rate is the greatest at this point. Conversely, the minimum value for total cost rate of product occurs at design point A. Design point C is the optimal situation if exergy efficiency is a single objective function, while design point A is the optimum condition if total cost rate of product is a single objective function. In multi-objective optimization, a process of decision-making for selection of the final optimal solution from the available solutions is required. The process of decision-making is usually performed with the aid of a hypothetical point, called the ideal point in Figure 10.11, at which both objectives have their optimal values independent of the other objectives. It is clear that it is impossible to have both objectives at their optimum point simultaneously and, as shown in Figure 10.11, the ideal point is not a solution located on the Pareto frontier. The closest point of the Pareto frontier to the ideal point might be considered as a desirable final solution. However, in this case, the Pareto optimum frontier exhibits weak equilibrium, that is, a small change in exergy efficiency due to the expected variation of operating parameters causes a large

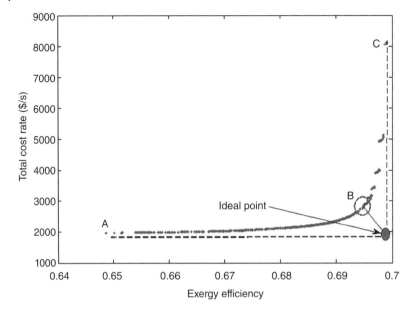

Figure 10.11 Pareto frontier showing best trade-off values for the objective functions for the gas turbine CHP system.

variation in the cost rate of product. Therefore, the equilibrium point cannot be utilized for decision-making in this problem. In selection of the final optimum point, it could be desired to achieve a better magnitude for each objective than its initial value of the base case problem. Because of this, as the optimized points in the B–C region have a maximum exergy efficiency increment of about 1% and a minimum total cost rate increment of 5000 relative to the design C, this region was eliminated from the Pareto curve, leaving only the region of A–B as shown in Figure 10.12.

Since each point on the Pareto solution from multi-objective optimization can be the optimized point, the selection of the optimum solution is dependent on the preferences and criteria of most interest to each decision maker. Hence, each decision maker may select a different point as the optimum solution. To provide a helpful tool for the optimal design of the gas turbine cycle, the following equation was derived for the Pareto optimal points curve (Figure 10.11):

$$\dot{C}_{total} = \frac{-3267\psi^2 + 4247\psi - 1371}{\psi^3 - 9573\psi^2 + 2549\psi - 825.8} \times 1000 \qquad (10.18)$$

This equation is valid in the range of $0.64 < \psi < 0.70$.

In this part of the multi-objective optimization, another two objective functions (total cost rate and unit CO_2 emissions) are considered in the optimization study. The result of this multi-objective optimization is shown in Figure 10.13. It can be seen that if we aim to reduce the CO_2 emissions of the cycle, which are mainly associated with thermodynamic characteristics of the cycle components (e.g., compressor and gas turbine isentropic efficiencies), the purchase cost of each item of equipment in the cycle should be selected to be as high as possible to increase the efficiency and output power. Therefore, the total cost rate increases leading to high efficiency components. But, it is clear that, by selecting the best component as well as using a low fuel mass flow rate in the

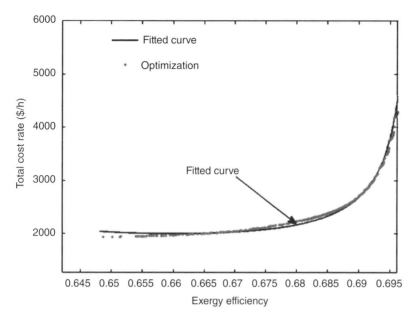

Figure 10.12 Selecting the optimal solution from Pareto Frontier.

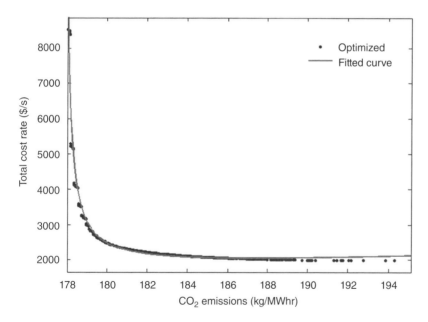

Figure 10.13 Pareto Frontier showing best trade-off values for the objective functions (CO_2 emissions and total cost rate).

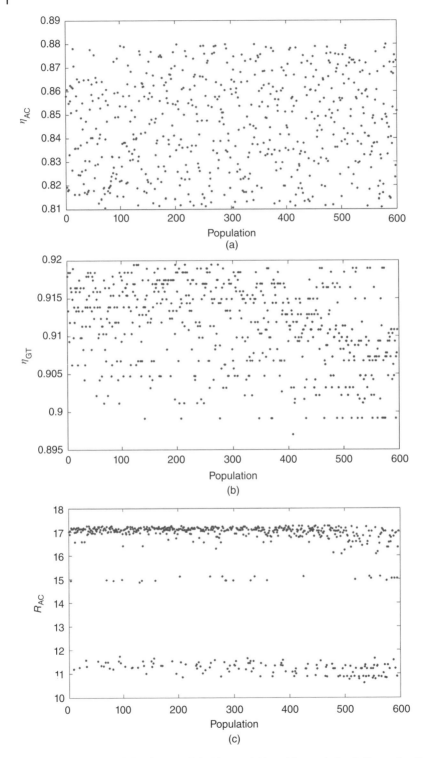

Figure 10.14 Scattered distribution of decision variables with population in Pareto frontier in Figure 10.11: (a) compressor isentropic efficiency, (b) gas turbine isentropic efficiency, (c) compressor pressure ratio, (d) T_3, (e) GTIT.

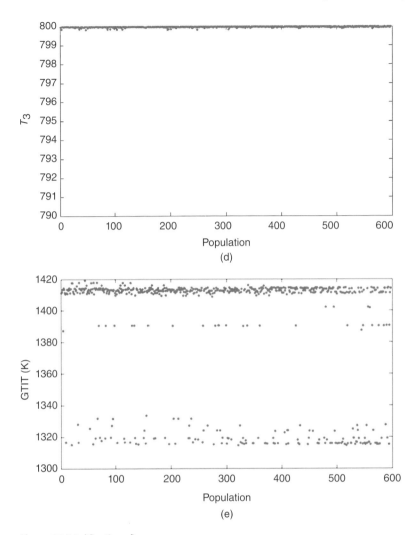

Figure 10.14 (*Continued*)

combustion chamber, the environmental impacts decrease. Hence, to provide the trend of this curve, the equation is fitted to all the points obtained by multi-objective optimization, yielding the following:

$$\dot{C}_{\text{total}} = \frac{0.02\varepsilon^4 - 5.4\psi\varepsilon^3 - 331.4\varepsilon^2 + 6.4\varepsilon + 0.83}{\varepsilon^3 - 178\varepsilon^2 - 3.36\varepsilon + 0.09} \times 1000 \qquad (10.19)$$

where ε is the CO_2 emissions per net output power ($\text{kg}_{CO2}/\text{MWh}$).

As we know, relevant design parameters need to be selected for each optimization problem, and this has been done in previous sections and chapters. However, the scattering distribution of these design parameters is important. In general, it can be found that each design parameter tends to its upper bound, its lower bound, or the range between these two bounds. The distribution of each of the design parameters considered in the present optimization is shown in Figure 10.14. It is seen there that η_{GT} and T_3 reach the upper bound, meaning that an increase in these parameters leads

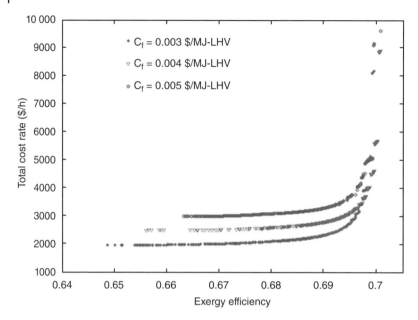

Figure 10.15 Sensitivity of Pareto optimum solution for exergy efficiency and cost rate to specific fuel cost, when $i = 11\%$.

to better optimization results. However, due to physical limitations it is not allowable to extend the region. But other design parameters have the scattered distributions in the ranges between the bounds. This confirms a reasonable optimization problem with proper decision variables.

To improve understanding of the insights attained through an optimization study, a sensitivity analysis is helpful. Therefore, a sensitivity analysis of the Pareto optimum solution is performed for the fuel specific cost, the interest rate, and the pinch point temperature. Figure 10.15 shows the sensitivity of the Pareto optimal frontier to variations of the specific fuel cost for a fixed interest rate. This figure shows that the Pareto frontier shifts upward as the specific fuel cost increases. At a lower exergetic efficiency, for which the weight of thermoenvironomic objective is higher, the sensitivity of the optimal solutions to fuel cost is much greater than is the case for the location of the Pareto frontier with the lower weight of the thermoenvironomic objective. In fact the exergetic objective does not have a significant effect on the sensitivity to economic parameters such as the fuel cost and interest rate. Moreover, at higher exergy efficiency, the purchase cost of equipment in the plant is increased so the cost rate of the plant increases too. Further, for a constant exergy efficiency, increasing the fuel cost leads to a rise in the total cost rate of product due to the fact that the fuel price plays a significant role in this objective function.

Figure 10.16 presents the sensitivity analysis of the Pareto optimum solution to the CO_2 emission and total cost rate due to changes in the fuel cost rate. It is observed that, to have a cycle that produces less CO_2, one may select components that have higher thermodynamic characteristics like isentropic efficiency. Such a selection leads to an

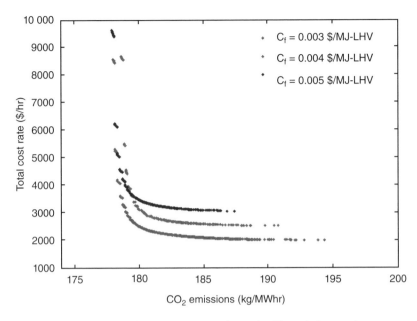

Figure 10.16 Sensitivity of Pareto optimum solution for CO_2 emissions and cost rate to the specific fuel cost, when $i = 11\%$.

increase the purchase cost of the equipment, however. Nonetheless, by increasing the fuel cost, the total cost rate of the product is increased because of the importance of fuel cost in the objective function.

Similar behavior is observed when examining the sensitivity of the Pareto optimal solution to variations in interest rate (see Figures 10.17 and 10.18). The final optimal solution that was selected in this case belongs to the region of Pareto frontier with significant sensitivity to the cost parameter. However, the region with lower sensitivity to the cost parameter is not reasonable for the final optimum solution due to the weak equilibrium of the Pareto frontier. This implies that a small change in plant exergy efficiency due to variations of operating parameters risks increasing the product cost rate drastically. Figure 10.19 shows the sensitivity of the Pareto optimum solution to the pinch point (PP) temperature in the HRSG of the CHP plant. It is found that, to keep total cost rate constant, an increase in the pinch point temperature leads to a decrease in the total exergy efficiency of the plant. This is observed because an increase in the pinch point temperature results in a lower energy recovery from the exhaust gas of the gas turbine. However, a decrease in PP temperature results in a decrease in the HRSG exergy destruction rate as well as an increase in its exergy efficiency.

Figure 10.20 shows the variation of the Pareto curve with changes in the pinch point temperature. It is seen that varying the PP temperature does not affect significantly the CO_2 emission of the plant. Thus, the CO_2 emission of the CHP plant is a direct function of the fuel injected into the combustion chamber and any parameters which alter the mass flow rate. In addition, the PP temperature is a design parameter of the bottoming cycle which has no significant effect on the power production of the system. This is why for CHP systems we need to be able to conduct a proper sensitivity analysis.

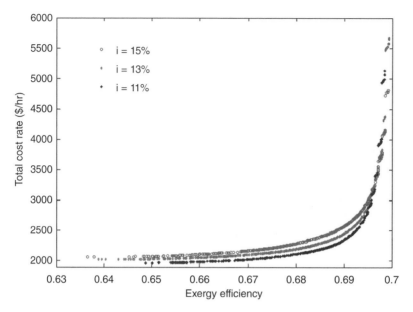

Figure 10.17 Sensitivity of Pareto optimum solution for exergy efficiency and total cost rate to interest rate, when $C_f = 0.003$ $/MJ (based on LHV).

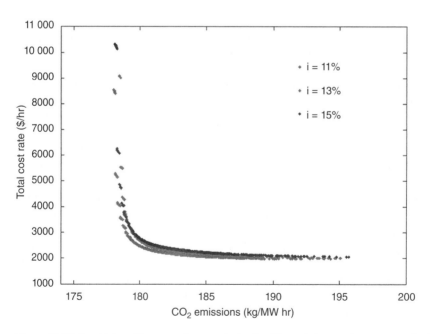

Figure 10.18 Sensitivity of Pareto optimum solution for CO_2 emissions and cost rate to the interest rate, when $C_f = 0.003$ $/MJ (based on LHV).

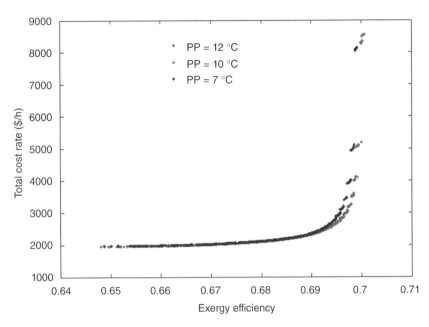

Figure 10.19 Sensitivity of Pareto optimum solution for exergy efficiency and cost rate to pinch point temperature, when $C_f = 0.003$ \$/MJ (based on LHV) and $i = 0.11$.

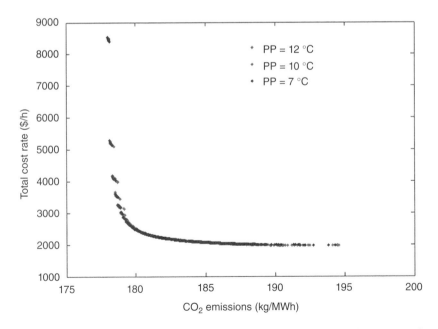

Figure 10.20 Sensitivity of Pareto optimum solution for CO_2 emissions and cost rate to the pinch point temperature when $C_f = 0.003$ \$/MJ (based on LHV) and $i = 0.11$.

10.3   Internal Combustion Engine (ICE) Cogeneration Systems

Another type of CHP system is based on the internal combustion engine (ICE), with the energy of the exhaust gases recovered to produce useful outputs.

A schematic of an ICE-based CHP system is shown in Figure 10.21 and the corresponding *T-s* diagram in Figure 10.22. This system consists of three major subsystems: an internal combustion engine, an organic Rankine cycle (ORC) and an ejector refrigeration cycle (ERC). The ORC and ERC are proposed as bottoming cycles to recover the waste heat of the ICE exhaust gas. In other words, the high temperature exhaust gases drive the bottoming cycles. A portion of the condensed flow from the condenser enters the ORC cycle at point 1 for power production. It then enters a pump, which increases the pressure to the desired operating pressure of the ORC. The flow then enters the heat recovery vapor generator (HRVG) at point 2. Exhaust gases from the engine enter the HRVG at point *a* to produce high temperature vapor at point 3, which rotates the expander to produce electrical power. Part of the stream extracted from expander is the primary flow for the ejector (point 4) and the rest is expanded through an expander to condenser pressure (point 5). The outlet flows of the ejector (point 6) and expander are mixed in the mixing chamber and then enter the condenser at point 7. Part of the condensed flow enters the refrigeration cycle (point 8). An expansion valve is used to reduce the pressure for cooling purposes at point 9.

The evaporator absorbs the heat of the building and the flow from it enters the ejector as the secondary flow, as a saturated vapor (point 10). The ejector operates based on the interaction of two different flows with different energy levels. The higher energy stream (stream 4) and the lower energy stream (stream 10) are called the mover, or primary, flow and secondary flow, respectively. The interaction of these two fluids during the process in the ejector leads to a pressure rise without using moving components. Note that the remainder of the refrigeration cycle is the same as a conventional compression refrigeration cycle.

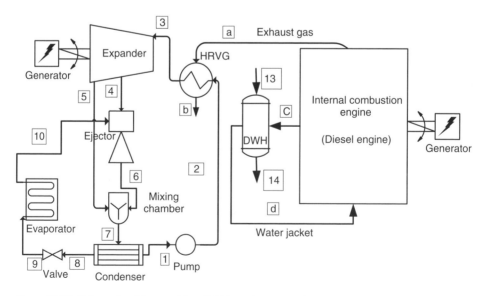

Figure 10.21 Schematic of the considered ICE-based cogeneration system.

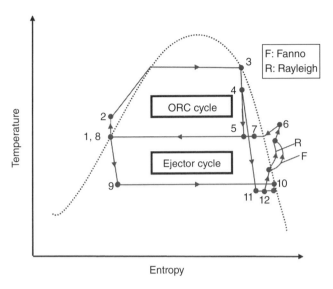

Figure 10.22 Temperature-entropy diagram of the ICE-based cogeneration system in Figure 10.21.

As shown in Figure 10.23, the motive flow is expanded through a converging–diverging nozzle and exits as a supersonic stream (point 11) while the pressure decreases significantly. This pressure reduction causes a suction condition at the inlet of the mixing chamber, causing the outlet flow of the evaporator to enter the ejector. Then, the primary and secondary flows are mixed in the mixing chamber at point 12. The mixed flow, which is still supersonic, enters a constant area duct where the flow becomes subsonic via a normal shock. The properties of the flow after the normal shock can be determined by finding the intersection of the Fanno and Rayleigh lines (see Figure 10.22). Finally, the flow enters a diffuser and its pressure increases to the condenser pressure (point 6). To provide the required direct hot water for the residential application, a heat exchanger absorbs the waste heat through the water jacket. Cold water enters the heat exchanger (point 13) at 20°C and exits at the desired temperature of 60°C (point 14).

10.3.1 Selection of Working Fluids

Working fluids play a significant role for low-grade heat recovery and refrigeration systems. Various working fluids exist, and they are usually classified into three categories according to the slope of the saturation vapor line in the T-s diagram. The three categories of working fluids are a dry fluid which has a positive slope, a wet fluid which has a negative slope, and an isentropic fluid which has an infinitely large slope. Lior et $al.$ [16] derive an expression to calculate the slope of the saturation vapor curve on a T-s diagram (dT/ds). Defining $\xi = ds/dT$, they find:

$$\xi = \frac{C_p}{T_{\mathrm{H}}} - \frac{((n.T_{\mathrm{H}})/(1 - T_{\mathrm{rH}})) + 1}{T_{\mathrm{H}}^2} \Delta H_{\mathrm{H}} \tag{10.20}$$

where n is suggested to be 0.375 or 0.38 [17], T_{rH} represents $T_{\mathrm{H}}/T_{\mathrm{C}}$, and ΔH_{H} denotes the enthalpy of evaporation. With this nomenclature, the working fluids can be classified in the three mentioned types: $\xi > 0$ for a dry fluid, $\xi \approx 0$ for an isentropic fluid, and $\xi < 0$ for a wet fluid. According to the application, different working fluids may be suitable for waste heat recovery or other applications.

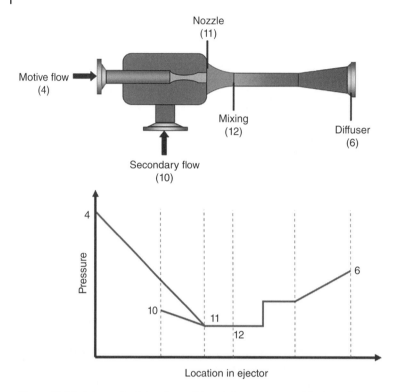

Figure 10.23 Schematic of the ejector [8] and the pressure profile through it.

Selecting a proper fluid is important and is governed by numerous factors. A working fluid should be selected to suit the thermodynamic conditions and environmental standards, and should have good availability and be inexpensive. Several factors are important from a thermodynamic perspective, suggesting the working fluid should exhibit a high critical temperature and pressure, a high vapor density, and an appropriate *T-s* curve in order to prevent formation of liquid drops in expander outlet. A working fluid should also be environmentally benign, including having a low ozone depletion potential and a low global warming potential. Non-flammability and non-toxicity are also considered desirable selection criteria for a working fluid.

In the analyses described here, R-134a, R-123, R-11, and R600 are adopted as the system working fluid. These can be categorized in isentropic (R-11), dry (R-123 and R-600), and wet (R-134a), according to the Equation 10.16. Table 10.6 lists characteristics of some widely used fluids. Also, the *T-s* diagram of the considered working fluids is presented in Figure 10.24.

10.3.2 Thermodynamic Modeling and Analysis

The considered system is composed of three major subsections: an ICE, an ORC, and an ERC. The ICE is treated as the prime mover to produce electricity via a generator, while the ORC and ERC recover the ICE wasted heat to produce additional electricity and cooling, respectively. The modeling and thermodynamic analysis of the system and its components are described in this section.

Table 10.6 Physical and thermodynamic properties of the considered working fluids [18, 19].

Parameter	Fluid		
	R-134a	R-600	R-11
Chemical formula	$C_3H_3F_5$	C_4H_{10}	CCl_3F
Type	Wet	Dry	Isentropic
Critical temperature (°C)	101.06	152	197.96
Critical pressure (kPa)	4059	3796	4408
Ozone depletion potential (−)	0.00	0	1
Global warming potential (−)	1430	4	4750
Flammability	No	No	No

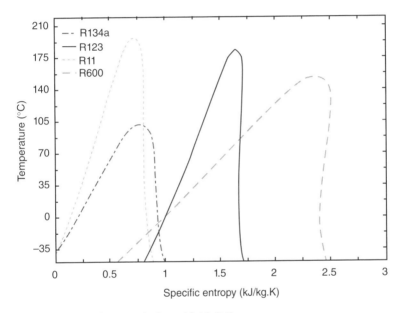

Figure 10.24 *T-s* diagram of selected fluids [19].

10.3.2.1 Internal Combustion Engine

In a typical ICE, one third of the fuel chemical energy is converted into useful electrical power and the remainder is rejected as waste heat. Cogeneration is of interest because it offers a way to benefit from the waste heat by producing useful outputs such as electricity, heating, and cooling. In a diesel engine, a commonly used ICE, the thermal efficiency and recoverable heat rate are a function of load, and are sensitive to part load operation, which represents the fraction of the nominal load. An increase in the operating load raises both the electrical power and the recoverable heat rate. As a result, the optimum part load value should be determined in the optimization process; consequently the engine specifications as a function of load are needed. There are several empirical methods to estimate the electrical power generation, the recoverable heat and the fuel mass flow rate. Here, we use the relation provided in reference [20].

The nominal diesel fuel consumption can be expressed as:

$$\dot{m}_{\text{f,nom}} = \frac{\dot{W}_{\text{Diesel}}}{\eta_{\text{Diesel,nom}} LHV} \tag{10.21}$$

The fuel mass flow rate at the part load operating condition can be calculated as follows [15]:

$$\frac{\dot{m}_{\text{f,PL}}}{\dot{m}_{\text{f,nom}}} = -0.02836 \exp(0.03254(PL)) + 0.2556 \exp(0.01912(PL)) \tag{10.22}$$

where PL is the partial load of the engine. The part load power output can be estimated as:

$$\frac{\dot{W}_{\text{Diesel,PL}}}{\dot{m}_{\text{f,PL}} LHV} = 1.07 \exp(-0.0005736(PL)) - 1.259 \exp(-0.05367(PL))\eta_{\text{Diesel,nom}} \tag{10.23}$$

The waste heat associated with the exhaust gas and water jacket can be calculated respectively as follows:

$$\frac{\dot{Q}_{\text{exh,PL}}}{\dot{m}_{\text{f,PL}} LHV} = 0.001016(PL)^2 - 0.1423(PL) + 31.72 \tag{10.24}$$

$$\frac{\dot{Q}_{\text{wj,PL}}}{\dot{m}_{\text{f,PL}} LHV} = 24.01 \exp(-0.0248(PL)) + 15.35 \exp(0.002822(PL)) \tag{10.25}$$

10.3.2.2 Organic Rankine Cycle

The waste heat of the exhaust gas can be recovered by an ORC system, which is modeled and assessed thermodynamically here by writing an energy balance equation for a control volume around each of its components:

Pump

$$\dot{W}_{\text{p}} = \dot{m}_{\text{p}}(h_{2s} - h_1)/\eta_{\text{p}} \tag{10.26}$$

where η_{p} denotes the isentropic efficiency of the pump.

HRVG

$$\dot{Q}_{\text{exh}} = \dot{m}_3(h_3 - h_{2,\text{ac}}) \tag{10.27}$$

Expander

$$\dot{W}_{\text{expd}} = [\dot{m}_3 h_3 - \dot{m}_4 h_{4,s} - (\dot{m}_3 - \dot{m}_4)h_{5,s}]\eta_{\text{expd}} \tag{10.28}$$

where η_{expd} denotes the isentropic efficiency of the expander.

Condenser

$$\dot{Q}_{\text{cond}} = \dot{m}_7(h_7 - h_1) \tag{10.29}$$

10.3.2.3 Ejector Refrigeration Cycle (ERC)

The ERC consists of an ejector, an evaporator, a valve, and a condenser. The components are modeled as follows:

Ejector Several assumptions are made here to render the analysis more tractable:

- The velocity at the inlet and outlet of the ejector are assumed to be negligible.
- Friction and mixing losses in the nozzle, mixing chamber, and diffuser are taken into account by means of each component's efficiency.
- The mixing chamber operates at constant pressure.
- The ejector does not exchange heat with the surroundings.

Considering the velocity at the ejector inlet as negligible, the outlet velocity of the nozzle can be expressed as:

$$u_{11} = \sqrt{(2\eta_n(h_4 - h_{11})}$$

(10.30)

The entertainment ratio is defined as:

$$\mu = \frac{\dot{m}_{sf}}{\dot{m}_{mf}}$$

(10.31)

The conservation of the mass balance in the mixing chamber leads to:

$$\dot{m}_{mf}u_{11} + \dot{m}_{sf}u_{10} = (\dot{m}_{mf} + \dot{m}_{sf})u_{12s}$$

(10.32)

Neglecting the secondary flow velocity (u_{10}), the isentropic velocity of the mixing outlet flow can be calculated as follows:

$$u_{12s} = \frac{u_{11}}{1 + \mu}$$

(10.33)

The mixing efficiency can be defined as:

$$\eta_m = \frac{u_{12}^2}{u_{12s}^2}$$

(10.34)

The outlet velocity from the mixing section flow can be determined as:

$$u_{12} = u_{11}\frac{\sqrt{\eta_m}}{1 + \mu}$$

(10.35)

and the outlet specific enthalpy of the mixing section can be expressed as:

$$h_{12} = \frac{h_4 + \mu h_{10}}{1 + \mu} - \frac{u_{12}^2}{2}$$

(10.36)

Kinetic energy is converted to pressure in the diffuser. Neglecting the outlet velocity of the diffuser, the outlet specific enthalpy can be evaluated as follows:

$$h_6 = h_{12} + \frac{h_{6s} - h_{12}}{\eta_d}$$

(10.37)

where h_{6s} represents the isentropic specific enthalpy of the diffuser outlet flow.

The entertainment ratio μ can be derived as follows:

$$\mu = \sqrt{\eta_n\eta_m\eta_d\frac{(h_4 - h_{11s})}{(h_{6s} - h_{12})} - 1}$$

(10.38)

As shown in Figure 10.25, the entertainment ratio calculations in the ejector involve a trial-and-error process. First, the entertainment ratio is guessed and, after the

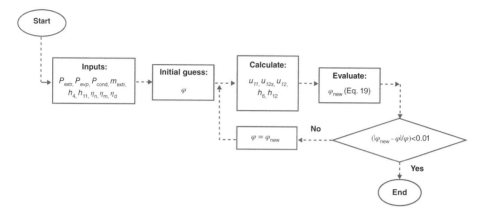

Figure 10.25 Flowchart of the ejector calculations.

estimation of some parameters, it is calculated using Equation 10.34. If the relative error exceeds the stopping criterion, the previous value of the entertainment ratio is automatically replaced with a new one and the procedure is repeated until the stopping criterion is satisfied.

Valve The process occurring in the valve is assumed isenthalpic. Hence,

$$h_8 = h_9 \tag{10.39}$$

Evaporator The outlet flow of the evaporator is assumed to be a saturated vapor. The heat rejected from the building can be calculated as follows:

$$\dot{Q}_{eva} = \dot{m}_9(h_{10} - h_9) \tag{10.40}$$

10.3.3 Exergy Analysis

The exergy analysis concepts and relations have already been detailed in previous chapters. Here we develop expressions for the overall exergy efficiency of the CHP system and the exergy destruction rates for each component in the system. The exergy destruction for this CHP system occurs in various components (heat recovery devices, expander, condenser, pump, etc.). Expressions for the component exergy destruction rates for the entire system are listed in Table 10.7.

Exergy efficiency is a significant indicator that frequently provides a finer understanding than energy efficiency. The exergy efficiency indicates how well the system performs compared with the ideal one. The exergy efficiency is generally defined as the exergy of the desired output(s) divided by the exergy of the primary input(s). The exergy efficiency of the CHP system considered here can be expressed as follows:

$$\Psi_{CCHP} = \frac{\dot{W}_{Diesel,PL} + \dot{W}_{expd} - \dot{W}_p + \dot{Ex}_{DHW} + \dot{Ex}_{cooling}}{\dot{Ex}_f} \tag{10.41}$$

10.3.4 Optimization

Different objective functions can be considered according to our needs. In this optimization study, the main objective functions considered are the exergy efficiency of the

Table 10.7 Expressions for component exergy destruction rates for of the ICE cogeneration system.

Component	Exergy destruction rate
Heat recovery vapor generator	$\dot{Ex}_{D,hrvg} = \dot{Ex}_2 - \dot{Ex}_3 - \dot{Ex}_{Q,hrvg}$
Expander	$\dot{Ex}_{D,expd} = \dot{Ex}_3 - \dot{Ex}_4 - \dot{Ex}_5 + \dot{W}_{expd}$
Condenser	$\dot{Ex}_{D,cond} = \dot{Ex}_7 - \dot{Ex}_8 - \dot{Ex}_1 - \dot{Ex}_{Q,cond}$
Pump	$\dot{Ex}_{D,pmp} = \dot{Ex}_1 - \dot{Ex}_2 + \dot{W}_{pmp}$
Evaporator	$\dot{Ex}_{D,eva} = \dot{Ex}_9 - \dot{Ex}_{10} - \dot{Ex}_{Q,eva}$
Expansion valve	$\dot{Ex}_{D,ev} = \dot{Ex}_8 - \dot{Ex}_9.$
Ejector	$\dot{Ex}_{D,ej} = \dot{Ex}_4 - \dot{Ex}_{10} - \dot{Ex}_6$
Mixing chamber	$\dot{Ex}_{D,mch} = \dot{Ex}_6 + \dot{Ex}_5 - \dot{Ex}_7$
Domestic water heater	$\dot{Ex}_{D,DWH} = \dot{Ex}_{13} - \dot{Ex}_{14} - \dot{Ex}_{Q,DHW}$
Diesel engine	$\dot{Ex}_{D,Diesel} = \dot{Ex}_f - \dot{Ex}_{Q,hrvg} - \dot{Ex}_{W,Diesel} - \dot{Ex}_{Q,DWH}$

CHP system, which is to be maximized, and the total cost rate of the system, which is to be minimized, while satisfying several reasonable constraints. These objective functions are a function of design parameters of the system and can be expressed as follows:

$$\Psi_{CHP} = \frac{\dot{W}_{net} + \dot{Ex}_{DHW} + \dot{Ex}_{cooling}}{\dot{Ex}_f} \tag{10.42}$$

$$\dot{C}_{total} = \sum_i \dot{Z}_i + \dot{C}_f + \dot{C}_{env} \tag{10.43}$$

Here, $\dot{Z}_i$ is the annual investment cost rate of the ith component of the system, which includes the cost rate of annualized capital investment and the annual operating and maintenance cost. The total capital investment cost includes two parts: the direct capital cost, which is the equipment purchase cost in this study, and the indirect capital cost. The purchase cost of each component should be defined as a function of the system design parameters. For this CHP system the purchase cost of each component is listed in Table 10.8.

The total capital investment of each component is taken to be 6.32 times of the purchase equipment cost of the same component [23]. Also, the cost of operation and

Table 10.8 Purchase cost of components [21, 22].

Component	Equipment purchase cost (EPC)
Expander	$\log_{10}(EPC_{expd}\,(\$)) = 2.6259 + 1.4398 \log(\dot{W}_{expd}) - 0.1776\,[\,\log_{10}(\dot{W}_{expd})]^2$
Heat exchanger	$\log_{10}(EPC_{he}(\$)) = 4.6656 - 0.1557 \log_{10}(A) + 0.1547\,[\log_{10}(A)]^2$
Pump	$\log_{10}(EPC_{pmp}\,(\$)) = 3.3892 + 0.0536 \log_{10}(\dot{W}_{pmp}) + 0.1538[\log_{10}(\dot{W}_{pmp})]^2$
Ejector	$EPC_{ej}(\$) = 1000 \times 16.14 \times 0.989 \times \left(m_4 \times \left(\frac{T_4}{P_4 \times 0.001} \right)^{0.05} \right) \times (P_6^{-0.75})$
Diesel	$EPC_D(\$) = \left(\frac{-138.71}{2} \times \ln(\dot{W}_{PL,diesel}) + 1727.1 \right) \times \dot{W}_{PL,Diesel}$

maintenance is almost 20% of the component purchase cost. In Equation 10.43, $\dot{C}_f$ and $\dot{C}_{env}$ represent the cost rate of fuel consumed by the ICE and the cost rate of environmental impact, respectively, which can be defined as:

$$\dot{C}_f = \psi_f \dot{m}_f \tag{10.44}$$

$$\dot{C}_{env} = \psi_{em} \dot{m}_{CO_2} \tag{10.45}$$

Here ψ_f and ψ_{em} are unit fuel cost and unit pollutant emission cost, respectively, and are given in reference [3].

10.3.4.1 Decision Variables

For this optimization problem, different decision variables are selected for different subsystems (i.e., ICE, ORC, and ERC). The decision variables and their intervals for various working fluids are listed in Table 10.9.

A multi-objective genetic algorithm is selected as the optimization method here for the determination of the optimal design parameters. The input parameters used in the multi-objective optimization are provided in Table 10.10.

10.3.4.2 Multi-objective optimization

For this CHP system, multi-objective optimization is applied considering both exergy efficiency and total cost rate as the objective functions, in order to determine the optimal design parameters of the CCHP system for a residential application. Note that the exergy efficiency is to be maximized, while the total cost rate is to be minimized, subject to satisfying reasonable constraints. In order to determine the optimal

Table 10.9 Decision variables and their range of variation for the optimization.

Decision variable	Interval			Reason
	R-134a	R-600	R-11	
$\dot{W}_{Diesel}$ (kW)		[50 to 200]		Commercial availability
PL (%)		[50 to 100]		Commercial availability
φ (−)		[0.1 to 1]		Commercial availability
F_{extr} (−)		[0.5 to 1]		Commercial availability
P_{hrvg} (kPa)	[2800 to 3975]	[2000 to 3114]	[500 to 1775]	Pinch point limitation
P_{cond} (kPa)	[600 to 770]	[208 to 284]	[80 to 125]	Temperature limitation
P_{eva} (kPa)	[180 to 290]	[56 to 103]	[20 − 40]	Temperature limitation

Table 10.10 Multi-objective genetic algorithm input parameters.

Parameter	Value
Population size	50
Crossover fraction	0.8
Mutation fraction	0.03
Objective function tolerance	0.0001

design parameters, seven decision variables with the highest influence of the system performance are selected, whose ranges were already given (Table 10.9). These design parameters are as follows: diesel engine nominal power ($\dot{W}_D$), engine part load fraction (*PL*), expander inlet pressure (P_{hrvg}), pressure of extraction fraction (F_{extr}), extraction ratio (φ), condenser pressure (P_{cond}), and evaporator pressure (P_{eva}). A genetic algorithm is applied as the optimization algorithm for multi-objective optimization. For this optimization it is assumed that the diesel nominal efficiency, $\eta_{D,nom}$, is 35%. The fuel lower heating value (LHV) and its chemical exergy factor are equal to 47.82 MJ/kg and 1.03, respectively [3]. The isentropic efficiency of the pump and the expander of the ORC are assumed to be 75% and 82%, respectively. As mentioned in the ERC analysis sub-section, the ejector has three subsystems, namely a converging–diverging nozzle, a mixing chamber, and a diffuser. For each, the efficiency is defined and its value is assumed to be 90%, 85%, and 85%, respectively [23]. In the economic analysis, some input economic data are provided. The interest rate, i_{eff}, is fixed at 11%. The nominal escalation rate, r_n, is assumed to be 5%. The economic life, n, and annual operating hours, τ, are considered to be 20 years and 7000 hours, respectively. The fuel cost, ψ_{fuel}, and pollutant emission cost, ψ_{em}, are assumed to be 0.168 and 0.02086 \$/kg, respectively [22]. In the DWH subsystem, water enters at 20°C and the exit temperature is set at 60°C, respectively.

The Pareto frontier for the three working fluids considered is presented in Figure 10.26. As mentioned, all of the points in the Pareto frontier have the possibility of being selected as the final optimal points. However, in some studies, there are some criteria that are invoked for the final point selection. In this study, the selection is based on equilibrium point theory, which aims at both objective functions taking on their "best" values, called the ideal or best point. However, it is usually not advantageous to attain this point in practice, which is why in multi-objective optimization the closest point to the ideal points are usually claimed to yield superior results compared to other points on the Pareto front. As shown for R134a in Figure 10.26, the ideal point is not located on the Pareto curve. Consequently, the final decision should be a trade-off point rather than the best point. As shown in Figure 10.26, the trade-off point is defined as the closest point of the Pareto frontier to the ideal point [8]. The trade-off points for all three cases are shown in this figure as points *A* to *C*. As is evident, R-11 is the most beneficial working fluid in terms of minimum total cost rate and maximum exergy efficiency, with values of 3.19 \$/hr for total cost rate and 66.85% efficiency, respectively. In addition, the lowest exergy efficiency, which is equal to 62.18%, occurs with R-134a as working fluid. In this instance, the total cost rate attains its highest value, 3.46 \$/hr. A brief inspection of Figure 10.26 allows one to conclude that for each working fluid there is a trade-off between total cost rate and exergy efficiency. Or, more specifically, the greater the exergy efficiency, the higher is the total cost rate. But, relative to different working fluids, it is determined that in this CHP system the fluid with the higher critical temperature leads to better performance from both exergy efficiency and total cost rate points of view. Therefore, as shown in Figure 10.26, R-11 with the highest critical temperature attains the best Pareto frontire, with the best total cost rate and exergy efficiency. Next come R600 and R134a.

Figure 10.27 presents an energy balance of the bottoming cycle, including the ORC, ERC, and DWH units. The input energy to the system should equal the output energy. As shown, calculation errors do not exceed 1.5%. Figures 10.28 to 10.34 present the variations of various decision variables linked to the optimum points of the Pareto frontiers

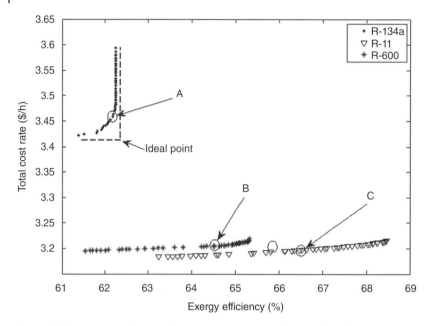

Figure 10.26 Pareto frontier showing best trade-off values for the objective functions.

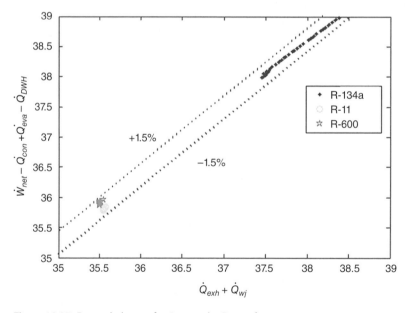

Figure 10.27 Energy balance of points on the Pareto front.

for all working fluids considered. Figure 10.28 illustrates the variation of diesel engine capacity with respect to population. As is shown for all of the working fluids, to achieve an optimum condition, the diesel engine capacity should be at its minimum value of 50 kW. Figure 10.29 shows the variation of diesel engine part load fraction for optimum points of the Pareto frontier. From Figure 10.29 one can find that, for all working fluids

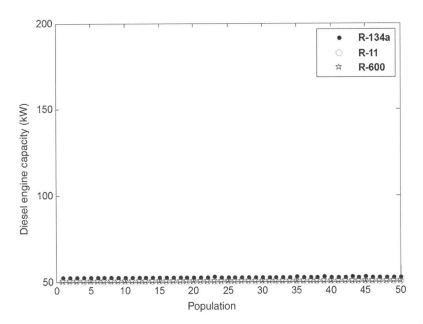

Figure 10.28 Scattered distribution of diesel engine capacity with population in Pareto frontier.

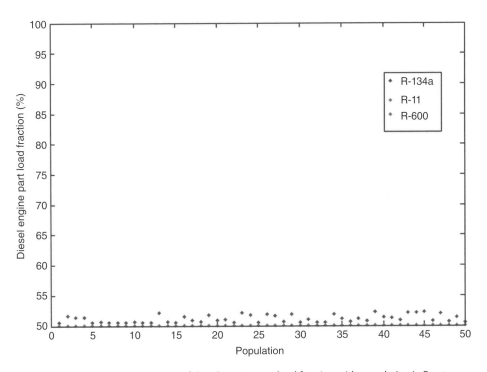

Figure 10.29 Scattered distribution of diesel engine part load fraction with population in Pareto frontier.

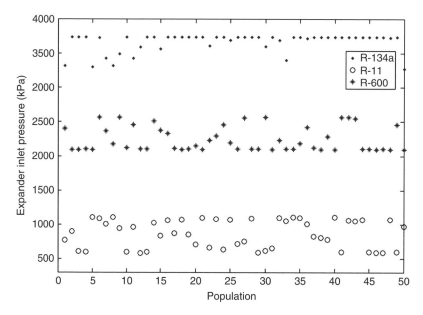

Figure 10.30 Scattered distribution of expander inlet pressure with population in Pareto frontier.

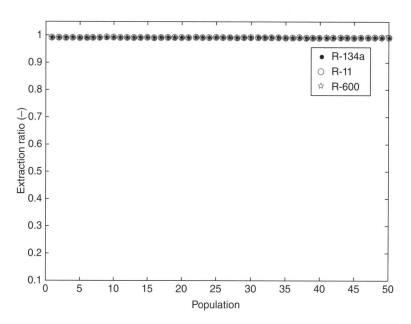

Figure 10.31 Scattered distribution of extraction ratio with population in Pareto frontier.

and all points of the Pareto frontier, the diesel engine partial load takes on its minimum value of 50%. This means that in order to receive a better benefit from the exergy efficiency and total cost rate, it is necessary to install ICEs with lower capacities at half load. Figure 10.30 shows the variation of expander inlet pressure for the optimum points on the Pareto frontier. Figure 10.31 shows the variation of the motive flow pressure fraction for various optimum points of the Pareto frontier. As depicted, for all working fluids and optimum points, this quantity takes on a maximum value of 1, which means that an increase in the pressure of the motive flow is beneficial from both exergy efficiency and total cost rate points of view. Figure 10.32 shows the variation of the extraction ratio for several points of the Pareto frontier.

As is shown here, for all working fluids the maximum value in the range is selected. In other words, to achieve an optimum operating condition with maximum exergy efficiency, minimum total cost rate or any other combination of the two, all of the wasted heat with the exhaust gas should be used in the ERC for cooling production rather than in the ORC for electrical power production. Figure 10.33 shows the variation of condenser pressure for several optimum points of the Pareto frontier. As is evident, all considered working fluids attain their minimum available values for the ranges in which a higher pressure condenser leads to lower ORC electrical power production and ERC cooling capacity. Figure 10.34 illustrates the variation of evaporator pressure for several points on Pareto frontier. As depicted in this figure, for all working fluids the evaporator pressure takes on its maximum available value since a higher evaporator pressure leads to a higher cooling capacity and exergy efficiency.

The optimal values of the design parameters for desirable points A to C are listed in Table 10.11. Moreover, information about these points such as net output power, cooling capacity, exergy efficiency, total cost rate, sustainability index, and cost ratio, are presented in Table 10.12. The cost ratio is defined as the total cost rate per unit of produced

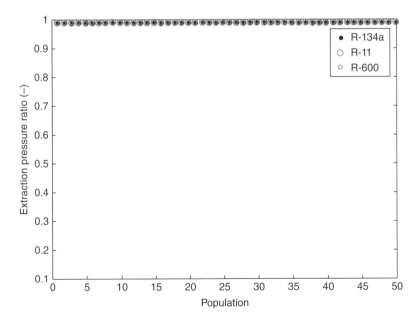

Figure 10.32 Scattered distribution of extraction pressure ratio with population in Pareto frontier.

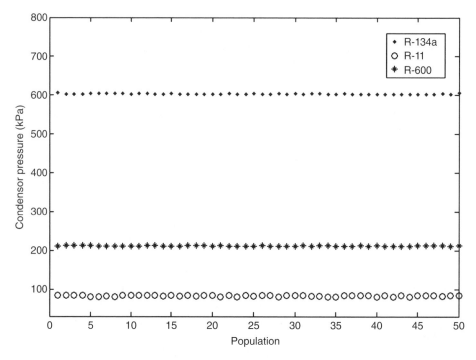

Figure 10.33 Scattered distribution of condenser pressure with population in Pareto frontier.

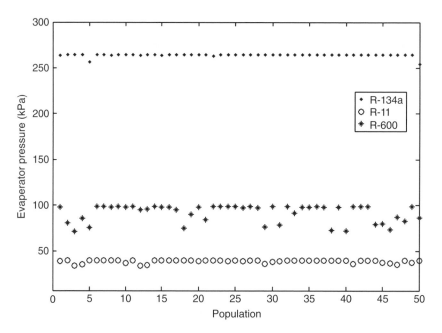

Figure 10.34 Scattered distribution of evaporator pressure with population in Pareto frontier.

Table 10.11 Decision variables of optimum points A to C.

Point	$\dot{W}_{Diesel}$ (kW)	PL (%)	P_{hrvg} (kPa)	F_{extr} (—)	φ (—)	P_{cond} (kPa)	P_{eva} (kPa)
A	52.48	50.59	3692.1	0.99	0.99	603.85	264.13
B	50.00	50.02	2106.90	0.99	0.99	212.13	97.14
C	50.20	50.00	1128.70	0.99	0.99	71.85	29.93

Table 10.12 Parameters of optimum points A to D.

Point	$\dot{W}_{net}$ (kW)	$\dot{Q}_{cool}$ (kW)	$\dot{m}_{DWH}$ (kg/s)	Ψ_{CHP} (%)	Total cost rate ($/h)	SI (—)	Cost ratio ($/kWh)
A	26.06	5.05	0.11	62.18	3.45	1.53	0.0.132
B	24.72	6.39	0.10	64.50	3.21	1.49	0.129
C	24.89	7.35	0.10	65.83	3.20	1.46	0.128

electrical power. Since the quality of "cold energy," "heat energy," and electrical power are different, the sum of heat, cold, and electrical energy is not the output capacity of the system. Thus, two factors, α and β, are introduced to define the price difference of cold, heat and electrical power and presented as: $\dot{W}_{net} + \alpha\dot{Q}_{heat} + \beta\dot{Q}_{cool}$. In this study, the values of α and β are taken to be 0.5 and 0.8, respectively [24]. As is evident from Table 10.12, although the highest value of net electrical power output is exhibited at point A, both the maximum exergy efficiency and minimum total cost rate are exhibited at point C. It is beneficial to consider such factors as SI and cost ratio, which are quantitative measures of the environmental impact of the system and the total cost rate per total produced electrical power, respectively. The best (highest) and worst SI values result from operating the system with the conditions at points A and C, respectively. Also, the most beneficial operating conditions from a cost ratio perspective results from operating at point C. The final decision on which operating point to use depends on many factors including costs, available space, and system efficiency. In this study, point C is selected based on its exergy efficiency, total cost rate, and cost ratio, as the best overall performance of the system for residential application is observed at this point.

To enhance understanding of the optimization of the CHP system, the effect of different decision variables on the objective functions is examined. For this analysis, each fluid is considered at its optimum point and the effect of varying a single decision variable is observed. Figure 10.35 shows the effect of varying diesel engine capacity on both objective functions. It is seen that an increase in this decision variable results in a significant increase in the total cost rate, while the exergy efficiency remains constant. This observation implies that the increments of the supplied and useful exergy rates are equal. This trend is consistent with the results in Figure 10.28; where all the Pareto points are considered, the diesel engine capacity takes on the minimum available value (50 kW). Figure 10.36 shows the impact of varying engine part load fraction on both objective functions, demonstrating that an increase in this parameter leads to a higher total cost rate and lower exergy efficiency. Therefore, the lower the diesel engine

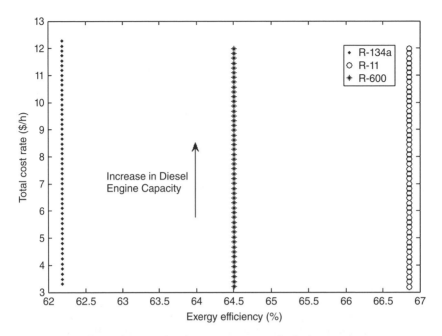

Figure 10.35 Effects of varying diesel engine capacity on both objective functions.

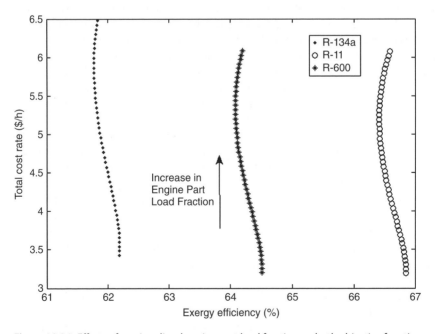

Figure 10.36 Effects of varying diesel engine part load fraction on both objective functions.

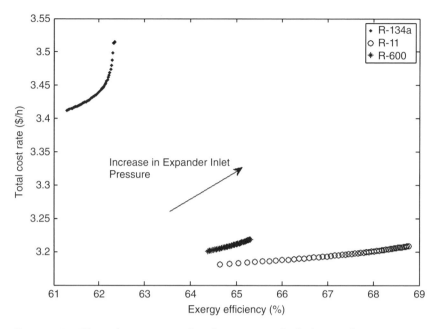

Figure 10.37 Effects of varying expander inlet pressure on both objective functions.

part load fraction, the better are both the objective functions. Figure 10.37 shows the trade-off between total cost rate and exergy efficiency when expander inlet pressure increases. The trend in Figure 10.37 provides useful insight into the scattered behavior of Figure 10.30. Specifically, the optimum value of the total cost rate occurs for a lower expander inlet pressure, while the optimum value of the exergy efficiency occurs at a higher expander inlet pressure.

The effect of the expander extraction ratio on the objective functions is shown in Figure 10.38, where an increase in expander extraction ratio is seen to improve both objective functions. This explains why all optimum points in the scattered distribution in Figure 10.31 occur for the maximum value of this variable. A similar trend is observed for extraction ratio in Figure 10.39. Figure 10.40 shows the effect of condenser pressure variations on both objective functions. An increase in condenser pressure leads to an almost negligible improvement in the total cost rate of the CHP system, while it significantly decreases the system exergy efficiency. This is why in Figure 10.33, where the scattered distribution of condenser pressure is given, this variable takes on the maximum available value for all optimum points. Figure 10.41 shows the effect varying evaporator pressure on both objective functions, demonstrating that an increase in evaporator pressure increases the exergy efficiency significantly for all cases and also increases the total cost rate. When the evaporator pressure increases, the cooling capacity of the system rises, which in turn leads to an increase in the numerator of the exergy efficiency expression. But the higher evaporator pressure requires more electricity for the pumps, which increases the system cost.

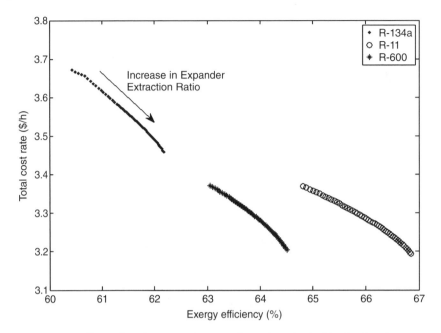

Figure 10.38 Effects of varying expander extraction ratio on both objective functions.

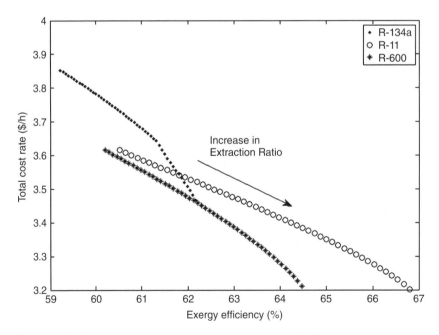

Figure 10.39 Effects of varying extraction ratio on both objective functions.

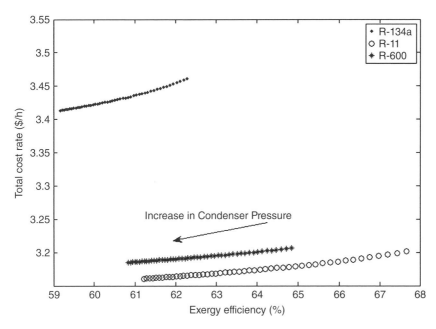

Figure 10.40 Effects of varying condenser pressure on both objective functions.

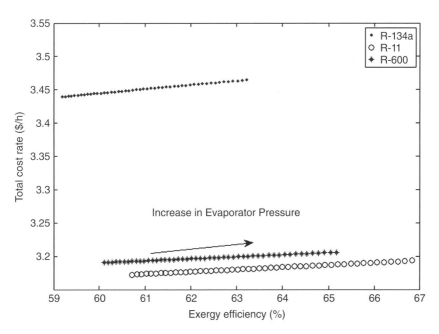

Figure 10.41 Effects of varying evaporator pressure on both objective functions.

10.4   Micro Gas Turbine Trigeneration System

Fossil fuel depletion and global warming are two important concerns for the sustainability of energy systems in the future. Demand for energy has been steadily rising despite limited availability of non-renewable fuel resources. Hence, efforts to develop more efficient energy systems are becoming increasingly significant. The efficiency of conventional power plants, which are usually based on single prime movers, is usually less than 40%. Thus, most of the input energy is lost, much as waste heat. The integration of systems to provide cooling, heating, and hot water with conventional plant can increase the overall plant efficiency to up to 70% [25]. Cogeneration systems represent one of the strategic technologies to increase the efficiency of energy usage. Among the ways of achieving cogeneration for combined heating and power (CHP), the use of micro-turbines is considered an attractive option.

Micro-turbines (MTs) are small size combustion turbines, with powers ranging between 20 and 10 MW [26]. From the environmental point of view, the use of MTs is of special interest nowadays due to the fact that the levels of CO_2 and NO_x emissions of MTs are significantly lower than those of reciprocating engines of similar capacity. In addition, MTs offer fuel flexibility [27]. One application of cogeneration systems is the coupling of MTs with absorption systems, both for single and double-effect units. The residual heat of the MT is used to drive the refrigeration system. The term "trigeneration" is applied in such an instance since an additional benefit, cooling, is obtained. A trigeneration system is shown in Figure 10.42, and is composed of a compressor, a combustion chamber, a gas turbine, a dual pressure heat recovery steam generator, a single effect absorption chiller, and a domestic water heater.

10.4.1   Thermodynamic Modeling

Relevant energy balances and governing equations for the main sections of the trigeneration plant shown in Figure 10.42 are described, and broken down into the following subsections: topping cycle (Brayton cycle), bottoming cycle (steam turbine and heat recovery steam generator), single effect absorption chiller, and domestic water heater.

10.4.1.1   Topping Cycle (Brayton Cycle)

We model a gas turbine cycle using the first law of thermodynamics. As seen in Figure 10.42, air at ambient conditions enters the air compressor at point 1 and, after compression (point 2), is supplied to the combustion chamber (CC). Fuel is injected in the combustion chamber and hot combustion gases exit (point 3) and pass through a gas turbine to produce power. The hot gas expands in the gas turbine to point 4. This flue gas is utilized in the dual pressure heat recovery steam generator (HRSG) to generate low pressure (LP) and high pressure (HP) steam. Energy balances and governing equations for the gas turbine cycle are given elsewhere [1, 28].

10.4.1.2   Bottoming Cycle

Energy balances and governing equations for the components of the bottoming cycle (steam turbine cycle and HRSG) are provided here.

Dual Pressure HRSG   A dual pressure HRSG with two economizers (LP and HP) and two evaporators (LP and HP) is used in the trigeneration cycle to provide both low- and high-pressure steam. The LP steam is used to drive the absorption chiller and the

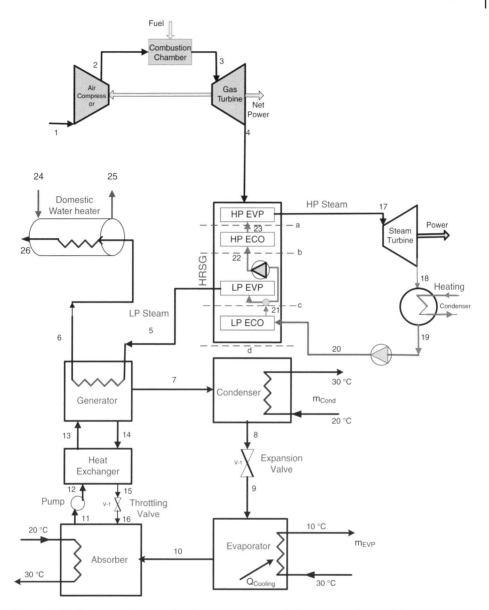

Figure 10.42 Schematic diagram of a trigeneration system for heating, cooling, and electricity generation.

HP steam to generate electricity. The temperature profile in the HRSG is shown in Figure 10.43, where the pinch point is defined as the difference between the temperature of the gas at the entrance of the evaporator (economizer side) and the saturation temperature. The dual pressure HRSG has two pinch points (PP_{HP} and PP_{LP}).

The temperature differences between the water leaving the economizers (T_{20} and T_{22}) and the saturation temperature (T_5 and T_{17}) are the approach points (AP_{HP} and AP_{LP}), which depend on the economizer's tube layout. Note that the pinch point and approach temperatures are considered constant here.

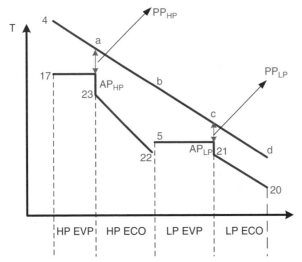

Figure 10.43 Temperature profile of HRSG.

Point number	Remarks
4	Hot gases entering HRSG
a	Hot gases exiting high pressure evaporator (HP EVP)
b	Hot gases exiting high pressure economizer (HP ECO)
c	Hot gases exiting low pressure evaporator
d	Hot gases exiting HRSG
20	Cold water entering HRSG
21	Hot water exiting low pressure economizer (LP ECO)
5	Saturated water exiting low pressure evaporator
22	Hot water entering high pressure economizer (HP ECO)
23	Hot water exiting high pressure economizer (HP ECO)
17	Saturated water exiting high pressure evaporator

Energy balances for each element of the HRSG are expressed as follows:

$$\dot{m}_{w,HP}(h_{17} - h_{23}) = \dot{m}_g C_{Pg}(T_4 - T_a) \tag{10.46}$$

$$\dot{m}_w(h_{23} - h_{22}) = \dot{m}_g C_{Pg}(T_a - T_b) \tag{10.47}$$

$$\dot{m}_{w,LP}(h_5 - h_{21}) = \dot{m}_g C_{Pg}(T_b - T_c) \tag{10.48}$$

$$\dot{m}_w(h_{21} - h_{20}) = \dot{m}_g C_{Pg}(T_c - T_d) \tag{10.49}$$

Steam Turbine An energy balance for the steam turbine in Figure 10.42 and its isentropic efficiency are written as follows:

$$\dot{m}_w h_{17} = \dot{W}_{ST} - \dot{m}_w h_{18} \tag{10.50}$$

$$\eta_{ST} = \frac{\dot{W}_{ST,act}}{\dot{W}_{ST,is}} \tag{10.51}$$

Condenser An energy balance for the condenser follows:

$$\dot{m}_{18} h_{18} = \dot{Q}_{cond} - \dot{m}_{19} h_{19} \tag{10.52}$$

Pump An energy balance for the pump and its isentropic efficiency can be expressed as follows:

$$\dot{m}_w h_{19} + \dot{W}_{pump} = \dot{m}_w h_{20} \tag{10.53}$$

$$\eta_{pump} = \frac{w_{is}}{w_{act}} \tag{10.54}$$

10.4.1.3 Absorption Chiller

The principle of mass conservation and the first and second laws of thermodynamics are applied to each component of the single effect absorption chiller. Each component is considered as a control volume with inlet and outlet streams, and heat transfer and work interactions are considered. Mass balances are applied for the total mass and each material of the working fluid solution. The governing and conservation equations for the total mass and each material of the solution for a steady state and steady flow system are expressed as follows [29]:

$$\sum \dot{m}_i = \sum \dot{m}_o \tag{10.55}$$

$$\sum (\dot{m}x)_i = \sum (\dot{m}x)_o \tag{10.56}$$

where $\dot{m}$ is the working fluid mass flow rate and x is mass concentration of LiBr in the solution. For each component of the absorption system, a general energy balance is written:

$$\dot{Q} - \dot{W} = \sum \dot{m}_o h_o - \sum \dot{m}_i h_i \tag{10.57}$$

The cooling load of the absorption chiller is defined as:

$$\dot{Q}_{cooling} = \dot{m}(h_{10} - h_9) \tag{10.58}$$

Further information about the thermodynamic modeling and energy balances for each component is given in [29].

10.4.1.4 Domestic Water Heater

The hot gases from the heat recovery heat exchanger enter the water heater to warm domestic hot water to 60°C. Water enters this heater at a pressure and temperature of 3 bar and 15°C, respectively. The energy balance for this component can be written as follows:

$$\dot{m}_{w,LP}(h_6 - h_{26}) = \dot{m}_w(h_{25} - h_{24}) \tag{10.59}$$

10.4.2 Exergy Analysis

The exergy of each flow in the plant is calculated and exergy destructions are determined for each major component. The source of exergy destruction (or irreversibility) in the combustion chamber is mainly chemical reaction and thermal losses. However, the exergy destruction in the heat exchanger of the system, that is condenser and HRSG, is due to the large temperature difference between the hot and cold fluids. The exergy destruction rate for each component of this trigeneration energy system is shown in Table 10.13.

Table 10.13 Expressions for exergy destruction rates for components of the trigeneration system.

Component	Exergy destruction rate expressions
Compressor	$\dot{E}x_1 + \dot{W}_{AC} = \dot{E}x_2 + \dot{E}x_D$
Combustion chamber (CC)	$\dot{E}x_2 + \dot{E}x_f = \dot{E}x_3 + \dot{E}x_D$
Gas turbine (GT)	$\dot{E}x_3 = \dot{W}_{GT} + \dot{E}x_4 + \dot{E}x_D$
HRSG	$\dot{E}x_4 + \dot{E}x_{20} = \dot{E}x_5 + \dot{E}x_{17} + \dot{E}x_d + \dot{E}x_D$
Steam turbine (ST)	$\dot{E}x_{17} = \dot{W}_{ST} + \dot{E}x_{18} + \dot{E}x_D$
Steam condenser	$\dot{E}x_{18} = \dot{E}x_{19} + \dot{E}x_Q + \dot{E}x_D$
Absorption generator	$\dot{E}x_5 + \dot{E}x_{13} = \dot{E}x_6 + \dot{E}x_{14} + \dot{E}x_7 + \dot{E}x_D$
Absorption heat exchanger	$\dot{E}x_{12} + \dot{E}x_{14} = \dot{E}x_{13} + \dot{E}x_{15} + \dot{E}x_D$
Absorber	$\dot{E}x_{10} + \dot{E}x_{16} = \dot{E}x_{11} + \dot{E}x^Q + \dot{E}x_D$
Expansion valve	$\dot{E}x_8 = \dot{E}x_9 + \dot{E}x_D$
Condenser	$\dot{E}x_7 = \dot{E}x_8 + \dot{E}x_Q + \dot{E}x_D$
Pump	$\dot{E}x_{11} + \dot{W}_{Pump} = \dot{E}x_{12} + \dot{E}x_D$
Domestic water heater	$\dot{E}x_6 + \dot{E}x_{24} = \dot{E}x_{25} + \dot{E}x_{26} + \dot{E}x_D$

The exergy efficiency, defined as the product exergy output divided by the exergy input, can be expressed for the overall trigeneration system as follows:

$$\psi_{tri} = \frac{\dot{W}_{net,GT} + \dot{W}_{net,ST} + \dot{E}x_{heating} + \dot{E}x_{cooling} + \dot{E}x_{hot\ water}}{\dot{E}x_f} \tag{10.60}$$

where $\dot{E}x_f$ is the fuel exergy flow rate, $\dot{E}x_Q$ is the exergy rate associated with heating and cooling, and $\dot{W}_{net,GT}$ and $\dot{W}_{net,ST}$ are the net output work rates of the gas turbine (GT) and steam turbine (ST) cycles.

10.4.3 Optimization

A multi-objective optimization method based on an evolutionary algorithm is applied to the trigeneration system for heating, cooling, electricity, and hot water to determine the best design parameters for the system. Objective functions, design parameters and constraints, and the overall optimization process are described in this section.

10.4.3.1 Definition of Objectives

Two objective functions are considered here for multi-objective optimization: exergy efficiency (to be maximized) and total cost rate of product (to be minimized). The cost of pollution damage is assumed to be added directly to the expenditures that must be paid, making the second objective function the sum of thermodynamic and environmental objectives. Consequently, the objective functions in this analysis can be expressed as follows:

Exergy Efficiency

$$\psi = \frac{\dot{W}_{net,GT} + \dot{W}_{net,ST} + \dot{E}x_{heating} + \dot{E}x_{cooling} + \dot{E}x_{hot\ water}}{\dot{E}x_f} \tag{10.61}$$

Total Cost Rate

$$\dot{C}_{\text{tot}} = \dot{C}_D + \dot{C}_f + \dot{C}_{\text{env}} + \sum_k \dot{Z}_k \tag{10.62}$$

where $\dot{Z}_K$ is the purchase cost of component k and $\dot{C}_D$ is the cost of exergy destruction, which is explained well elsewhere [6]. The cost rates of environmental impact and fuel can be expressed as:

$$\dot{C}_{\text{env}} = C_{CO}\dot{m}_{CO} + C_{NOx}\dot{m}_{NOx} \tag{10.63}$$

Here, $\dot{m}_{CO}$ and $\dot{m}_{Nox}$ can be expressed as follows:

$$m_{NO_x} = \frac{0.15 \times 10^{16} \tau^{0.5} \exp\left(-\frac{71\,100}{T_{PZ}}\right)}{P_3^{0.05}\left(\frac{\Delta P_3}{P_3}\right)^{0.5}} \tag{10.64}$$

$$m_{CO} = \frac{0.179 \times 10^9 \ \tau^{0.5} \exp\left(\frac{7800}{T_{PZ}}\right)}{P_3^2 \tau \left(\frac{\Delta P_3}{P_3}\right)^{0.5}} \tag{10.65}$$

10.4.3.2 Decision Variables

The following decision variables (design parameters) are selected for this study: compressor pressure ratio (R_{AC}), compressor isentropic efficiency (η_{AC}), gas turbine isentropic efficiency (η_{GT}), gas turbine inlet temperature (*GTIT*), high pressure pinch point temperature (PP_{HP}) difference, low pressure pinch point temperature (PP_{LP}) difference, high pressure steam pressure (P_{HP}), low pressure steam pressure (P_{LP}), steam turbine isentropic efficiency (η_{ST}), pump isentropic efficiency (η_p), and evaporator temperature (T_{EVP}). Although the decision variables may be varied during optimization, they are normally required to be within reasonable ranges; these constraints are listed in Table 10.14, based on earlier reports [4, 20].

Table 10.14 Optimization constraints and their rationales.

Constraint	Reason
$GTIT < 1550$ K	Material temperature limit
$P_2/P_1 < 22$	Commercial availability
$\eta_{AC} < 0.9$	Commercial availability
$\eta_{GT} < 0.9$	Commercial availability
$P_{HP} < 40$ bar	Commercial availability
$P_{LP} < 5.5$ bar	Commercial availability
$10\,°C < PP_{HP} < 22°C$	Heat transfer limit
$12\,°C < PP_{LP} < 22°C$	Heat transfer limit
$\eta_{ST} < 0.9$	Commercial availability
$\eta_p < 0.9$	Commercial availability
$2°C < T_{EVP} < 6°C$	Cooling load limitation

10.4.3.3 Evolutionary Algorithm: Genetic Algorithm

Genetic algorithms apply an iterative, stochastic search strategy to find an optimal solution and imitate in a simplified manner principles of biological evolution [30]. A characteristic of an evolutionary algorithm is a population of individuals, where an individual consists of the values of the decision variables (structural and process variables here) and is a potential solution to the optimization problem [31]. More details about genetic algorithms and their procedures are given elsewhere [31, 32].

10.4.4 Optimization Results

The results of the optimization are described. Figure 10.44 shows the Pareto frontier solution for the trigeneration system for the objective functions in Equations 10.61 and 10.62 after multi-objective optimization. It can be seen in this figure that the total cost rate of products increases moderately as the total exergy efficiency of the cycle increases to about 56%. Increasing the total exergy efficiency from 56% to 58% increases the cost rate of product significantly. It is shown in Figure 10.44 that the maximum exergy efficiency exists at design point *A* (57.64%), while the total cost rate of products is the greatest at this point (1394 $/hr). On the other hand, the minimum value for the total cost rate of product occurs at design point *E*, which is about 1365.7 $/hr. Design point *A* is the optimal situation when efficiency is the sole objective function, while design point *E* is the optimum point when total cost rate of product is the sole objective function.

In multi-objective optimization, a process of decision-making for selection of the final optimal solution from the available solutions is required. The process of decision-making is usually performed with the aid of a hypothetical point in Figure 10.44 (the equilibrium point), at which both objectives have their optimal

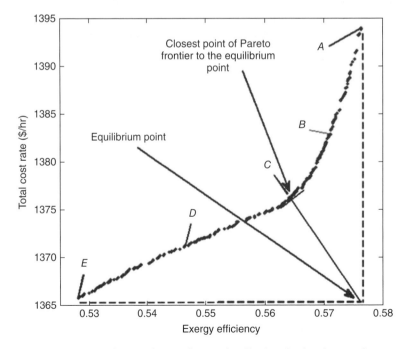

Figure 10.44 Pareto frontier showing best trade-off values for the objective functions.

Table 10.15 Optimized values for design parameters of the system based on multi-objective optimization.

Design parameter	A	B	C	D	E
η_{AC}	0.88	0.88	0.88	0.88	0.88
η_{GT}	0.9	0.9	0.9	0.9	0.9
R_{AC}	19.977	19.999	19.996	19.999	20
$GTIT$ (K)	1499.6	1499.5	1499.7	1499.7	1499.8
P_{LP} (bar)	5.6796	5.6405	2.02	2.043	2.9621
P_{HP} (bar)	39.949	39.955	39.924	24.719	15.021
PP_{HP} (°C)	8.096	16.233	19.939	19.941	19.897
PP_{LP} (°C)	24.778	24.65	24.981	24.83	24.971
T_{EVP} (°C)	4.1983	4.293	4.3565	5.9145	7.3016
η_{FWP}	0.77333	0.78133	0.738	0.79267	0.858
η_{ST}	0.88945	0.87902	0.76592	0.75	0.75769

values independent of the other objectives. It is clear that it is impossible to have both objectives at their optimum point simultaneously and, as shown in Figure 10.44, the equilibrium point is not a solution located on the Pareto frontier. The closest point of the Pareto frontier to the equilibrium point might be considered as a desirable final solution. Nevertheless, in this case, the Pareto optimum frontier exhibits weak equilibrium, that is, a small change in exergy efficiency from varying the operating parameters causes a large variation in the total cost rate of product. Therefore, the equilibrium point cannot be utilized for decision-making in this problem. In selection of the final optimum point, it is desired to achieve a better magnitude for each objective than its initial value for the base case problem. Because of this the optimized points in the $C–E$ region have the maximum exergy efficiency increment of about 1% and minimum total cost rate increment of about 20 $/hr relative to the design point, C. Therefore, design point C can be a good candidate for the multi-objective optimization. Note that in multi-objective optimization and the Pareto solution, each point can be utilized as the optimized point. Therefore, the selection of the optimum solution depends on the preferences and criteria of the decision maker, suggesting that each may select a different point as for the optimum solution depending on his/her needs. Table 10.15 shows all the design parameters for points $A–E$.

As shown in Figure 10.44, the optimized values for exergy efficiency on the Pareto frontier range between 53% and 58%. To provide a good relation between exergy efficiency and total cost rate, a curve is fitted on the optimized points obtained from the evolutionary algorithm. This fitted curve is shown in Figure 10.45 and can be expressed as follows:

$$\dot{C}_{total} = \frac{-17.42\psi^5 - 23.75\psi^4 - 19.24\psi^3 + 5.2\psi^2 + 37.6\psi - 15.94}{\psi^5 - 3.5\psi^4 + 4.65\psi^3 - 3.01\psi^2 + 0.97\psi - 0.12} \tag{10.66}$$

To see the variation of thermodynamic characteristics, three different points on the Pareto frontier are considered (A, C, and E). Table 10.16 shows the total cost rate of the system, the total exergy destruction rate, the system efficiency, and the sustainability index. From point A to point E in this table, both total cost rate of the system and exergy

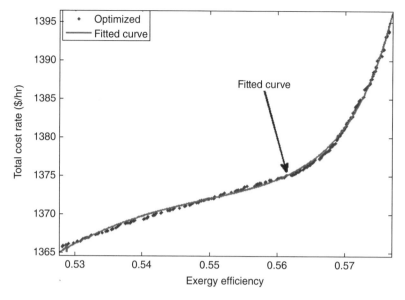

Figure 10.45 Fitted curve of the optimal values in the Pareto frontier.

Table 10.16 Thermodynamic characteristics of three different points on the Pareto frontier.

Point	$\dot{Ex}_{D,tot}$ (MW)	$\dot{C}_{tot}$ ($/h)	Ψ	Z_k ($)	SI	CO_2 emission (kg/MWh)
A	35.4	1393.4	57.58	8.704×10^6	2.19	130.14
C	37.3	1375.5	56.28	7.928×10^6	2.09	134.97
E	38.4	1365.8	52.80	7.470×10^6	2.01	143.75

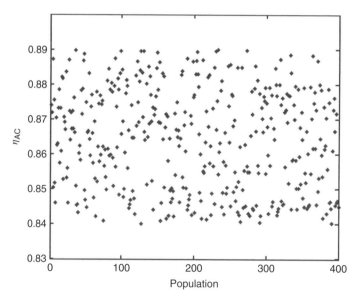

Figure 10.46 Scatter distribution of compressor isentropic efficiency with population in Pareto frontier.

efficiencies decrease. As already stated, point A is preferred when exergy efficiency is a single objective function and design point E when total cost rate is a single objective function. Design point C exhibits better results for both objectives. Other thermodynamic properties correctly confirm this trend. For instance, from point A to E, the total exergy destruction increases when the exergy efficiency decreases.

To better understand the variations of all design parameters, the scattered distributions of the design parameters are shown in Figures 10.46 to 10.56. The results show that gas turbine inlet temperature, compressor pressure ratio, and gas turbine isentropic efficiency tend to become as high as possible. This observation means that an increase in these parameters leads to better optimization results. Due to physical limitations, however, it is not allowed to extend this region. The reason for the compressor pressure ratio behavior is that an increase in this parameter increases the outlet temperature and decreases the mass flow rate injected to the combustion chamber. As the first term in the total cost rate (Equation 10.62) is directly associated with the mass flow rate of the fuel, any decrease in this term results in a decrease in the objective function. This is why in the scattered distribution for the compressor pressure ratio achieves a maximum value in its range. Similar results are seen for the gas turbine isentropic efficiency. In addition, the maximum value for the gas turbine inlet temperature ($GTIT$) is selected based on the evolutionary algorithm. The higher the $GTIT$, the higher the achieved exergy efficiency, since one of the objective functions is supposed to be maximized. It can be concluded that compressor pressure ratio and gas turbine inlet temperature have positive effects on both objective functions since they attain high values.

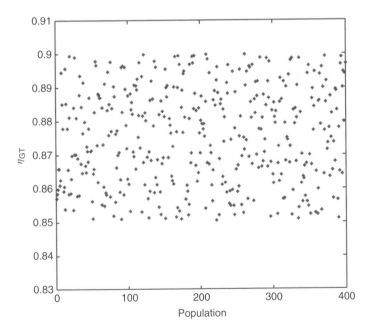

Figure 10.47 Scatter distribution of gas turbine isentropic efficiency with population in Pareto frontier.

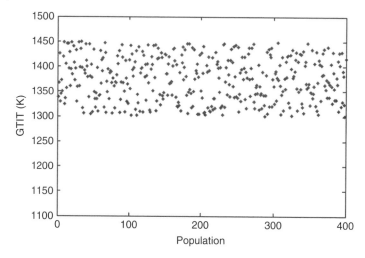

Figure 10.48 Scatter distribution of *GTIT* with population in Pareto frontier.

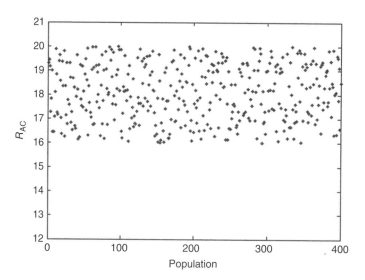

Figure 10.49 Scatter distribution of compressor pressure ratio with population in Pareto frontier.

10.4.5 Sensitivity Analysis

To provide better insight into the study, a sensitivity analysis is conducted to determine the variation of each objective function with some of the main design parameters. The design parameters that have major effects on both objectives are considered: air compressor efficiency, compressor presser ratio, gas turbine efficiency, gas turbine inlet temperature (*GTIT*), high pressure steam pressure (*HP*), low pressure steam pressure (*LP*), high and low pressure pinch point temperatures (PP$_{HP}$, PP$_{LP}$), and steam turbine efficiency. Therefore, the variations of objective functions for five considered points shown in the Pareto curve (Figure 10.44) are shown in Figures 10.57 to Figure 10.65. The variations of other points on the Pareto curve exhibit the same trends as these five points.

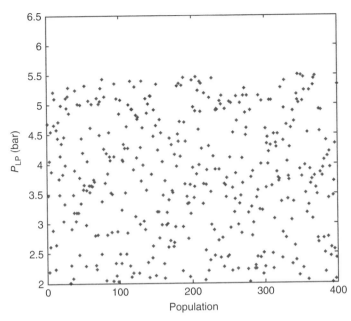

Figure 10.50 Scatter distribution of low pressure steam pressure with population in Pareto frontier.

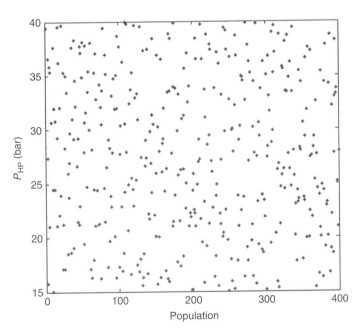

Figure 10.51 Scatter distribution high pressure steam pressure with population in Pareto frontier.

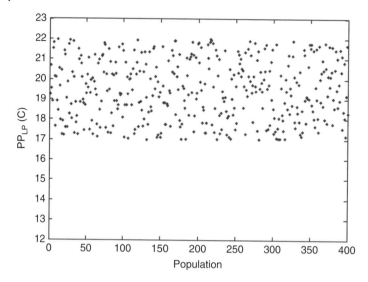

Figure 10.52 Scatter distribution of low pressure pinch point with population in Pareto frontier.

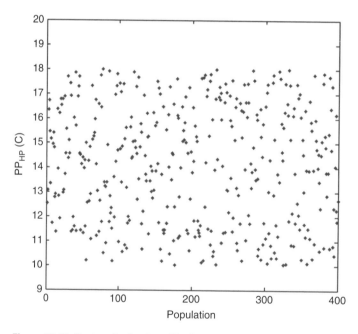

Figure 10.53 Scatter distribution of high pressure pinch point with population in Pareto frontier.

Figure 10.57 shows the variation of compressor pressure ratio for the five selected optimal points on the Pareto curve. As shown in this figure, an increase in compressor pressure ratio has a positive effect on both objective functions. Increasing this design parameter first leads to an increase in exergy efficiency of the cycle, due to the fact that an increase in pressure ratio results in an increase in compressor outlet temperature. In this case, air enters the combustion chamber with higher temperature; hence the total mass

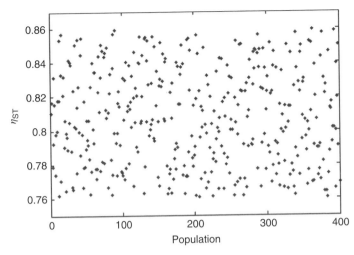

Figure 10.54 Scatter distribution of steam turbine isentropic efficiency with population in Pareto frontier.

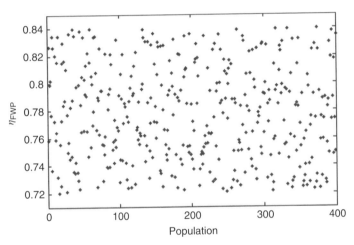

Figure 10.55 Scatter distribution of feed water pump isentropic efficiency with population in Pareto frontier.

flow rate injected into the combustion chamber decreases. According to Equation 10.61, a reduction in combustion chamber fuel mass flow rate leads to an increase in cycle exergy efficiency. On the other hand, a reduction in total fuel mass flow rate has a significant effect on the total cost rate of the cycle. Equation 10.62 shows that a decrease in total fuel mass flow rate leads to a reduction in total cost of fuel which is an important part of this objective function. In addition, this reduction leads to a decrease in greenhouse gas emissions of the combustion chamber. Therefore, this reduction leads to a decrease in cost of environmental impacts. Also, an increase in pressure ratio results in a reduction of the cost of exergy destruction; this is due to a reduction of exergy destruction rate of the combustion chamber attributable to the fact that air passes through the combustion chamber at a higher temperature. Therefore, the temperature difference between the

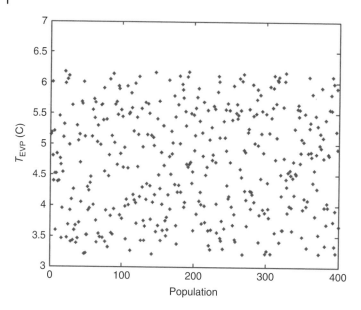

Figure 10.56 Scatter distribution of evaporator temperature with population in Pareto frontier.

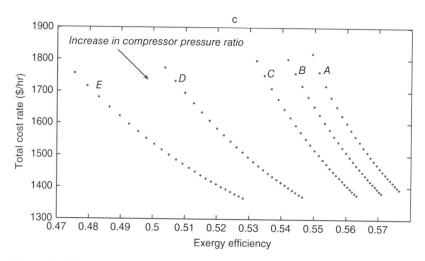

Figure 10.57 Variation with compressor pressure ratio of exergy efficiency and total cost rate.

working fluid and flame temperature decreases, which leads to a reduction in entropy generation. Finally, these three effects result in a significant reduction in total cost of the cycle, as shown in Figure 10.57.

Figure 10.58 shows the variation of both objective functions with compressor efficiency. As shown in this figure, an increase in this efficiency has a positive effect on exergy efficiency and total cost rate. An increase in compressor efficiency results in the better compressor performance and leads to a decrease in irreversibility and an increase in the gas turbine exergy efficiency. But this increase also has an effect on first and last term of the second objective function ($\dot{C}_D$ and $\dot{Z}_{AC}$).

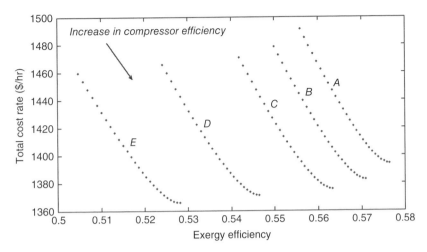

Figure 10.58 Variation with compressor efficiency of exergy efficiency and total cost rate.

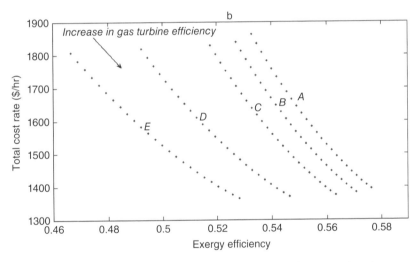

Figure 10.59 Variation with gas turbine efficiency of exergy efficiency and total cost rate.

Since a higher compressor efficiency results in a decrease in exergy destruction, the cost of exergy destruction which is directly associated with exergy destruction decreases. Also, to have higher compressor efficiency, a compressor with higher cost is required. But in this case the first effect is dominant. Therefore, an increase in compressor efficiency leads to a decrease in total cost of the system. The same result is observed for the variation of gas turbine efficiency on both objective functions (see Figure 10.59). An increase in gas turbine efficiency results in an increase in cycle exergy efficiency and a decrease in total cost of the system. Another important parameter which has a significant effect on the performance of the gas turbine cycle is gas turbine inlet temperature (*GTIT*).

The effects of varying this temperature on both objective functions are shown in Figure 10.60. An increase in *GTIT* leads to an increase in gas turbine output power

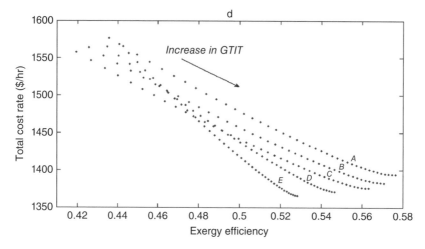

Figure 10.60 Variation with gas turbine inlet temperature of exergy efficiency and total cost rate.

which is a major part of the first objective function. In addition, increasing the *GTIT* leads to an increase in gas turbine outlet temperature. Therefore, flue gases pass through the heat recovery steam generator with higher temperature and the heating load of the system increases. The effect of an increase in *GTIT* on total cost of the system is significantly affected by the first term of Equation 10.62. Since an increase in *GTIT* leads to drastic decrease in exergy destruction, the cost of exergy destruction decreases. Note that although raising this temperature results in an increase in flame temperature which in turn increases the mass flow rate of NO_x according to Equation 10.64, this increase can be negligible compared to the decrease in exergy destruction. Hence, the combination of these two effects leads to a decrease in total cost of the system.

Figure 10.61 shows the effect of boiler high pressure steam on both objective functions at five selected points on the Pareto curve. The effect of raising this parameter

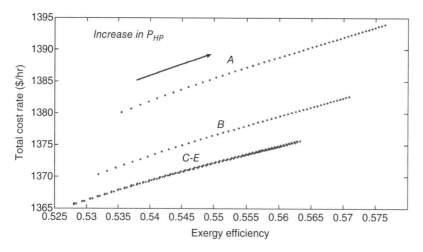

Figure 10.61 Variation with gas turbine inlet temperature of exergy efficiency and total cost rate.

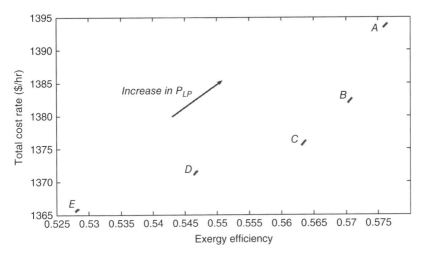

Figure 10.62 Variation with gas turbine inlet temperature of exergy efficiency and total cost rate.

has a positive effect on exergy efficiency of the cycle, mainly due to the fact that an increase in this pressure leads to an increase in the enthalpy of the main steam entering the steam turbine, which directly increases the steam turbine output power. Since the action just changes the steam turbine work, the variation of exergy efficiency is limited within 5%. The effect of varying the high pressure steam pressure on total cost of the plant is also shown in Figure 10.61. It is observed that an increase in this parameter has a negative effect on the cost of the plant. The reason is that increasing this temperature requires a larger pump with higher work and larger pipes in the HRSG, which lead to higher investment costs. Therefore, this is a good parameter to be selected as a design parameter because it will lead to a conflict between two objective functions.

Figure 10.62 shows the effect of varying the low pressure steam on both objective functions. A similar result is obtained when examining the high pressure steam. Note that the variation range of exergy efficiency for this parameter is within 3%.

Another two important design parameters in the HRSG are the high and low pressure pinch point temperatures. The pinch point temperature is defined as the smallest temperature difference between the hot gases and the steam being raised; here, the pinch point occurs at the hot gases exiting the evaporator and the saturated steam at the high or low drum pressure [33]. Figure 10.63 shows the variation of high pressure pinch point (PP_{HP}) on both objective functions. As shown in this figure, an increase in PP_{HP} has a negative effect on exergy efficiency while it has a positive effect on total cost of the plant. When the pinch point temperature increases, the hot gases cannot transfer most of their energy to the working fluid which leads to a decrease in HRSG efficiency and an increase in HRSG exergy destruction. The first effect can affect the quality of the main steam entering the HRSG and lower the main steam temperature. This reduction in both quality and temperature of the main steam reduces the steam turbine output work. Therefore, the exergy efficiency of the cycle decreases. But, since the heat transfer rate of hot flue gases over the evaporators is not changing, an increase in pinch point temperature leads to a decrease in heat transfer surface area which reduces the evaporator cost. Hence, increasing this temperature leads to a decrease in total cost of the plant. Figure 10.64 shows the effect of variation of low pressure pinch point temperature

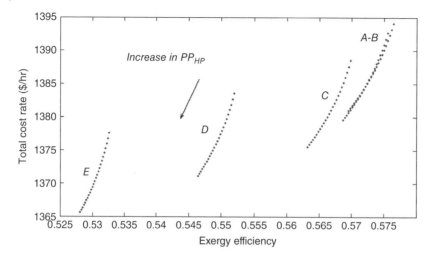

Figure 10.63 Variation with gas turbine inlet temperature of exergy efficiency and total cost rate.

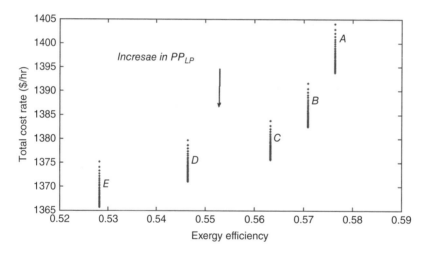

Figure 10.64 Variation with gas turbine inlet temperature of exergy efficiency and total cost rate.

on both objective functions. It is observed that an increase in low pressure pinch point does not have a significant effect on the exergy efficiency, but it can decrease the total cost of the plant due to the reduction of heat transfer surface area for the evaporators. The same trend is observed for the high pressure pinch point, as seen in Figure 10.63.

Figure 10.65 shows the effect of steam turbine isentropic efficiency on both objective functions. An increase in this efficiency has a positive effect on exergy efficiency of the cycle whereas it has a negative effect on total cost rate of the plant. As shown in this figure, an increase in steam turbine isentropic efficiency results in an increase in exergy efficiency. According to Equation 10.51, an increase in this efficiency leads to an increase in steam turbine actual work and a reduction in irreversibility. But increasing the steam turbine isentropic efficiency raises the steam turbine initial cost because a steam turbine with higher isentropic efficiency requires higher cost.

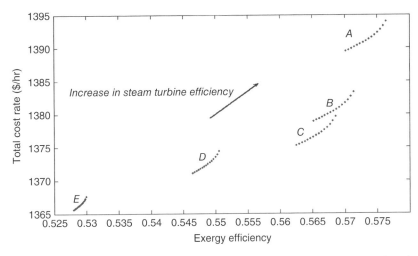

Figure 10.65 Variation with gas turbine inlet temperature of exergy efficiency and total cost.

10.5 Biomass Based Trigeneration System

Renewable energy is derived from natural resources such as sunlight, wind, rain, tides, waves, geothermal heat, and biomass. These are naturally replenished when used. About 16% of global final energy consumption comes from renewable resources, with 10% coming from traditional biomass, mainly for heating, and 3.4% from hydroelectricity [34]. New renewables (small hydro, modern biomass, wind, solar, geothermal, and bio-fuels) account for another 3% and are growing rapidly. Biomass, as a renewable energy source, is biological material from living, or recently living, organisms [35]. Comprehensively, biomass comprises all the living matter present on earth and, as an energy source, can either be used directly, or converted into other energy products such as biofuels [35]. Biomass resources are currently mainly used for heating, cooling, and electricity generation [36]. Direct combustion of biomass with coal is the most common method of conversion and provides the greatest potential for large scale utilization of biomass energy in the near term [37]. Other thermochemical conversion technologies such as gasification and pyrolysis are technically feasible and potentially efficient, compared to combustion, for power generation. However, these technologies either lack maturity and reliability or are not economically viable for large scale utilization [38].

A schematic of a biomass based trigeneration system is shown in Figure 10.66. This trigeneration produces electricity, heating, and fresh water.

The biomass gasifier produces syngas from woody material. The produced syngas is burned in a combustion chamber, which results in an increase in the hot air outlet temperature from the compressor. High temperature gas products enter a ceramic heat exchanger and exchange heat with pressurized air at the exit of compressor. A heat recovery steam generator (HRSG) is used to produce high pressure steam to feed the multi-effect desalination thermal vapor compression (MED-TVC). At the point 8, combustion products enter the organic evaporator at a temperature of 207.8 °C to evaporate the organic fluid to generate additional electrical power.

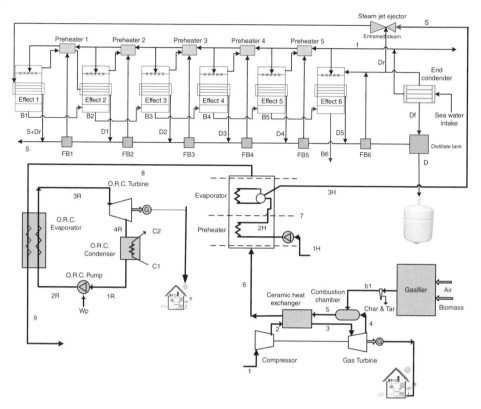

Figure 10.66 Schematic of biomass based trigeneration for electricity, heating, and fresh water production.

10.5.1 Thermodynamic Modeling

For the thermodynamic modeling of the trigeneration system, we focus on the gasifier and multi-effect desalination (MED) parts, since other parts have already been explained in this chapter.

10.5.1.1 Gasifier

To simulate the gasification process taking place in the gasifier, a thermodynamic equilibrium model is used. The general form of chemical reaction in the gas producer is assumed as follows [39]:

$$CH_xO_yN_z + wH_2O + m(O_2 + 3.76N_2) \rightarrow x_1H_2 + x_2CO + x_3H_2O$$
$$+ x_4CO_2 + x_5CH_4 + x_6N_2 \qquad (10.67)$$

Here, $CH_xO_yN_z$ is the biomass chemical formula and w is the amount of water per kmol of biomass. The coefficients x_1 to x_6 can be obtained from atomic balances and using equilibrium constant equations. The atomic balances can be expressed as:

$$x_2 + x_4 + x_5 = 1 \qquad (10.68)$$

$$2x_1 + 2x_3 + 4x_5 = x + 2w \qquad (10.69)$$

$$x_2 + 2x_3 + x_4 = y + w + 2m \qquad (10.70)$$

$$x_2 + x_3 + 2x_4 = y + 2m + w \tag{10.71}$$

The other equations can be obtained from equilibrium reactions. Pyrolysis products before reaching the reduction zone are fired and prior to exiting the gasifier achieve equilibrium state, so the reactions can be written as follows:

$$C + CO_2 \rightarrow 2CO \tag{10.72}$$

$$C + H_2O \rightarrow CO + H_2 \tag{10.73}$$

Zainal *et al.* [40] and Higman and Vander Burgt [41] show that Equations 10.72 and 10.73 can be combined to give the water-gas shift reaction:

$$CO + H_2O \rightarrow CO_2 + H_2 \tag{10.74}$$

Another equilibrium reaction is the methane reaction, which can be expressed as:

$$C + 2H_2 \rightarrow CH_4 \tag{10.75}$$

The equilibrium constants for the water-gas shift reactions respetively can be written as follows [42]:

$$K_1 = \frac{x_1 \times x_4}{x_2 \times x_3} \tag{10.76}$$

$$K_2 = \frac{x_5}{x_1^2} \times n_{total} \tag{10.77}$$

The equilibrium constants can be obtained using the Gibbs function change for each reaction as follows [43]:

$$LnK = -\frac{\Delta G_T^\circ}{\overline{R}T} \tag{10.78}$$

$$\Delta G_T^\circ = \sum_i v_i \Delta \overline{g}_{f,T,i}^\circ \tag{10.79}$$

Here, ΔG_T° is the standard Gibbs free energy of reaction, $\Delta \overline{g}_{f,T,i}^\circ$ is the standard Gibbs function of formation at temperature T for the gas species i, and $\overline{R}$ is the universal gas constant, $8.314 \, kJ/(kmol.K)$. Finally, for the calculation of m the energy equation is applied as:

$$\overline{h}_{f,biomass}^\circ + w\overline{h}_{f,H_2O}^\circ = x_1(\overline{h}_{f,H_2}^\circ + \Delta\overline{h})x_2(\overline{h}_{f,CO}^\circ + \Delta\overline{h}) + x_3(\overline{h}_{f,H_2O}^\circ + \Delta\overline{h})$$
$$+ x_4(\overline{h}_{f,CO_2}^\circ + \Delta\overline{h}) + x_5(\overline{h}_{f,CH_4}^\circ + \Delta\overline{h}) + x_6(\overline{h}_{f,N_2}^\circ + \Delta\overline{h}) \tag{10.80}$$

where $\overline{h}_f^\circ$ is the formation enthalpy (in kJ/kmol), and $\Delta\overline{h}^\circ$ is enthalpy difference between the given state and the reference state.

10.5.1.2 Multi-effect Desalination Unit

As shown in Figure 10.66, the MED-TVC consists of six evaporators, six feed water preheaters, six flash boxes, a condenser, and a thermo-vapor compressor. Assuming two-phase flow within each effect and using conservation laws, the process can be modeled mathematically. The main conservation laws used are mass, salinity, and energy balances, which are presented in Table 10.17.

Table 10.17 Conservation laws for multi-effect desalination unit.

Part of process	Balance equations	Description
First effect	$B_1 = F_1 - D_1$	Mass
	$x_{sw}F_1 = x_{B_1}B_1$	Salinity
	$D_1L_1 + F_1C_p(T_1 - T_{f1}) = (D_r + s)L_o$	Energy
Second to sixth effects	$B_i = F_i + B_{i-1} - D_i - (i-1) \cdot F_{i-1} \cdot y_{i-1} + \left[y_{i-1} \left(D_r + \sum_{m=1}^{i-2} D_k \right) \right]$	Mass
	$x_{sw}F_i + x_{B_{i-1}}B_{i-1} = x_{B_i}B_i$	Salinity
	$D_iL_i + F_iC_p(T_i - T_{fi}) = D_{i-1}L_{i-1} - (i-1) \cdot F_{i-1}y_{i-1}L_{i-1} +$	
	$\quad y_{i-1}\left(D_r + \sum_{m=1}^{i-2} D_m \right) \cdot L_{i-1} + B_{i-1}C(T_{i-1} - \overline{T}_i)$	Energy
Condenser	$D_f = D_6 - D_r + y_6(D_r + D_1 + D_2 + D_3 + D_4 + D_5)$	Mass
	$D_fL_6 = \dot{m}_{sw}C_p(T_f - T_{sw})$	Energy
	$F = F_1 + F_2 + F_3 + F_4 + F_5 + F_6$	Mass
Fresh water flow rate	$86.4\,D$	

Source: Ettouney 2002. Reproduced with permission of Elsevier.

In addition, heat transfer coefficients, the logarithmic mean temperature difference, and the heat transfer area for each effect, the preheater, and the condenser are presented in Table 10.18.

In order to obtain values for the above-mentioned parameters, the temperature profile is required. Table 10.19 shows thermodynamic characteristics necessary for the simulation process.

The flashing vapor condensation temperatures for effects 1 to n are:

$$\overline{T}_1 = T_{oc} + NEA_1 \quad \text{for effect 1} \tag{10.81}$$

$$\overline{T}_i = T_{vi} + NEA_i \quad \text{for effects 2 to n} \tag{10.82}$$

Here, the non-equilibrium allowance (NEA) in effects can be defined as

$$NEA_i = \frac{0.33(T_{i-1} - T_i)^{0.55}}{Tv_i} \tag{10.83}$$

The flashing brines temperature of effects 2 to n is:

$$\dot{T}_i = T_i + NEA_i \tag{10.84}$$

Also, the vapor condensation temperature and saturated vapor temperature of each effect can be written respectively as follows:

$$T_{c_i} = T_{vi} - (\Delta T_p)_i \tag{10.85}$$

$$T_{vi} = T_i - BPE \tag{10.86}$$

Here, ΔT_p is the temperature drop due to pressure loss and BPE is the boiling point elevation. More detail about MED-TVC thermodynamic modeling can be found in References [46, 47].

Table 10.18 Formulation of heat transfer coefficient, heat transfer area and logarithmic mean temperature difference.

Description	Balance equations
Heat transfer coefficient	
Effect 1 Effects 2 to n	$U_{e_1} = 1.9394 + 1.40562 \times 10^{-3}T_{oc} - 2.07525 \times 10^{-5}T_{oc}^2 + 2.3186 \times 10^{-6}T_{oc}^3$ $U_{ei} = 1.9394 + 1.40562 \times 10^{-3}T_{vi-1} - 2.07525 \times 10^{-5}T_{vi-1}^2 + 2.3186 \times 10^{-6}T_{vi-1}^3$
Preheaters 1 to n-2 Preheater n-1	$U_{H_i} = 14.18251642 + 0.011383865T_{vi} + 0.013381501T_{f_{i+1}}$ $U_{H_{n-1}} = 14.18251642 + 0.011383865T_{v_{n-1}} + 0.013381501T_f$
Condenser	$U_{con.} = 1.6175 + 1.537 \times 10^{-4}T_{v_n} - 1.825 \times 10^{-4}T_{v_n}^2 + 8.026 \times 10^{-8}T_{v_n}^3$
Logarithmic mean temperature difference	
Preheaters 1 to n-2	$LMTD_{H_i} = \dfrac{(T_{f_i} - T_{f_{i+1}})}{\ln \dfrac{T_{v_i} - T_{f_{i+1}}}{T_{v_i} - T_{f_i}}}$
Preheater n-1	$LMTD_{H_{n-1}} = \dfrac{(T_{f_{n-1}} - T_f)}{\ln \dfrac{T_{v_{n-1}} - T_f}{T_{v_{n-1}} - T_{f_{n-1}}}}$
Condenser	$LMTD_{Cond} = \dfrac{(T_f - T_{sw})}{\ln \dfrac{T_{v_n} - T_{sw}}{T_{v_n} - T_f}}$
Heat transfer area	
Effect 1	$A_{e_1} = \dfrac{(D_r + S)L_o}{U_{e_1}(T_{oc} - T_1)}$
Effects 2 to n	$A_{e_m} = \dfrac{\left(D_{i-1} - (i-1)y_{i-1}F_i + \left(D_r + \sum_{m=1}^{i-2} D_m\right)y_{i-1}\right)L_{i-1}}{U_{ei}(T_{v_{i-1}} - T_i)}$
Preheaters 1 to n-1	$A_{H_i} = \dfrac{(i\,F_i(T_{f_{i+1}} - T_{f_{i+1}}))}{U_{H_i}LMTD_{H_i}}$
Preheater n	$A_{H_n} = \dfrac{(nF_n(T_{f_n} - T_f))}{U_{H_n}LMTD_{H_n}}$
Condenser	$A_{Cond} = \dfrac{\left(D_f + (D_r + \sum_{m=1}^{n-1} D_m)y_n\right)L_n}{U_{Cond}LMTD_{Cond}}$

Source: Esfahani 2012. Reproduced with permission of Elsevier.

Table 10.19 Thermodynamic properties of
multi-effect desalination unit.

Parameter	Value	Unit
Sea water temperature	25	°C
Sea water pressure	101	kPa
Salinity of sea water	36 000	ppm
Salinity of last effect brine	70 000	ppm
Boiling point elevation (BPE)	0.8	—
Top brine temperature	69	°C
Last brine temperature	45.2	°C

10.5.2 Exergy Analysis

Exergy analysis is a powerful tool for evaluating efficiencies and identifying inefficient processes in a system. The specific exergy of a material quantity or flow is usually composed of two parts: physical exergy and chemical exergy. That is,

$$ex = ex_{ph} + ex_{ch} \tag{10.87}$$

The specific physical exergy of each state depends on its pressure and temperature, and can be expressed as:

$$ex_{ph} = (h - h_o) - T_o(s - s_o) \tag{10.88}$$

In general, the specific chemical exergy for a gas mixture can be written as follows [48]:

$$ex_{ch} = \sum_i x_i ex_{o,i}^{ch} + \overline{R}T_o \sum_i x_i \ln x_i \tag{10.89}$$

where x_i is the molar fraction of the ith component, and $ex_{o,i}^{ch}$ is the standard exergy of the ith pure material. In order to calculate the exergy for MED-TVC streams the seawater specific enthalpy and specific entropy are determined as follows [49]:

$$h = m_{f,s} h_s + m_{f,w} h_w \tag{10.90}$$
$$s = m_{f,s} s_s + m_{f,w} s_w \tag{10.91}$$

The seawater properties at the dead state are evaluated at $T = 298$ K, P=101 kPa and a salinity of 0.036%. At this condition the salt properties are: $C_{p_s} = 0.8368$ kJ/kg · K, $h_{s_o} = 20.92$ kJ/kg, $s_{s_o} = 0.0732978$ kJ/kg · K. Then, the enthalpy and entropy of salt at the temperature T can be written as:

$$h_s = h_{s_o} + C_{ps}(T - T_0) = 20.92 + 0.836(T - T_0) \tag{10.92}$$
$$s_s = s_{s_o} + C_{ps} \ln(T/T_0) = 0.073 + 0.836 \ln(T/T_0) \tag{10.93}$$

Considering saline water as an ideal solution, the entropy of a component and saline water can be determined respectively as follows [50]:

$$\overline{s}_i = \overline{s}(P, T)_{i,pure} - R_u \ln x_i \tag{10.94}$$
$$\overline{s} = x_s \overline{s}_s + x_w \overline{s}_w = x_s [\overline{s}_{s,pure}(PT) - R_u \ln x_s] + x_w [\overline{s}_{w,pure}(PT) - R_u \ln x_w]$$
$$= x_s \overline{s}_{s,pure}(P, T) + x_w \overline{s}_{w,pure}(P, T) - R_u(x_s \ln x_s + x_w \ln x_w) \tag{10.95}$$

The chemical exergy of the biomass fuel can be expressed as the product of the lower heating value and the coefficient β [38]. That is,

$$\text{ex}_{\text{biomass}} = \beta \text{LHV}_{\text{wood}} \tag{10.96}$$

$$\beta = \frac{1.044 + 016\frac{Z_H}{Z_C} - .34493\frac{Z_O}{Z_C}\left(1 + .0531\frac{Z_H}{Z_C}\right)}{1 - 0.4124\frac{Z_O}{Z_C}} \tag{10.97}$$

$$\text{LHV} = 349.1\text{C} + 1178.3\text{H} + 100.5\text{S} - 103.4\text{O} - 15.1\text{N} - 21.1\text{ Ash} \tag{10.98}$$

where Z_O, Z_H, and Z_C are the mass components of oxygen, hydrogen, and carbon in biomass. For the biomass considered here and the above equation, the higher heating value is 19 980 kJ/kg. The exergy destruction rates of the main system components are given in Table 10.20.

10.5.3 Optimization

Two output parameters, namely the total cost rate and the exergy efficiency of the trigeneration system, are considered as objective functions. These can be defined as

$$\psi = \frac{\dot{\text{Ex}}_{\text{Q,domestic}} + \dot{W}_{\text{net,ORC}} + \dot{W}_{\text{net,GT}} + (\dot{\text{Ex}}_{\text{steam,out,HRSG}} - \dot{\text{Ex}}_{\text{D,MED-TVC}})}{\dot{\text{Ex}}_{\text{biomass}}} \tag{10.99}$$

$$\dot{C}_{\text{total}} = \dot{Z}_{\text{total}} + \dot{C}_{\text{biomass}} \tag{10.100}$$

$$\dot{Z}_{\text{total}} = \dot{Z}_{\text{gasif}} + \dot{Z}_{\text{C+}}\dot{Z}_{\text{GT}} + \dot{Z}_{\text{HE}} + \dot{Z}_{\text{CC}} + \dot{Z}_{\text{DWH}} + \dot{Z}_{\text{P,R}} + \dot{Z}_{\text{HRSG}} + \dot{Z}_{\text{Eva,R}} + \dot{Z}_{\text{Tur,R}} + \dot{Z}_{\text{cond,R}} + \dot{Z}_{\text{MED-TVC}} \tag{10.101}$$

Here, $\dot{C}_{\text{biomass}}$ is the biomass cost rate which mainly depends on the type of material, collection, and processing cost. The fuel cost can be written as a function of lower heating value as given in [38]. More details about the purchase cost of each piece of equipment are given elsewhere [3]. To model the trigeneration system, Matlab simulation code is used. The parts of the trigeneration system (gas turbine unit, gasifier, HRSG, ORC and MED-TVC, hot water heater) are first modeled including all outputs and exergy flow rates. Several simplifying assumptions are made to make the model tractable while providing adequate accuracy:

- All elements operate at steady state.
- Pressure drops in the ORC components are negligible.
- Heat losses from the pipeline and other auxiliary components are negligible.
- The feed flow rates in each MED effect are equal.

A modified thermodynamic model of the biomass gasification process is used to determine the composition of the produced syngas. To ensure and validate the accuracy of the developed code, the gasifier simulation is compared with data from literature. With a gasification temperature of 1100 K, the model exhibits good agreement with experimental data and results from previous modeling. Experimental data are available from Jayah *et al.* [54] for a moisture content range of 12.5% to 18%, and for various chip sizes and air to fuel ratios. Table 10.21 compares the results of present analysis and other works.

Table 10.20 Exergy destruction rates and exergy efficiencies of components of system.

Component	Exergy destruction rate	Exergy efficiency
ORC pump	$\dot{Ex}_D = \dot{Ex}_{1R} - \dot{Ex}_{2R} + \dot{W}_{pump,ORC}$	$\psi = \dfrac{\dot{Ex}_{2R} - \dot{Ex}_{1R}}{\dot{W}_{pump,ORC}}$
ORC evaporator	$\dot{Ex}_D = \dot{Ex}_{3R} - \dot{Ex}_{2R} + \dot{Ex}_8 - \dot{Ex}_9$	$\psi = \dfrac{\dot{Ex}_{3R} - \dot{Ex}_{2R}}{\dot{Ex}_8 - \dot{Ex}_9}$
ORC turbine	$\dot{Ex}_D = \dot{Ex}_{3R} - \dot{Ex}_{4R} - \dot{W}_{Tur,ORC}$	$\psi = \dfrac{\dot{W}_{Tur,ORC}}{\dot{Ex}_{3R} - \dot{Ex}_{4R}}$
ORC condenser	$\dot{Ex}_D = \dot{Ex}_{4R} - \dot{Ex}_{1R} + \dot{Ex}_{C1} - \dot{Ex}_{C2}$	$\psi = \dfrac{\dot{Ex}_{4R} - \dot{Ex}_{1R}}{\dot{Ex}_{C1} - \dot{Ex}_{C1}}$
Compressor	$\dot{Ex}_D = \dot{Ex}_1 - \dot{Ex}_2 + \dot{W}_{comp}$	$\psi = \dfrac{\dot{Ex}_{4R} - \dot{Ex}_{1R}}{\dot{Ex}_{c_2} - \dot{Ex}_{c_1}}$
Gas turbine	$\dot{Ex}_D = \dot{Ex}_3 - \dot{Ex}_4 - \dot{W}_{GT}$	$\psi = \dfrac{\dot{W}_{GT}}{\dot{Ex}_3 - \dot{Ex}_4}$
Combustion chamber	$\dot{Ex}_D = \dot{Ex}_4 - \dot{Ex}_{b1} - \dot{Ex}_5$	$\psi = \dfrac{\dot{Ex}_5 - \dot{Ex}_4}{\dot{Ex}_{b1}}$
Heat exchanger	$\dot{Ex}_D = \dot{Ex}_2 - \dot{Ex}_3 + \dot{Ex}_5 - \dot{Ex}_6$	$\psi = \dfrac{\dot{Ex}_3 - \dot{Ex}_2}{\dot{Ex}_5 - \dot{Ex}_6}$
Domestic water heater	$\dot{Ex}_D = \dot{Ex}_{D1} - \dot{Ex}_{D2} + \dot{Ex}_9 - \dot{Ex}_{10}$	$\psi = \dfrac{\dot{Ex}_{D2} - \dot{Ex}_{D1}}{\dot{Ex}_9 - \dot{Ex}_{10}}$
Gasifier	$\dot{Ex}_D = \dot{Ex}_{biomass} - \dot{Ex}_{b1}$	$\psi = \dfrac{\dot{Ex}_{b1}}{\dot{Ex}_{biomass}}$
HRSG	$\dot{Ex}_D = \dot{Ex}_{1H} - \dot{Ex}_{3H} + \dot{Ex}_3 - \dot{Ex}_8$	$\psi = \dfrac{\dot{Ex}_{3H} - \dot{Ex}_{1H}}{\dot{Ex}_3 - \dot{Ex}_8}$
MED-TVC	Exergy destruction rate	

Effect 1

$$\dot{Ex}_D = \dot{m}_o(h_{v_o} - h_{l_o}) - T_D(s_{v_o} - s_{l_o}) - F_1 C\left(T_1 - T_{f,1} - T_D \ln\left(\frac{T_1}{T_{f,1}}\right)\right)$$
$$-D_1 L_1\left(1 - \frac{T_D}{T_{v,1}}\right)$$

Effects 2 to 6

$$\dot{Ex}_D = (D_n + y_{n-1}(D_1 + \ldots + D_{n-1} + D_r) - (n-1)y_{n-1}F_n)L_n\left(1 - \frac{T_D}{T_{n-1}}\right)$$
$$-F_n C\left(T_n - T_{f,n} - T_D \ln\left(\frac{T_n}{T_{f,n}}\right)\right) + B_{n-1} C\left(\Delta T - T_D \ln\left(\frac{T_{n-1}}{T_n}\right)\right)$$
$$-D_n L_n\left(1 - \frac{T_D}{T_{v,n}}\right)$$

End condenser

$$\dot{Ex}_D = (D_f + y_6(D_1 + \ldots + D_6 + D_r))L_6\left(1 - \frac{T_D}{T_6}\right)$$
$$-\dot{m}_{sw} C\left(T_f - T_{sw} - T_D \ln\left(\frac{T_f}{T_{sw}}\right)\right)$$

Recycled fresh water

$$\dot{Ex}_D = \dot{m}_o C\left((T_{v_6} - T_{sw}) - T_D \ln\left(\frac{T_{v_6}}{T_{sw}}\right)\right)$$

Rejected brine water

$$\dot{Ex}_D = B_6 C\left(T_6 - T_{sw} - T_D \ln\left(\frac{T_6}{T_{sw}}\right)\right)$$

Steam jet ejector

$$\dot{Ex}_D = \dot{m}_{1H}(h_{3H} - h_{o,c} - T_D(s_{3H} - s_{o,c}) - D_r(h_{o,c} - h_{D,r} - T_D(s_{o,c} - s_{D,r}))$$

Source: Adapted from Khanmohammadi 2015.

Table 10.21 Comparison between result of present analysis and the other experimental and numerical data for MC = 0.16 and $T_{gasif} = 1100$ K.

Species	Present analysis	Experimental [52]	Model [42]	Model [42]
H_2	21.3	17	18.04	21.06
CO	25.43	18.4	17.86	19.6
CH_4	1.48	1.3	0.11	0.64
CO_2	10.3	10.6	11.84	12.01
N_2	41.5	52.7	52.15	46.68

Table 10.22 Thermodynamic specifications of various points in the trigeneration system.

Point	$\dot{m}$(kg/s)	P (kPa)	T (K)	h (kJ/kg)	s (kJ/kg K)	ex (kJ/kg)
b1	1.892	—	—	655.2	8.65	5560
1	7.17	101.3	298	305.6	6.81	1.85
2	7.17	909.9	600.7	622.9	6.90	291.7
3	7.17	909.9	1250	1370	7.74	789.1
4	7.17	101.8	848.8	785.4	7.78	191.3
5	9.062	101.3	1350	1544	8.45	768
6	9.062	101.3	877	953.2	7.91	337.3
7	9.062	101.3	591.3	622.5	7.42	280.4
8	9.062	101.3	480.4	500.6	7.23	88.33
9	9.062	101.3	423	438.3	7.09	67.18
10	9.062	101.3	393	406.3	7.01	58.52
1R	4.375	1000	313	74.57	0.271	36.53
2R	4.375	3000	337.3	76.42	0.277	36.66
3R	4.375	3000	367	212.2	0.654	60.1
4R	4.375	960	313	196.5	0.6613	42.44
H1	1.539	101.3	303	125.8	0.436	0.184
H2	1.539	1519.5	3030.1	127.6	0.436	1.98
H3	1.539	1519.5	471.3	844.9	6.44	875.8
D1	1.541	101.3	303	125.8	0.436	0.184
D2	1.541	202.6	348	314	1.016	15.83

The results of the exergy analysis of the trigeneration system are now considered. It is assumed that the inlet air composition is 71% N_2 and 29% O_2. Table 10.22 shows selected thermodynamic specifications of the trigeneration system. Also, Table 10.23 presents various important parameters of the trigeneration system for its initial operation conditions.

Table 10.23 Performance parameters of trigeneration system for initial state.

Parameter	Unit	Value
Fuel (biomass) flow rate	kg/s	0.6
Exergy efficiency	%	27.9
Steam mass flow rate of HRSG	kg/s	1.54
Output pressure of HRSG	kPa	1500
Gasification air mass flow rate	kg/s	1.09
Combustion air mass flow rate	kg/s	7.17
Total exergy destruction rate	kW	13329
ORC turbine output	kW	163.3
Gas turbine output	kW	1915
Domestic hot water flow rate	kg/s	1.54
Fresh water flow rate	m³/day	1142
Gasifier purchase cost	$/h	5.9
Compressor purchase cost	$/h	4.47
Gas turbine cost rate	$/h	2.23
Heat exchanger cost rate	$/h	6.38
Combustion chamber cost rate	$/h	0.44
HRSG cost rate	$/h	0.88
Domestic water heater cost rate	$/h	0.03
Organic Rankine cycle cost rate	$/h	2.38
Multi-effect desalination unit cost rate	$/h	99.06
Biomass fuel cost	$/GJ	2

10.5.3.1 Decision Variables

The decision variables considered in this study are gasification temperature (T_{gasif}), compressor pressure ratio (r_{p}), combustion temperature (T_{comb}), gas turbine inlet temperature ($GTIT$), maximum pressure of organic Rankine cycle (P_{3R}), compressor isentropic efficiency (η_{comp}), gas turbine isentropic efficiency (η_{GT}), and temperature difference between MED-TVC effects ($\Delta T_{\text{effects}}$). Reasonable ranges are permitted for each.

10.5.4 Optimization Results

The results of multi-objective optimization of the trigeneration system, considering the two objective functions in Equations 10.99 and 10.100, are shown in Figure 10.67.

Figure 10.67 shows the non-dominated optimum solution in the plane of exergy efficiency–total cost rate. As indicated in this figure, improvements to one of the objective functions make the other deteriorate accordingly. Points A and C represent the lowest total cost rate (100 $/h) and highest exergy efficiency (42%) of the trigeneration system, respectively. The values of objective functions at these points are very close to the value obtained from single objective optimization. A decision-making process is necessary for selecting the optimal condition from available optimal solutions in a multi-objective optimization. All points in Figure. 10.67 are optimum solutions for the

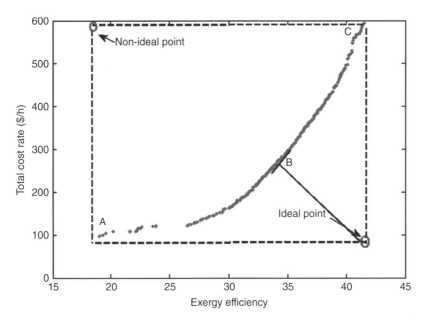

Figure 10.67 Pareto frontier showing best trade-off values for the objective functions.

Table 10.24 Values of decision variables and objective functions at points A, B, and C on the Pareto frontier curve.

Point	$\Delta T_{effects}$	r_{ac}	GTIT	T_{gasif}	T_{comb}	$P_{max,ORC}$	MC	η_{comp}	η_{GT}
A	4.9	5.7	1209.1	950.01	1270.66	2642.59	0.34	0.889	0.89
B	4.94	10.91	1198.06	952.11	1445.71	1346.28	0.1	0.847	0.91
C	4.818	10.89	1249.86	952.1	1399.4	1199.17	0.101	0.88	0.90

trigeneration system. The concept of a hypothetical (ideal) point on the Pareto curve at which two objective functions are at their optimum values regardless of the other objective can help the decision-making process. In reality, the ideal point in Figure 18 does not exist but the closest point to this one can be considered as the best solution. So, it can be assumed that the perpendicular line to the Pareto curve from ideal point, which has a minimum distance from the ideal point, represents the optimum solution for the optimization problem (Point B). Table 10.24 represents values of the decision variables at three points on the Pareto curve.

Note that in multi-objective optimization and the Pareto solution, each point can be the optimized point. Therefore, the selection of the optimum solution depends on the preferences and criteria of each decision maker. Hence, each decision maker may select a different point as the optimum solution so as to better suit his/her requirements. To provide a helpful tool for the optimal design of the gas turbine cycle, the following equation can be derived for the Pareto optimal points curve (Figure 10.68):

$$\dot{C}_{total} = \frac{-37.77\psi^3 + 5419\psi^2 + 571\psi + 41.9}{\psi^3 - 112.2\psi^2 + 3213\psi - 588.5} \tag{10.102}$$

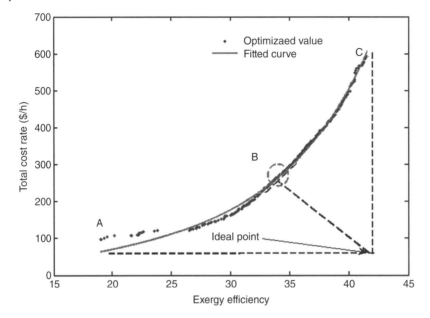

Figure 10.68 Fitted curve of the Pareto optimal points.

10.6 Concluding Remarks

This chapter began with some introductory information about CHP and trigeneration energy systems followed by a comprehensive modeling and optimization of several common CHP and trigeneration systems. The model and corresponding simulation computer code were verified with actual data to ensure accuracy. The results were seen to exhibit good agreement with actual operating data. Various objective functions were defined for each system and appropriate constraints were considered. A genetic algorithm was applied to determine the best optimal design parameters, and sensitivity analyses were conducted to ascertain how each design parameter affects the objective functions. Several concluding remarks can be drawn from the material covered in the chapter:

- The sensitivity analyses of the gas turbine CHP system showed that an increase in compressor isentropic efficiency results in a decrease in the compressor power consumption, and that by increasing the fuel unit cost, the compressor pressure ratio increases in order to decrease the objective function. In addition, it was observed that an increase in the unit cost of fuel causes the combustion chamber inlet temperature to decrease due to the fact that increasing the combustor inlet temperature reduces the exergy destruction in the combustion chamber. The results also showed that higher compressor and gas turbine isentropic efficiencies reduce the exergy destruction in the compressor and the turbine, and reduce the net cycle fuel consumption and operating cost. Further, an increase in gas turbine inlet temperature (GTIT) was seen to decrease the exergy destruction in the combustion chamber (and HRSG) and to lower fuel consumption as well.
- The optimization of the IC cogeneration system provided useful insights for determining a suitable working fluid. The results showed that cooling production is

preferable to electrical power production in waste heat recovery from both total cost rate and exergy efficiency viewpoints. The higher the critical temperature, the better the waste heat recovery performance. In addition, R-11 is the most beneficial working fluid among those considered for the waste heat recovery in terms of maximum exergy efficiency and total cost rate.

- The results for the micro gas turbine trigeneration system showed that an increase in compressor pressure ratio, compressor efficiency, gas turbine efficiency, and gas turbine inlet temperature (GTIT) have a positive effect on both objective functions. Raising GTIT has a significant effect on total cost rate because it leads to a decrease in cost of exergy destruction which is an important portion of the objective function. In addition, an increase in pinch point temperature lowers the exergy efficiency of the cycle but positively affects the total cost rate of the plant. An increase in the pinch point temperature causes a reduction in heat transfer surface area which leads to a decrease in the HRSG cost.

- The optimization of the biomass based trigeneration system demonstrated that the total cost rate of system can decrease significantly to reach 100 $/h, regardless of the exergy efficiency as an objective function. It was also observed that variations in each of the decision variables within reasonable ranges have different effects on exergy efficiency and total cost of system. Selecting the optimal point and considering the effect of varying performance parameters on the optimal point are important and should be considered in energy systems design.

References

1 Ahmadi, P. and Dincer, I. (2010) Exergoenvironmental analysis and optimization of a cogeneration plant system using Multimodal Genetic Algorithm (MGA). *Energy* **35**:5161–5172.

2 Al-Sulaiman, F. A., Dincer, I. and Hamdullahpur, F. (2010) Exergy analysis of an integrated solid oxide fuel cell and organic Rankine cycle for cooling, heating and power production. *Journal of Power Qources* **195**:2346–2354.

3 Ahmadi, P. (2013) Modeling, analysis and optimization of integrated energy systems for multigeneration purposes. PhD thesis. Faculty of Engineering and Applied Science, University of Ontario Institute of Technology.

4 Horlock, J. H. (2003) *Advanced gas turbine cycles*. Pergamon Press.

5 Dincer, I. and Rosen, M. A. (2012) *Exergy: Energy, Environment and Sustainable Development*, Newnes.

6 Bejan, A., Tsatsaronis, G. and Moran, M. J. (1996) *Thermal Design and Optimization*, John Wiley & Sons, Inc., New York.

7 Ahmadi, P., Dincer, I. and Rosen, M. A. (2013) Thermodynamic modeling and multi-objective evolutionary-based optimization of a new multigeneration energy system. *Energy Conversion and Management* **76**:282–300.

8 Ahmadi, P., Dincer, I. and Rosen, M. A. (2015) Multi-objective optimization of an ocean thermal energy conversion system for hydrogen production. *International Journal of Hydrogen Energy* **40**:7601–7608.

9 Roosen, P., Uhlenbruck, S. and Lucas, K. (2003) Pareto optimization of a combined cycle power system as a decision support tool for trading off investment vs. operating costs. *International Journal of Thermal Sciences* **42**:553–560.

10 Ahmadi, P., Dincer, I. and Rosen, M. A. (2011) Exergy, exergoeconomic and environmental analyses and evolutionary algorithm based multi-objective optimization of combined cycle power plants. *Energy* **36**:5886–5898.

11 Ahmadi, P. and Dincer, I. (2011) Thermodynamic and exergoenvironmental analyses, and multi-objective optimization of a gas turbine power plant. *Applied Thermal Engineering* **31**:2529–2540.

12 Valero, A., Lozano, M. A., Serra, L., Tsatsaronis, G., Pisa, J., Frangopoulos, C. and Von Spakovsky, M. (1994) CGAM problem: definition and conventional solution. *Energy* **19**(3):279–286.

13 Frangopoulos, C. A. (1994) Application of the thermoeconomic functional approach to the CGAM problem. *Energy* **19**:323–342.

14 Tsatsaronis, G. and Pisa, J. (1994) Exergoeconomic evaluation and optimization of energy systems—application to the CGAM problem. *Energy* **19**:287–321.

15 Valero, A., Lozano, M., Serra, L. and Torres, C. (1994) Application of the exergetic cost theory to the CGAM problem. *Energy* **19**:365–381.

16 Lior, N. and Zhang, N. (2007) Energy, exergy, and second law performance criteria. *Energy* **32**:281–296.

17 Reid, R. C., Prausnitz, J. M. and Poling, B. E. (1987) The properties of gases and liquids.

18 Javan, S., Mohamadi, V., Ahmadi, P. and Hanafizadeh, P. (2016) Fluid selection optimization of a combined cooling, heating and power (CCHP) system for residential applications. *Applied Thermal Engineering* **96**:26–38.

19 Mago, P. J., Chamra, L. M., Srinivasan, K. and Somayaji, C. (2008) An examination of regenerative organic Rankine cycles using dry fluids. *Applied Thermal Engineering* **28**:998–1007.

20 Hajabdollahi, Z., Hajabdollahi, F., Tehrani, M. and Hajabdollahi, H. (2013) Thermo-economic environmental optimization of Organic Rankine Cycle for diesel waste heat recovery. *Energy* **63**:142–151.

21 Shengjun, Z., Huaixin, W. and Tao, G. (2011) Performance comparison and parametric optimization of subcritical Organic Rankine Cycle (ORC) and transcritical power cycle system for low-temperature geothermal power generation. *Applied Energy* **88**:2740–2754.

22 Mabrouk, A. A., Nafey, A. and Fath, H. (2007) Thermoeconomic analysis of some existing desalination processes. *Desalination* **205**:354–373.

23 Ahmadi, P., Dincer, I. and Rosen, M. A. (2013) Performance assessment and optimization of a novel integrated multigeneration system for residential buildings. *Energy and Buildings* **67**:568–578.

24 Wang, M., Wang, J., Zhao, P. and Dai, Y. (2015) Multi-objective optimization of a combined cooling, heating and power system driven by solar energy. *Energy Conversion and Management* **89**:289–297.

25 Ahmadi, P., Rosen M. A. and Dincer, I. (2012) Multi-objective exergy-based optimization of a polygeneration energy system using an evolutionary algorithm. *Energy* **46**:21–31.

26 Tom, K. (2008) *Combined Heating and Power and Emissions Trading: Options for Policy Makers*, International Energy Agency.

27 Bruno, J. C., Ortega-López, V. and Coronas, A. (2009) Integration of absorption cooling systems into micro gas turbine trigeneration systems using biogas: case study of a sewage treatment plant. *Applied Energy* **86**:837–847.

28 Ahmadi, P., Dincer, I. and Rosen, M. A. (2012) Exergo-environmental analysis of an integrated organic Rankine cycle for trigeneration. *Energy Conversion and Management* **64**:447–453.

29 Dincer, I. and Kanoglu, M. (2011) *Refrigeration systems and applications.* John Wiley & Sons, Ltd., UK.

30 Scrinivas, N. D. (1994) Multiobjective Optimization Using Non Dominated Sorting in Genetic Algorithms. *Evolutionary Computation* **2**:221.

31 Ghaffarizadeh, A. (2006) *Investigation On Evolutionary Algorithms Emphasizing Mass Extinction.* Shiraz University of Technology, Shiraz, Iran.

32 Deb, K. (2001) *Multi-Objective Optimization Using Evolutionary Algorithms*, John Wiley & Sons Ltd, UK.

33 Hajabdollahi, H., Ahmadi, P. and Dincer, I. (2011) An exergy-based multi-objective optimization of a heat recovery steam generator (HRSG) in a combined cycle power plant (CCPP) using evolutionary algorithm. *International Journal of Green Energy* **8**:44–64.

34 Dresselhaus, M. and Thomas, I. (2001) Alternative energy technologies. *Nature* **414**:332–337.

35 Cohce, M., Dincer, I. and Rosen, M. (2011) Energy and exergy analyses of a biomass-based hydrogen production system. *Bioresource technology* **102**:8466–8474.

36 Anselmo Filho, P. and Badr, O. (2004) Biomass resources for energy in North-Eastern Brazil. *Applied Energy* **77**:51–67.

37 Hughes, E. E. and Tillman, D. A. (1998) Biomass cofiring: status and prospects 1996. *Fuel Processing Technology* **54**:127–142.

38 Lian, Z., Chua, K. and Chou, S. (2010) A thermoeconomic analysis of biomass energy for trigeneration. *Applied Energy* **87**:84–95.

39 Barman, N. S., Ghosh, S. and De, S. (2012) Gasification of biomass in a fixed bed downdraft gasifier – A realistic model including tar. *Bioresource Technology* **107**:505–511.

40 Zainal, Z., Ali, R., Lean, C. and Seetharamu, K. (2001) Prediction of performance of a downdraft gasifier using equilibrium modeling for different biomass materials. *Energy Conversion and Management* **42**:1499–1515.

41 Higman, C. and Van der Burgt, M. (2011) *Gasification*, Gulf professional publishing.

42 Jarungthammachote, S. and Dutta, A. (2007) Thermodynamic equilibrium model and second law analysis of a downdraft waste gasifier. *Energy* **32**:1660–1669.

43 Khanmohammadi, S., Atashkari, K. and Kouhikamali, R. (2015) Exergoeconomic multi-objective optimization of an externally fired gas turbine integrated with a biomass gasifier. *Applied Thermal Engineering* **91**:848–859.

44 Ettouney, H. (2002) *Fundamentals of Salt Water Desalination*, Elsevier Science, B. V., New York.

45 Esfahani, I. J., Ataei, A., Shetty, V., Oh, T., Park, J. H. and Yoo, C. (2012) Modeling and genetic algorithm-based multi-objective optimization of the MED-TVC desalination system. *Desalination* **292**:87–104.

46 Alamolhoda, F., KouhiKamali, R. and Asgari, M. (2016) Parametric simulation of MED–TVC units in operation. *Desalination and Water Treatment* **57**(22):1–14.

47 Kamali, R., Abbassi, A., Vanini, S. S. and Avval, M. S. (2008) Thermodynamic design and parametric study of MED-TVC. *Desalination* **222**:596–604.

48 Khanmohammadi, S. and Azimian, A. R. (2015) Exergoeconomic Evaluation of a Two-Pressure Level Fired Combined-Cycle Power Plant. *Journal of Energy Engineering* **141**(3).

49 Hosseini, S. R., Amidpour, M. and Behbahaninia, A. (2011) Thermoeconomic analysis with reliability consideration of a combined power and multi stage flash desalination plant. *Desalination* **278**:424–433.

50 Kahraman, N. and Cengel, Y. A. (2005) Exergy analysis of a MSF distillation plant. *Energy Conversion and Management* **46**:2625–2636.

51 Khanmohammadi, S., Atashkari, K. and Kouhikamali, R. (2015) Performance assessment and multi-objective optimization of a trigeneration system with a modified biomass gasification model. *Modares Mechanical Engineering* **15**:209–222.

52 Jayah, T., Aye, L., Fuller, R. and Stewart, D. (2003) Computer simulation of a downdraft wood gasifier for tea drying. *Biomass and Bioenergy* **25**:459–469.

Study Questions/Problems

1   For the gas turbine based CHP system shown in Figure 10.3, repeat the multi-objective optimization by considering exergy efficiency and normalized CO_2 emissions as the two objective functions and discuss the results.

2   For the ICE cogeneration system shown in Figure 10.21, conduct a sensitivity analysis for different unit costs of fuel and sketch the Pareto curve. Discuss the results.

3   In gas turbine CHP systems, the combustion chamber usually has the highest exergy destruction compared to other components. In order to reduce this exergy

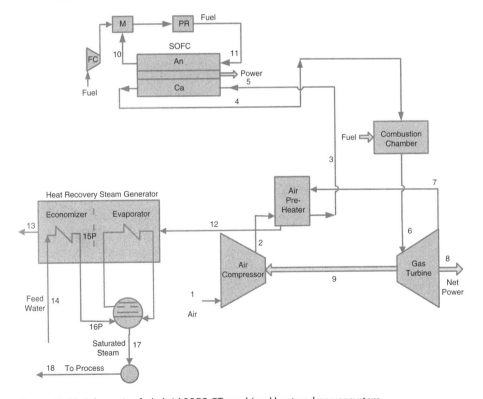

Figure 10.69 Schematic of a hybrid SOFC-GT combined heat and power system.

destruction, a SOFC can be used between the air preheater and the combustion chamber. A schematic of a typical SOFC-based CHP system is shown in Figure 10.69. Using the data provided for the gas turbine CHP system, apply an optimization to this CHP system. Assume typical inputs for the fuel cell.

4 For the CHP system shown in Figure 10.70, determine the optimal design parameters when the total cost rate of the system is the only objective function.

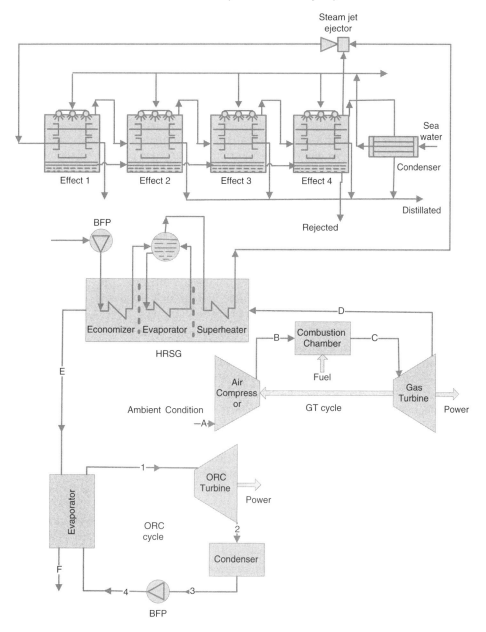

Figure 10.70 Schematic diagram of gas turbine based CHP system for electricity, heating, and fresh water production.

11

Modeling and Optimization of Multigeneration Energy Systems

11.1 Introduction

Energy drives processes and is essential to life. Energy exists in several forms, such as light, heat, and electricity. Concerns exist regarding limitations on easily accessible supplies of energy resources and the contribution of energy processes to global warming, as well as other environmental concerns such as air pollution, acid precipitation, ozone depletion, forest destruction, and radioactive emissions [1]. There are various alternatives to fossil fuels, including solar, geothermal, hydropower, wind, and nuclear energy. The use of many of the available natural energy resources is limited due to their reliability, quality, and energy density. Nuclear energy has the potential to contribute a significant share of large scale energy supply without contributing to climate change. Advanced technologies aimed at mitigating global warming are being proposed and tested in many countries. Among these technologies, multigeneration processes can make important contributions due to their potential for high efficiencies as well as low operating costs and pollution emissions per energy output.

The efficiency of multigeneration energy systems is often higher than that for either trigeneration or CHP, sometimes due to the additional products (e.g., hydrogen and potable and hot water). Figures 11.1 and 11.2 illustrate two multigeneration energy systems. The system in Figure 11.1 produces electricity, cooling, heating, hot water, and hydrogen. An electrolyzer is used for hydrogen production, driven by part of the electricity generated by a solar concentrating collector. Hot water enters the electrolyzer and is reacted electrochemically to split it into hydrogen and oxygen. The heating system is composed of two parts, one for hot water production and another for space heating. Heat rejected from the storage system enters an absorption cooling system to produce cooling and air conditioning. If the system is extended to produce potable water, a desalination system must be used – such as the multigeneration energy system shown in Figure 11.2. In this case, a portion of the heat produced by the solar concentrator is used by a desalination system, while part of the electricity generated by the power unit drives the pumps. Other parts of the system are the same as in Figure 11.1. These two figures are representative of typical multigeneration energy systems that use only solar energy as an input. Other configurations that combine renewable and conventional energy sources are also possible, and are discussed subsequently.

Various benefits are possible with multigeneration energy systems, including higher plant efficiency, reduced thermal losses and wastes, reduced operating costs, reduced

Optimization of Energy Systems, First Edition. Ibrahim Dincer, Marc A. Rosen, and Pouria Ahmadi.
© 2017 John Wiley & Sons Ltd. Published 2017 by John Wiley & Sons Ltd.

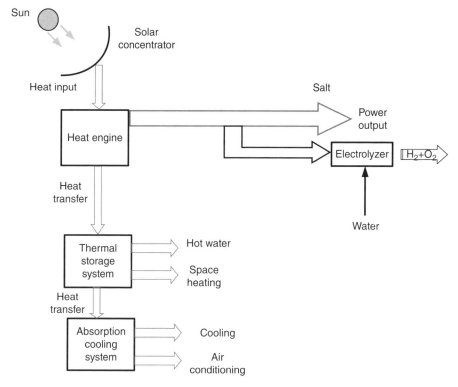

Figure 11.1 A multigeneration energy system for producing electricity, cooling, heating, hot water, and hydrogen. Source: Adapted from Dincer 2012.

greenhouse gas emissions, reduced use of resources, shorter transmission lines, fewer distribution units, multiple generation options, increased reliability, and less grid failure [2]. These benefits are discussed below.

Regarding efficiencies, for instance, consider the following. The overall efficiency of a conventional power plant that uses a fossil fuel with a single prime mover is usually less than 40%. That is, more than 60% of the heating value of the fuel entering the conventional power plant is lost. Furthermore, the overall efficiency of a conventional power plant that produces electricity and product heat separately is around 60% [3]. However, by utilizing the waste heat from the prime mover, the efficiency of the multigeneration plant can reach 80% [4]. In a multigeneration plant, the waste heat from the electricity generation unit is used to operate the cooling and heating systems without the need for extra fuel, unlike a conventional power plant that requires extra energy resources. Thus, a multigeneration plant uses less energy to produce the same output as a conventional plant, and has correspondingly lower operating costs.

Multigeneration can also reduce greenhouse gas (GHG) emissions. Since a multigeneration energy system often uses less fuel to produce the same output compared to a conventional power plant, a multigeneration plant typically emits less GHGs. Despite this advantage, there are some limitations to using multigeneration plants in a distributed manner because of their on site gas emissions. Another important benefit of using multigeneration energy systems is that they can reduce costs and energy losses because they often require fewer electricity transmission lines and distribution units.

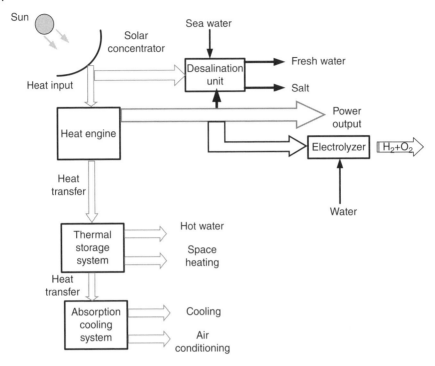

Figure 11.2 A multigeneration energy system for producing electricity, cooling, heating, hot water, hydrogen, and fresh water [2].

The conventional production of electricity is usually from a centralized plant that is generally located far from the end user. The losses from transmission and distribution of the electricity from a centralized system to the user can be about 9% [3]. These benefits have motivated researchers and designers to develop suitable multigeneration energy systems. The potential efficiency improvement is often the most significant factor in implementing a multigeneration energy system. Further assessments before designing or selecting multigeneration plants, such as evaluations of initial capital and operating costs, are needed to ensure efficient and economic multigeneration plant construction and performance [3].

Many analysis and optimization studies of multigeneration energy systems have been reported in recent years. Some of these systems are being considered as options for mitigating climate change. Note that various methods are available to achieve each purpose of a multigeneration energy system; hence the appropriate application of each subsystem is important in meeting the system's requirements. In this chapter we describe several multigeneration energy systems, ranging from non-renewable to renewable based systems and covering different locations. We also apply optimization in order to determine optimal design parameters. We utilize exergy and define various objective functions and constraints. An evolutionary algorithm based optimization is then applied to each system and the optimal design parameters are determined. In order to enhance understanding of the design criteria, sensitivity analyses are conducted, demonstrating how each objective function varies when small changes in selected design parameters are applied. Finally, some closing remarks are provided on the efficient design of multigeneration systems followed by some practical questions.

11.2 Multigeneration System Based On Gas Turbine Prime Mover

The gas turbine has proven to be a reliable and beneficial option as a prime mover for numerous reasons. Therefore, many multigeneration energy systems are based on this prime mover, including the one considered in this section. This system is composed of five subsystems, as shown in Figure 11.3. Electricity is produced by gas and steam turbines while cooling is produce based on two cycles, a single effect absorption chiller and an ejector refrigeration cycle. A PEM electrolyzer driven by electricity produced by the ejector produces hydrogen. Finlay a domestic hot water heater is incorporated, which uses thermal energy from the absorption generator. A complete explanation of each subsystem is given below.

The integrated multigeneration system in Figure 11.3 contains a compressor, a combustion chamber (CC), a gas turbine, a double pressure heat recovery steam generator (HRSG) to produce superheated steam, a single effect absorption chiller, a heat recovery vapor generator (HRVG) to produce ORC vapor that is driven by heat from flue gases from the HRSG, an organic Rankine cycle (ORC) ejector refrigeration system, a PEM electrolyzer for hydrogen production, and a heater for domestic hot water production. Air at ambient conditions enters the air compressor at point 1 and exits after compression (point 2). The hot air enters the combustion chamber (CC), into which fuel is injected, and hot combustion gases exit (point 3) and pass through the gas turbine to produce shaft power. The hot gas expands in the gas turbine to point 4. Hot flue gases enter the double pressure HRSG to provide high and low pressure steam at points 5 and 14. The high pressure steam enters the steam turbine to generate shaft power while the low pressure steam enters the generator of the absorption system to provide the cooling load of the system. The low pressure line leaving the generator has adequate energy for use in a domestic water heater that provides hot water at 50°C. Furthermore, flue gases leaving the HRSG at point C enter a heat recovery vapor generator to provide electricity and cooling. Since the flue gases have a low temperature, around 160°C, an ORC cycle is used, consisting of an ORC turbine to generate electricity and a steam ejector to provide the system cooling load. These flue gases enter the HRVG at point d to produce saturated vapor at point 29, which leaves the HRVG at point 28. Saturated vapor at point 29 enters the ORC turbine and work is produced.

The extraction turbine and ejector play important roles in this combined cycle. The high pressure and temperature vapor is expanded through the turbine to generate power, and the extracted vapor from the turbine enters the supersonic nozzle of the ejector as the primary vapor. The stream exiting the ejector (point 33) mixes with turbine exhaust (point 31) and is cooled in the preheater; it enters the condenser where it becomes a liquid by rejecting heat to the surroundings. Some of the working fluid leaving the condenser enters the evaporator after passing through the throttle valve (point 39), and the remainder flows back to the pump (point 37). The ORC pump increases the pressure (point 40), and high pressure working fluid is heated in the preheater (point 41) before entering the HRVG. The low pressure and temperature working fluid after the valve (point 39) enters the evaporator, providing a cooling effect for space cooling. Some of the electricity is considered for residential applications while some directly drives a PEM electrolyzer to produce hydrogen. In this analysis, waste heat is used as an input energy source for the multigeneration system and R123 is selected as the working fluid

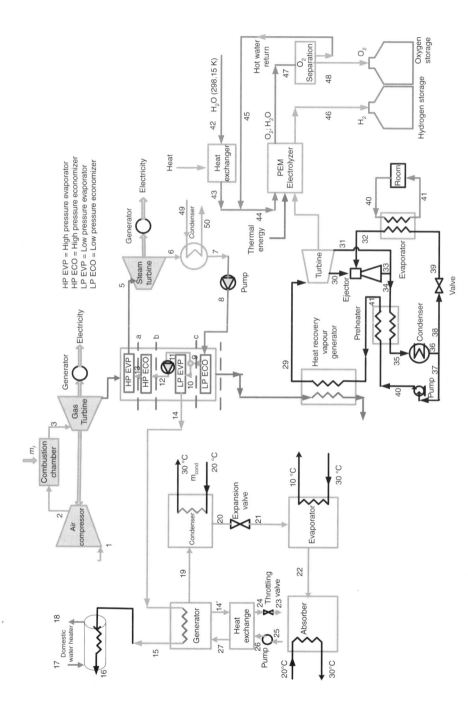

Figure 11.3 Schematic diagram of a multigeneration energy system based on a micro gas turbine, a dual pressure heat recovery steam generator, an absorption chiller, and an ejector refrigeration cycle.

because it is a non-toxic, non-flammable, and non-corrosive refrigerant with suitable thermophysical properties and characteristics.

11.2.1 Thermodynamic Modeling

For thermodynamic modeling, the multigeneration system in Figure 11.3 is divided into six main parts: gas turbine (Brayton) cycle, Rankine cycle with double pressure HRSG, single effect absorption chiller, organic Rankine cycle (ORC), domestic water heater, and PEM electrolyzer. The fuel injected into the combustion chamber is natural gas. We determine the temperature profile in the plant, the input and output enthalpy and exergy flow rates, the exergy destruction rates, and the energy and exergy efficiencies. Energy balances and governing equations for various multigeneration components (see Figure 11.3) are written, as described in the subsequent subsections.

11.2.1.1 Brayton Cycle
The Brayton cycle is composed of four main components, each of which is described below.

Air Compressor Air at ambient pressure and temperature T_1 enters the compressor. The compressor outlet temperature is a function of the compressor isentropic efficiency η_{AC}, the compressor pressure ratio r_{AC}, and the specific heat ratio γ_a, as follows:

$$T_2 = T_1 \times \left(1 + \frac{1}{\eta_{AC}}\left(r_{AC}^{\frac{\gamma_a-1}{\gamma_a}} - 1\right)\right) \tag{11.1}$$

The compressor work rate is a function of the air mass flow rate $\dot{m}_a$, the air specific heat, and the temperature difference across the compressor, and can be expressed as follows:

$$\dot{W}_{AC} = \dot{m}_a C_{pa}(T_2 - T_1) \tag{11.2}$$

where C_{pa} denotes the specific heat at constant pressure of air, and is treated as a function of temperature as follows [5]:

$$C_{pa}(T) = 1.048 - \left(\frac{3.83T}{10^4}\right) + \left(\frac{9.45T^2}{10^7}\right) - \left(\frac{5.49T^3}{10^{10}}\right) + \left(\frac{7.92T^4}{10^{14}}\right) \tag{11.3}$$

Combustion Chamber (CC) The outlet properties of the combustion chamber are a function of the air mass flow rate, the fuel lower heating value (LHV), and the combustion efficiency, which are related as follows:

$$\dot{m}_a h_2 + \dot{m}_f LHV = \dot{m}_g h_3 + (1 - \eta_{cc})\,\dot{m}_f LHV \tag{11.4}$$

where η_{cc} is the combustion efficiency. The combustion chamber outlet pressure is defined by considering a pressure drop across the combustion chamber ΔP_{cc} as follows:

$$\frac{P_3}{P_2} = 1 - \Delta P_{CC} \tag{11.5}$$

The combustion reaction and its species coefficients can be expressed as follows:

$$\lambda C_{x1}H_{y1} + (x_{O_2}O_2 + x_{N_2}N_2 + x_{H_2O}H_2O + x_{CO_2}CO_2 + x_{Ar}Ar) \rightarrow y_{CO_2}CO_2$$
$$+ y_{N_2}N_2 + y_{O_2}O_2 + y_{H_2O}H_2O + y_{NO}NO + y_{CO}CO + y_{Ar}Ar) \tag{11.6}$$

where

$$y_{CO_2} = (\lambda x_1 + x_{CO_2} - y_{CO})$$

$$y_{N_2} = (x_{N_2} - y_{NO})$$

$$y_{H_2O} = \left(x_{H_2O} + \frac{\lambda \times y_1}{2}\right)$$

$$y_{O_2} = \left(x_{O2} - \lambda \times x_1 - \frac{\lambda \times y_1}{4} - \frac{y_{CO}}{2} - \frac{y_{NO}}{2}\right)$$

$$y_{Ar} = x_{Ar}$$

$$\lambda = \frac{n_f}{n_{air}}$$

Gas Turbine The gas turbine outlet temperature can be written as a function of gas turbine isentropic efficiency η_{GT}, the gas turbine inlet temperature T_3, and the gas turbine pressure ratio P_3/P_4 as follows:

$$T_4 = T_3(1 - \eta_{GT}\left[1 - \left(\frac{P_3}{P_4}\right)^{\frac{1-\gamma_g}{\gamma_g}}\right]) \tag{11.7}$$

The gas turbine output power can be expressed as

$$\dot{W}_{GT} = \dot{m}_g C_{pg}(T_3 - T_4) \tag{11.8}$$

Here, $\dot{m}_g$ is the gas turbine mass flow rate, which can be written as

$$\dot{m}_g = \dot{m}_f + \dot{m}_a \tag{11.9}$$

The net output power of the gas turbine cycle can be expressed as

$$\dot{W}_{net} = \dot{W}_{GT} - \dot{W}_{AC} \tag{11.10}$$

where C_{pg} denotes the specific heat at constant pressure of the combustion gas, and is taken to be a function of temperature as follows:

$$C_{pg}(T) = 0.991 + \left(\frac{6.997T}{10^5}\right) + \left(\frac{2.712T^2}{10^7}\right) - \left(\frac{1.2244T^3}{10^{10}}\right) \tag{11.11}$$

11.2.1.2 Bottoming Cycle

Energy balances and governing equations for the components of the bottoming cycle (steam turbine cycle and HRSG) are provided here.

Dual Pressure HRSG A dual pressure HRSG with two economizers (LP and HP) and two evaporators (LP and HP) is used in the multigeneration cycle to provide both low- and high-pressure steam. The LP steam is used to drive the absorption chiller and the HP steam to generate electricity. The temperature profile in the HRSG is shown in Figure 11.4, where the pinch point is defined as the difference between the temperature of the gas at the entrance of the evaporator (economizer side) and the saturation temperature. The dual pressure HRSG has two pinch points (PP_{HP} and PP_{LP}). The temperature differences between the water leaving the economizers (T_{20} and T_{22}) and the saturation temperature (T_5 and T_{17}) are the approach points (AP_{HP} and AP_{LP}),

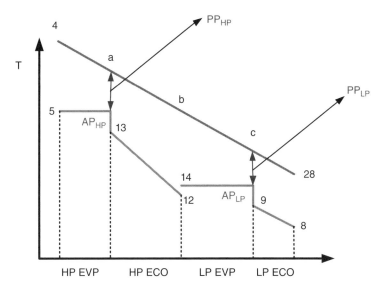

Figure 11.4 Temperature profile of HRSG.

which depend on the economizer's tube layout. Note that the pinch point and approach temperatures are considered constant here. Energy balances for each element of the HRSG are expressed as follows:

$$\dot{m}_{w,HP}(h_5 - h_{13}) = \dot{m}_g C_{Pg}(T_4 - T_a) \tag{11.12}$$

$$\dot{m}_{w,HP}(h_{13} - h_{12}) = \dot{m}_g C_{Pg}(T_a - T_b) \tag{11.13}$$

$$\dot{m}_{w,LP}(h_{11} - h_{10}) = \dot{m}_g C_{Pg}(T_b - T_c) \tag{11.14}$$

$$\dot{m}_w(h_9 - h_8) = \dot{m}_g C_{Pg}(T_c - T_d) \tag{11.15}$$

Steam Turbine   An energy balance for the steam turbine shown in Figure 11.3 and an expression for the steam turbine isentropic efficiency are written respectively as follows:

$$\dot{m}_w h_5 = \dot{W}_{ST} - \dot{m}_w h_6 \tag{11.16}$$

$$\eta_{ST} = \frac{\dot{W}_{ST,act}}{\dot{W}_{ST,is}} \tag{11.17}$$

Condenser   An energy balance for the condenser follows:

$$\dot{m}_6 h_6 = \dot{Q}_{cond} - \dot{m}_7 h_7 \tag{11.18}$$

Pump   An energy balance for the pump and an expression for its isentropic efficiency follows:

$$\dot{m}_w h_7 + \dot{W}_{pump} = \dot{m}_w h_8 \tag{11.19}$$

$$\eta_{pump} = \frac{\dot{W}_{is}}{\dot{W}_{act}} \tag{11.20}$$

11.2.1.3 Absorption Chiller

The principle of mass conservation and the first and second laws of thermodynamics are applied to each component of the single effect absorption chiller. In our analysis, each component is considered as a control volume with inlet and outlet streams, and heat and work interactions are considered. Mass balances are applied for the total mass and each material of the working fluid solution. The governing and conservation rate equations for total mass and each material of the solution for a steady state and steady flow case follow [6]:

$$\sum \dot{m}_i = \sum \dot{m}_o \tag{11.21}$$

$$\sum (\dot{m}x)_i = \sum (\dot{m}x)_0 \tag{11.22}$$

Here, $\dot{m}$ is the working fluid mass flow rate and x is mass concentration of LiBr in the solution. For each component of the absorption system, a general energy rate balance is written as

$$\dot{Q} - \dot{W} = \sum \dot{m}_o h_o - \sum \dot{m}_i h_i \tag{11.23}$$

The cooling load of the absorption chiller is defined as

$$\dot{Q}_{\text{cooling}} = \dot{m} \times (h_{22} - h_{21}) \tag{11.24}$$

Further information about the thermodynamic modeling and energy balances for each component was already given in the refrigeration chapter in this book (Chapter 5).

11.2.1.4 Domestic Hot Water Heater

The hot gases from the heat recovery heat exchanger enter the water heater to warm domestic hot water to 60°C. Water enters this heater at a pressure and temperature of 3 bar and 20°C, respectively. An energy rate balance for this component is as follows:

$$\dot{m}_{w,LP}(h_{15} - h_{16}) = \dot{m}_w(h_{18} - h_{17}) \tag{11.25}$$

11.2.1.5 Organic Rankine Cycle

The hot flue gases leaving the HRSG still have energy that can be utilized in a heat recovery vapor generator in an organic Rankine cycle to produce both cooling and electricity. Energy balances and governing equations for the various components of the ORC cycle (see Figure 11.3) are provided below.

Ejector An ejector, which is a type of pump, uses the Venturi effect of a converging–diverging nozzle to convert the mechanical energy (pressure) of a motive fluid to kinetic energy (velocity), creating a low pressure zone that draws in and entrains a suction fluid. After passing through the throat of the injector, the mixed fluid expands and the velocity is reduced, recompressing the mixed fluids by converting velocity back to pressure. The motive fluid may be a liquid, steam, or any other gas.

The process occurring in the ejector (Figure 11.5) is assumed to be steady state, one dimensional, and adiabatic, and no work is done during the process. The velocities at the inlet and outlet of the ejector can be considered negligible [7]. For simplicity the effect of losses in the nozzle, mixing section, and diffuser are accounted for by the efficiency for each section of the ejector. The primary motive flow enters the ejector at point 30, and

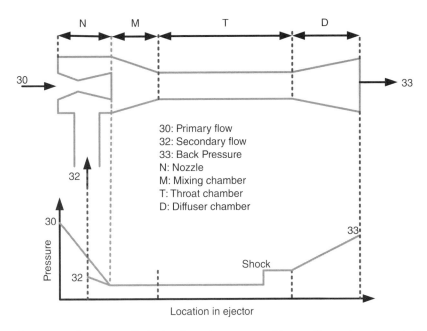

Figure 11.5 Pressure profile in the ejector for system multigeneration system shown in Figure 11.3. *Source:* Adapted from Wang 2009.

the suction flow exits the evaporator at point 32. The process in the ejector includes the expansion of the high pressure prime motive flow through the nozzle, mixing with the low pressure secondary flow in the mixing section at constant pressure, and diffusing to the outlet of the ejector (point 33) while the kinetic energy of the mixture is converted to pressure head. An important parameter for the secondary flow is the entrainment ratio, defined as

$$\omega = \frac{\dot{m}_{32}}{\dot{m}_{30}} \tag{11.26}$$

In the nozzle section in Figure 11.5, the inlet velocity of the primary flow $V_{\mathrm{pf,n1}}$ is negligible, so the exit specific enthalpy and velocity of primary flow can be expressed as

$$V_{pf,n_2} = \sqrt{2\eta_{\mathrm{noz}}(h_{pf,n_1} - h_{pf,n_2,s})} \tag{11.27}$$

where h_{pf,n_1} is the specific enthalpy at point 30 and $h_{pf,n_2,s}$ is the exit specific enthalpy of the primary flow for isentropic expansion and η_{noz} is the nozzle efficiency.

The momentum conservation equation for the mixing chamber area can be written as

$$\dot{m}_{30}V_{pf,n_2} + \dot{m}_{32}V_{sf,n_2} = (\dot{m}_{30} + \dot{m}_{32})V_{mf,m,s} \tag{11.28}$$

Neglecting the secondary flow velocity V_{sf,n_2} compared to the primary flow velocity V_{pf,n_2}, the exit velocity of mixed flow $V_{mf,m,s}$ can be expressed as

$$V_{mf,m,s} = \frac{V_{pf,n_2}}{1+\omega} \tag{11.29}$$

The mixing chamber efficiency can be expressed as

$$\eta_{\text{mix}} = \frac{V^2_{mf,m}}{V^2_{mf,m,s}} \tag{11.30}$$

Therefore, the actual velocity of the mixed flow is expressed as

$$V_{mf,m} = \frac{V_{pf,n2} \sqrt{\eta_{\text{mix}}}}{1 + \omega} \tag{11.31}$$

The energy equation for the mixing chamber gives

$$\dot{m}_{30}\left(h_{pf,n_2} + \frac{V^2_{pf,n_2}}{2}\right) + \dot{m}_{32}\left(h_{sf,n_2} + \frac{V^2_{sf,n_2}}{2}\right) = \dot{m}_{33}\left(h_{mf,m} + \frac{V^2_{mf,m}}{2}\right) \tag{11.32}$$

By simplifying this equation and using Equations 11.26 and 11.31, the specific enthalpy of mixed flow is obtained:

$$h_{mf,m} = \frac{h_{pf,n_1} + \omega h_{sf,n_2}}{1 + \omega} - \frac{V^2_{mf,m}}{2} \tag{11.33}$$

In the diffuser section, the mixed flow converts its kinetic energy to a pressure increase. Assuming the exit velocity of the mixed flow to be negligible and considering the diffuser efficiency, the actual exit specific enthalpy of the mixed flow can be written as

$$h_{33} = h_{mf,m} + (h_{mf,d,s} - h_{mf,m})/\eta_{\text{dif}} \tag{11.34}$$

where $h_{mf,d,s}$ is the ideal exit specific enthalpy of the mixed flow with isentropic compression, and η_{dif} is the diffuser efficiency.

Using these equations, the entrainment ratio is expressed as [7]:

$$\omega = \sqrt{\eta_{\text{noz}}\eta_{\text{mix}}\eta_{\text{dif}}\frac{h_4 - h_a}{h_5 - h_b}} - 1 \tag{11.35}$$

where η_{noz}, η_{mix} and η_{dif} are the nozzle, mixing chamber, and diffuser efficiencies. A flow chart for the ejector modeling is shown in Figure 11.6.

11.2.1.6 Heat Recovery Vapor Generator (HRVG)

As shown in Figure 11.3, R123 vapor is generated in the HRVG using the hot flue gases leaving the HRSG. An energy rate balance for this component is written as

$$\dot{m}_4 C_p(T_d - T_{28}) = \dot{m}_{ORC}(h_{29} - h_{41}) \tag{11.36}$$

ORC Turbine Saturated vapor at point 29 enters the ORC turbine and part exits the ORC turbine to drive the ejector. Writing an energy rate balance for a control volume around the ORC turbine gives

$$\dot{W}_{ORC,T} = \dot{m}_{29}h_{29} - \dot{m}_{30}h_{30} - \dot{m}_{31}h_{31} \tag{11.37}$$

Preheater The hot vapor leaving the ejector at point 34 enters a preheater to increase the temperature of the working fluid at point 40. An energy rate balance for this component can be written as

$$\dot{m}_{34}(h_{34} - h_{35}) = \dot{m}_{40}(h_{40} - h_{41}) \tag{11.38}$$

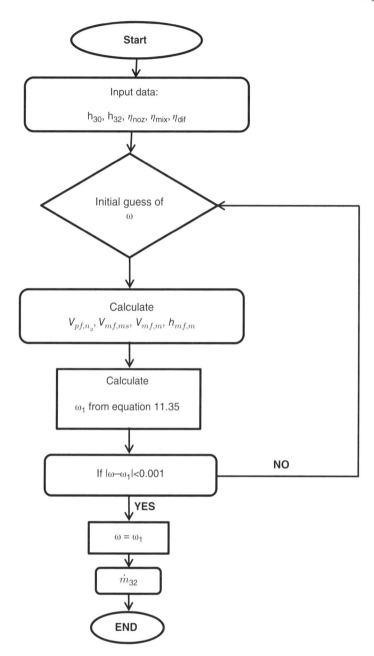

Figure 11.6 Ejector modeling flowchart.

Condenser The saturated vapor leaving the preheater at point 35 enters the condenser. The saturated liquid exiting the condenser is divided into two branches: one to an ORC pump to return to the ORC cycle and another to an expansion valve to provide the cooling capacity of the system. An energy rate balance for the condenser can be written as

$$\dot{Q}_{cond} = \dot{m}_{35}(h_{35} - h_{36}) \tag{11.39}$$

ORC Pump The ORC pump work can be expressed using an energy rate balance for a control volume around the ORC pump as follows:

$$\dot{W}_{ORC,pump} = \dot{m}_{ORC}(h_{40} - h_{37}) \tag{11.40}$$

Expansion Valve An energy balance for the expansion valve gives

$$h_{38} = h_{39} \tag{11.41}$$

Evaporator As shown in Figure 11.3, an evaporator is used for district cooling. An energy rate balance for this component is written as

$$\dot{Q}_{cooling} = \dot{m}_{32}(h_{32} - h_{39}) \tag{11.42}$$

PEM Electrolyzer Hydrogen as an energy carrier can facilitate sustainable energy systems. The development of sustainable carbon-neutral energy sources has become one of the most significant challenges in the world today. Hydrogen can be produced from various energy sources using methods such as biomass conversion, steam methane reforming, and water splitting. Hydrogen can be produced in a relatively environmentally benign manner (depending on the source of the input energy) via splitting water by photo-catalysis, thermochemical cycles, and electrolysis. Currently, both thermochemical and photocatalysis hydrogen production are not economically competitive. Water electrolysis is a mature technology for large scale hydrogen production. Hydrogen production by proton exchange membrane (PEM) electrolysis has numerous advantages, such as low environmental impact and ease of maintenance.

The PEM electrolyzer for H_2 production is illustrated on the right side of Figure 11.3. During electrolysis, electricity and heat are both supplied to the electrolyzer to drive the electrochemical reactions. As shown in Figure 11.3, liquid water is fed to the PEM electrolyzer at ambient temperature and enters a heat exchanger that heats it to the PEM electrolyzer temperature before it enters the electrolyzer. Leaving the cathode, the H_2 produced dissipates heat to the environment and cools to the reference environment temperature. The oxygen gas produced at the anode is separated from the water and oxygen mixture and then cooled to the reference environment temperature. The remaining water is returned to the water supply stream for the next hydrogen production cycle. The overall PEM electrolysis reaction is simply water splitting, that is, electricity and heat are used to separate water into hydrogen and oxygen. Hydrogen is stored in a tank for later usage.

Thermochemical modeling is carried out for the PEM electrolyzer, along with energy and exergy analyses. The total energy needed by the electrolyzer can obtained as

$$\Delta H = \Delta G + T\Delta S \tag{11.43}$$

where ΔG is the Gibbs free energy and $T\Delta S$ represents the thermal energy requirement. The values of G, S, and H for hydrogen, oxygen and water can be obtained from thermodynamic tables. The total energy need is the theoretical energy required for H_2O electrolysis without any losses. The catalyst used in PEM electrolysis provides an alternative path for the reaction with lower activation energy. The mass flow rate of hydrogen is determined by [8]:

$$\dot{N}_{H_2,out} = \frac{J}{2F} = \dot{N}_{H_2O,reacted} \tag{11.44}$$

Here, J is the current density and F is the Faraday constant. The PEM electrolyzer voltage can be expressed as

$$V = V_0 + V_{act,a} + V_{act,c} + V_{ohm} \tag{11.45}$$

where V_0 is the reversible potential, which is related to the difference in free energy between reactants and products and can be obtained with the Nernst equation as follows:

$$V_0 = 1.229 - 8.5 \times 10^{-4}(T_{PEM} - 298) \tag{11.46}$$

Here, $V_{act,a}$, $V_{act,c}$ and V_{ohm} are the activation overpotential of the anode, the activation overpotential of the cathode, and the Ohmic overpotential of the electrolyte, respectively. Ohmic overpotential in the proton exchange membrane (PEM) is caused by the resistance of the membrane to the hydrogen ions transported through it. The ionic resistance of the membrane depends on the degree of humidification and thickness of the membrane as well as the membrane temperature. The local ionic conductivity $\sigma(x)$ of the proton exchange membrane can be expressed as [9]:

$$\sigma_{PEM}[\lambda(x)] = [0.5139\lambda(x) - 0.326] \exp\left[1268\left(\frac{1}{303} - \frac{1}{T}\right)\right] \tag{11.47}$$

where x is the distance into the membrane measured from the cathode–membrane interface and $\lambda(x)$ is the water content at a location x in the membrane. The value of $\lambda(x)$ can be calculated in terms of the water content at the membrane–electrode edges:

$$\lambda(x) = \frac{\lambda_a - \lambda_c}{D} x + \lambda_c \tag{11.48}$$

Here, D is the membrane thickness, and λ_a and λ_c are the water contents at the anode–membrane and the cathode–membrane interfaces, respectively. The overall Ohmic resistance can thus be expressed as [9]:

$$R_{PEM} = \int_0^D \frac{dx}{\sigma_{PEM}[\lambda(x)]} \tag{11.49}$$

Based on Ohm's law, the following equation can be written for the Ohmic overpotential:

$$V_{ohm,PEM} = JR_{PEM} \tag{11.50}$$

The activation overpotential, V_{act}, caused by a deviation of net current from its equilibrium and an electron transfer reaction, must be differentiated from the concentration of the oxidized and reduced species. Then,

$$V_{act,i} = \frac{RT}{F} \sinh^{-1}\left(\frac{J}{2J_{0,i}}\right), \quad i = a, c \tag{11.51}$$

Here, J_o is the exchange current density, which is an important parameter in the activation overpotential. It characterizes the electrode's capabilities in the electrochemical reaction. A high exchange current density implies a high reactivity of the electrode, which results in a lower overpotential. The exchange current density for electrolysis can be expressed as [8]

$$J_{0,i} = J_i^{ref} \exp\left(-\frac{V_{act,i}}{RT}\right), \quad i = a, c \tag{11.52}$$

where J_i^{ref} is the pre-exponential factor and $V_{act,i}$ is the activation energy for the anode and cathode. Further details about PEM electrolysis modeling can be found elsewhere [8, 9].

11.2.2 Exergy Analysis

Much detail on exergy has been given in previous chapters but a few pertinent points are made here since they are germane to the present analysis. The exergy of each flow is calculated and the changes in exergy are determined for each major component. Expressions for the exergy destruction rates for all components in the multigeneration system are given in Table 11.1. Since in this multigeneration energy system, a combustion reaction occurs in the combustion chamber, it is important to determine chemical exergy where combustion occurs. Chemical exergy is equal to the maximum amount of work that can be obtained when a substance is brought from the reference environment state to the dead state by a process including heat transfer and exchange of substances only with the reference environment. The maximum work is attained when the process is reversible. Alternatively, chemical exergy can also be viewed as the exergy of a substance that is at the reference environment state.

Chemical exergy is also equivalent to the minimum amount of work necessary to produce a substance at the reference environment state from the constituents in the reference environment, where all are at the same temperature and pressure. Chemical

Table 11.1 Expressions for exergy destruction rates for components of the multigeneration system.

Component	Exergy destruction rate expression
Air compressor	$\dot{Ex}_{D,AC} = \dot{Ex}_1 - \dot{Ex}_2 + \dot{W}_{AC}$
Combustion chamber (CC)	$\dot{Ex}_{D,CC} = \dot{Ex}_2 + \dot{Ex}_f - \dot{Ex}_3$
Gas turbine (GT)	$\dot{Ex}_{D,GT} = \dot{Ex}_3 - \dot{Ex}_4 - \dot{W}_{GT}$
HRSG	$\dot{Ex}_{D,HRSG} = \dot{Ex}_4 + \dot{Ex}_8 - \dot{Ex}_5 - \dot{Ex}_c$
Steam turbine (ST)	$\dot{Ex}_{D,ST} = \dot{Ex}_5 - \dot{Ex}_6 - \dot{W}_{ST}$
Steam condenser	$\dot{Ex}_{D,cond} = \dot{Ex}_6 + \dot{Ex}_{49} - \dot{Ex}_7 - \dot{Ex}_{50}$
Pump	$\dot{Ex}_{D,P} = \dot{Ex}_7 - \dot{Ex}_8 + \dot{W}_P$
Heat recovery vapor generator	$\dot{Ex}_{D,HRVG} = \dot{Ex}_c + \dot{Ex}_{41} - \dot{Ex}_{28} - \dot{Ex}_{29}$
ORC turbine	$\dot{Ex}_{D,ORC\,T} = \dot{Ex}_{29} - \dot{W}_{ORC} - \dot{Ex}_{30} - \dot{Ex}_{31}$
Ejector	$\dot{Ex}_{D,ejector} = \dot{Ex}_{30} + \dot{Ex}_{32} - \dot{Ex}_{33}$
Preheater	$\dot{Ex}_{D,PRH} = \dot{Ex}_{34} + \dot{Ex}_{40} - \dot{Ex}_{35} - \dot{Ex}_{41}$
ORC pump	$\dot{Ex}_{D,ORC\,pump} = \dot{Ex}_{37} + \dot{W}_{ORC} - \dot{Ex}_{40}$
ORC condenser	$\dot{Ex}_{D,cond} = \dot{Ex}_{35} - \dot{Ex}_{36} - \dot{Ex}_{Q,cond}$
ORC evaporator	$\dot{Ex}_{D,EVP} = \dot{Ex}_{39} + \dot{Ex}_{40} - \dot{Ex}_{32} - \dot{Ex}_{41}$
ORC expansion valve	$\dot{Ex}_{D,EXV} = \dot{Ex}_{38} - \dot{Ex}_{39}$
Domestic water heater	$\dot{Ex}_{D,DWH} = \dot{Ex}_{15} + \dot{Ex}_{17} - \dot{Ex}_{16} - \dot{Ex}_{18}$
PEM electrolyzer	$\dot{Ex}_{D,PEM} = \dot{Ex}_{44} + \dot{W}_{PEM} - \dot{Ex}_{46} - \dot{Ex}_{47} + \dot{Ex}_Q$
Absorption condenser	$\dot{Ex}_{D,Cond} = \dot{Ex}_{19} - \dot{Ex}_{20} - \dot{Ex}_Q$
Absorption expansion valve	$\dot{Ex}_{D,EXV} = \dot{Ex}_{20} - \dot{Ex}_{21}$
Absorption evaporator	$\dot{Ex}_{D,EVP} = \dot{Ex}_{21} - \dot{Ex}_{22} + \dot{Ex}_Q$
Absorber	$\dot{Ex}_{D,Abs} = \dot{Ex}_{22} + \dot{Ex}_{23} - \dot{Ex}_{25} - \dot{Ex}_Q$
Absorption pump	$\dot{Ex}_{D,P} = \dot{Ex}_{25} + \dot{W}_P - \dot{Ex}_{26}$
Absorption heat exchanger	$\dot{Ex}_{D,HEX} = \dot{Ex}_{26} + \dot{Ex}_{14'} - \dot{Ex}_{24} - \dot{Ex}_{27}$
Absorption generator	$\dot{Ex}_{D,Gen} = \dot{Ex}_{14} + \dot{Ex}_{27} - \dot{Ex}_{15} - \dot{Ex}_{14'} - \dot{Ex}_{19}$

exergy has two main parts, reactive exergy resulting from the chemical reactions necessary to produce species that do not exist as stable components in the reference environment, and concentration exergy resulting from the difference between the chemical concentration of a species in a system and its chemical concentration in the reference environment [10]. The concentration part is related to the exergy of purifying or diluting a substance, such as separating oxygen from air.

11.2.2.1 Exergy Efficiency

The exergy efficiency, defined as the product exergy output divided by the exergy input, for the gas turbine, CHP, and the overall multigeneration system, can be expressed as follows:

$$\Psi_{power} = \frac{\dot{W}_{net,GT}}{\dot{Ex}_f} \tag{11.53}$$

$$\Psi_{CHP} = \frac{\dot{W}_{net,GT} + \dot{Ex}_{heating}}{\dot{Ex}_f} \tag{11.54}$$

$$\Psi_{multi} = \frac{\dot{W}_{net,GT} + \dot{W}_{net,ST} + \dot{W}_{net,ORC} + \dot{Ex}_{heating} + \dot{Ex}_{cooling,chiller} + \dot{Ex}_{cooling,ORC} + \dot{Ex}_{H_2} + \dot{Ex}_{18}}{\dot{Ex}_f} \tag{11.55}$$

Here,

$$\dot{Ex}_{heating} = \dot{Q}_{cond} \left(1 - \frac{T_0}{T_{cond}}\right) \tag{11.56}$$

$$\dot{Ex}_{cooling} = \dot{Q}_{cooling} \left(\frac{T_0 - T_{EVP}}{T_{EVP}}\right) \tag{11.57}$$

$$\dot{Ex}_{H_2} = \dot{m}_{H_2} ex_{H_2} \tag{11.58}$$

11.2.3 Economic Analysis

In order to perform the economic analysis and optimization of the multigeneration systems, we first determine the purchase cost of each component used in the system as a function of the main design parameters. In this section, the cost function of each component and some economic parameters are defined and explained.

The investment cost of equipment is most detailed and accurate when obtained from vendors. Comprehensive methods to express the variation of investment cost as a mathematical function of the variation of key parameter costs are useful and convenient. Alternatively, these complex data and mathematical cost functions can be presented in an approximate and compact form as described below.

11.2.3.1 Brayton Cycle

The investment cost of each component in the gas turbine cycle is given as follows:

Air Compressor The purchase cost of the air compressor is a function of air mass flow rate, compressor pressure ratio, and compressor isentropic efficiency is expressible as follows [11]:

$$Z_{AC}(\$) = C_{11} \dot{m}_{air} \frac{1}{C_{12} - \eta_{AC}} \left(\frac{P_2}{P_1}\right) \ln\left(\frac{P_2}{P_1}\right) \tag{11.59}$$

where $C_{11} = 44.71$ \$/(kg/s) and $C_{12} = 0.95$.

Combustion Chamber (CC) The combustion chamber, where fuel is burned to produce high temperature flue gases, is a major thermodynamic component in the gas turbine cycle. The purchase cost of the combustion chamber is a function of the air mass flow rate entering the chamber and the gas turbine inlet temperature (GTIT). The purchase cost of combustion chamber is expressed as

$$Z_{CC}(\$) = C_{21}\dot{m}_{\text{air}}\{1 + \exp[C_{22}(T_3 - C_{23})]\}\frac{1}{0.995 - \frac{P_3}{P_2}} \tag{11.60}$$

where $C_{21} = 28.98\ \$/(\text{kg/s})$, $C_{22} = 0.015\frac{1}{\text{K}}$ and $C_{23} = 1540$.

Gas Turbine (GT) The purchase cost of the gas turbine is a function of inlet gas mass flow rate, gas turbine pressure ratio, and gas turbine isentropic efficiency. The cost function can be defined as follows [11]:

$$Z_{GT}(\$) = C_{31}\dot{m}_{\text{gas}}\frac{1}{C_{32} - \eta_{GT}}\ln\left(\frac{P_3}{P_4}\right)\{1 + \exp[C_{33}(T_3 - 1570)]\} \tag{11.61}$$

where $C_{31} = 301.45\ \$/(\text{kg/s})$, $C_{23} = 0.025\frac{1}{\text{K}}$ and $C_{32} = 0.95$.

11.2.3.2 Steam Cycle

Exhaust gases exiting the gas turbine at point 4 still have sufficient thermal energy to produce vapor at point 5. The steam cycle in this multigeneration system consists of a dual pressure heat recovery steam generator (HRSG), a steam turbine, a condenser, and a pump. Expressions for the purchase cost of each component follow:

Heat Recovery Steam Generator (HRSG) The purchase cost of a double pressure HRSG is a function of several design parameters such as the high and low pressures, the high and low pressure mass flow rates, and the flue gas mass flow rate passing at each pressure level. The cost function of a double pressure HRSG can be expressed as follows [11]:

$$Z_{HRSG}(\$) = C_{41}\sum_i\left[f_{p,i}f_{T,\text{steam},i}f_{T,\text{gas},i}\left(\frac{\dot{Q}}{\Delta T_{ln,i}}\right)^{0.8}\right]$$
$$+ C_{42}\sum_j(f_{p,j}\dot{m}_{\text{steam},j}) + C_{43}\dot{m}_{\text{gas}}^{1.2} \tag{11.62}$$

where

$$f_{p,i} = 0.0971\frac{P_i}{30\ bar} + 0.9029 \tag{11.63}$$

$$f_{T,\text{steam},i} = 1 + \exp\left(\frac{T_{\text{out,steam},i} - 830}{500\ K}\right) \tag{11.64}$$

$$f_{T,\text{gas},i} = 1 + \exp\left(\frac{T_{\text{out,gas},i} - 990}{500\ K}\right) \tag{11.65}$$

Here, $C_{41} = 4138.85\ \$/\left(\frac{\text{kW}}{\text{K}}\right)^{0.8}$, $C_{42} = 13\,380\$/(\text{kg/s})$ and $C_{43} = 1489.7\ \$/(\text{kg/s})$.

Steam Turbine The purchase cost of the steam turbine is a function of turbine inlet temperature, steam turbine isentropic efficiency, and turbine work, and can be expressed as follows:

$$Z_{ST}(\$) = C_{51}(\dot{W}_{ST})^{0.7}\left[1 + \left(\frac{0.05}{1 - \eta_{ST}}\right)^3\right]\left\{1 + \exp\left(\frac{T_5 - 866\ K}{10.42\ K}\right)\right\} \quad (11.66)$$

where $C_{31} = 3880.5\frac{\$}{kW^{0.7}}$.

Condenser The purchase cost of the condenser can be expressed as follows:

$$Z_{cond}(\$) = C_{61}\frac{\dot{Q}_{cond}}{k\Delta T_{ln}} + C_{62}\dot{m}_{CW} \quad (11.67)$$

Where $C_{61} = 280.74\frac{\$}{m^2}$, $C_{62} = 746\$/(kg/s)$, $k = 2200\frac{W}{m^2 K}$

Here, $\dot{m}_{CW}$ is the cooling water mass flow rate and ΔT_{ln} is the log mean temperature difference.

Pump The purchase cost of the pump can be written as follows:

$$Z_{pump}(\$) = C_{71}(\dot{W}_P)^{0.71}\left(1 + \frac{0.2}{1 - \eta_P}\right) \quad (11.68)$$

where $C_{71} = 705.48\ \$/kW$.

11.2.3.3 ORC Cycle

The ORC cycle shown in Figure 11.3 has several components, for which the purchase costs are described below.

Heat Recovery Vapor Generator (HRVG) The purchase cost of the HRVG can be expressed as follows [12]:

$$Z_{HRVG}(\$) = 1010\ (A_{HRVG})^{0.8} \quad (11.69)$$

where

$$A_{HRVG} = \frac{\dot{m}_{28}\ Cp_g(T_C - T_{28})}{U_{HRVG}\Delta T_{ln}} \quad (11.70)$$

Here, U_{HRVG} is the overall heat transfer coefficient for the HRVG and has a value of $0.88\ \frac{kW}{m^2 K}$.

ORC Turbine The cost function of the ORC turbine follows [12]:

$$Z_T(\$) = 4750\ (\dot{W}_T)^{0.75} \quad (11.71)$$

Here, $\dot{W}_T$ is the work rate generated by the turbine (in kW).

Ejector The purchase cost of the ejector is a function of the motive mass flow rate, the inlet motive temperature and pressure, and the outlet pressure. The cost function is expressible as follows [13]:

$$Z_{ejector}(\$) = 1000 \times 15.96\dot{m}_{30}\left(\frac{T_{30} + 273.15}{\frac{P_{30}}{1000}}\right)^{0.05}\left(\frac{P_{33}}{1000}\right)^{-0.75} \quad (11.72)$$

where P denotes pressure in kPa and T temperature in °C.

Evaporator The purchase cost of the ORC evaporator can be expressed as follows [12]:

$$Z_{EVP}(\$) = 309.14(A_{EVP})^{0.85} \tag{11.73}$$

where

$$A_{EVP} = \frac{\dot{Q}_{EVP}}{U_{EVP}\Delta T_{\text{ln}}} \tag{11.74}$$

Here, U_{EVP} is the overall heat transfer coefficient for evaporator with the value of $0.2 \frac{\text{kW}}{\text{m}^2\text{K}}$.

Expansion Valve The purchase cost of expansion valve can be expressible as follows [12]:

$$Z_{EXV}(\$) = 37\left(\frac{P_{38}}{P_{39}}\right)^{0.68} \tag{11.75}$$

Preheater The cost function of a preheater in ORC cycle can be treated as a heat exchanger that can be calculated as [12]:

$$Z_{\text{preheater}}(\$) = 1000(A_{\text{preheater}})^{0.65} \tag{11.76}$$

where

$$A_{\text{preheater}} = \frac{\dot{Q}_{\text{preheater}}}{U_{\text{preheater}}\Delta T_{\text{ln}}} \tag{11.77}$$

Here, $U_{\text{preheater}}$ is the overall heat transfer coefficient for preheater and has a value of $0.85 \frac{kW}{m^2 K}$.

Condenser The purchase cost of the ORC condenser can be written as follows [12]:

$$Z_{\text{cond}}(\$) = 516.62(A_{\text{cond}})^{0.6} \tag{11.78}$$

where

$$A_{EVP} = \frac{\dot{Q}_{\text{cond}}}{U_{\text{cond}}\Delta T_{\text{ln}}} \tag{11.79}$$

Here, U_{cond} is the overall heat transfer coefficient for evaporator and has a value of $0.15 \frac{kW}{m^2 K}$.

Pump The cost of the ORC pump can be expressed as follows:

$$Z_{\text{pump}}(\$) = 200(\dot{W}_P)^{0.65} \tag{11.80}$$

11.2.3.4 Absorption Chiller

The purchase cost of the absorption chiller is a function of all of its design parameters. These can be compacted and approximated as function of the cooling load of the chiller as follows [14]:

$$Z_{\text{chiller}}(\$) = 1144.3(\dot{Q}_{EVP})^{0.67} \tag{11.81}$$

Here, $\dot{Q}_{EVP}$ is the cooling load of the absorption chiller (in kW). It can be determined using an energy balance for the control volume around the evaporator shown in Figure 11.3

11.2.3.5 PEM Electrolyzer

The purchase cost of the electrolyzer is a function of the electricity input to split water and can be expressed as [15]:

$$Z_{PEM}(\$) = 1000\dot{W}_{PEM} \tag{11.82}$$

11.2.3.6 Domestic Hot Water (DHW) Heater

The cost of the domestic water heater considered in this system can be expressed as follows [12]:

$$Z_{DWH}(\$) = 0.3 m_{DWH} \tag{11.83}$$

Here, m_{DWH} is the hot water production in a cubic meter that is calculated using the energy balance equation for a control volume around the DHW heater.

11.2.3.7 Capital recovery factor (CRF)

The capital recovery factor is the ratio of a constant annuity to the present value of receiving that annuity for a given length of time [16]. Using an interest rate i, the capital recovery factor is defined as

$$CRF = \frac{i \times (1+i)^n}{(1+i)^n - 1} \tag{11.84}$$

Here, i denotes the interest rate and n the total operating period of the system in years. The cost rate of each device is determined as

$$\dot{Z} = \frac{Z_k CRF\Phi}{N \times 3600} \tag{11.85}$$

where Z_k is the purchase cost of the kth component, N is the annual number of operation hours for the unit, and φ is the maintenance factor, which is often 1.06 [16].

11.2.4 Multi-objective Optimization

A multi-objective optimization method based on an evolutionary algorithm is applied to the multigeneration system for heating, cooling, electricity, hot water, and hydrogen to determine the best design parameters for the system. Objective functions, design parameters, and constraints, as well as the overall optimization process are described in this section.

11.2.4.1 Definition of Objectives

Two objective functions are considered here for multi-objective optimization: exergy efficiency (to be maximized) and total cost rate of product (to be minimized). The cost of pollution damage is assumed to be added directly to the expenditures that must be paid, making the second objective function the sum of thermodynamic and environmental objectives. Consequently, the objective functions in this analysis can be expressed as follows:

Exergy Efficiency

$$\Psi_{\text{multi}} = \frac{\dot{W}_{net,GT} + \dot{W}_{net,ST} + \dot{W}_{net,ORC} + \dot{Ex}_{\text{heating}} + \dot{Ex}_{\text{cooling,chiller}} + \dot{Ex}_{\text{cooling,ORC}} + \dot{Ex}_{H_2} + \dot{Ex}_{18}}{\dot{Ex}_f} \tag{11.86}$$

Total Cost Rate

$$\dot{C}_{tot} = \sum_k \dot{Z}_k + \dot{C}_f + \dot{C}_{env} \tag{11.87}$$

where $\dot{Z}_K$ is the purchase cost rate of each component, and the cost rates of environmental impact and fuel are expressed respectively as

$$\dot{C}_{env} = C_{Co}\dot{m}_{Co} + C_{NOx}\dot{m}_{NOx} + C_{Co_2}\dot{m}_{Co_2} \tag{11.88}$$

$$\dot{C}_f = c_f\dot{m}_f\ LHV \tag{11.89}$$

More details about equipment purchase cost can be found elsewhere [17]. The purchase cost of each component in this multigeneration system was explained in section 11.2.3. Also, C_f is the fuel cost which is taken to be 0.003 \$/MJ here. In this analysis, we express the environmental impact as the total cost rate of pollution damage (\$/s) due to CO, NO_x and CO_2 emissions by multiplying their respective flow rates by their corresponding unit damage costs, C_{CO}, C_{NOx} and C_{CO2}, values for which are taken to be 0.02086 \$/kg, 6.853 \$/kg and 0.0240 \$/kg, respectively [17]. The cost of pollution damage is assumed here to be added directly to other system costs.

11.2.4.2 Decision Variables

The following decision variables (design parameters) are selected for this study: compressor pressure ratio (r_{AC}), compressor isentropic efficiency (η_{AC}), gas turbine isentropic efficiency (η_{GT}), gas turbine inlet temperature ($GTIT$), high pressure pinch point temperature (PP_{HP}) difference, low pressure pinch point temperature (PP_{LP}) difference, high pressure value (P_{HP}), low pressure value (P_{LP}), steam turbine isentropic efficiency (η_{ST}), pump isentropic efficiency (η_p), condenser pressure (P_{Cond}), absorption chiller evaporator temperature (T_{EVP}), ORC turbine inlet pressure (P_{ORC}), ORC turbine extraction pressure ($P_{ex,ORC}$), and ORC evaporator pressure ($P_{EVP,ORC}$). Although the decision variables may be varied in the optimization procedure, each is normally required to be within a reasonable range and this is accomplished using constraints. The constraints applied here are based on earlier reports and listed in Table 11.2.

11.2.5 Optimization Results

The optimization results are now described. The genetic algorithm optimization is performed for 250 generations, using a search population size of $M = 100$ individuals, a crossover probability of $p_c = 0.9$, a gene mutation probability of $p_m = 0.035$, and a controlled elitism value $c = 0.55$. Figure 11.7 shows the Pareto frontier solution for the multi-objective optimization of the multigeneration system for the objective functions in Equations 11.86 and 11.87. It can be seen in this figure that the total cost rate of products increases moderately as the total exergy efficiency of the cycle increases to about 65%. Increasing the total exergy efficiency from 65% to 68% increases the cost rate of product significantly.

The Pareto optimal curve (best rank) is clearly visible in the lower part of the figure (solid line) which is separately shown in Figure 11.7. As shown there, the maximum exergy efficiency exists at design point D (67.89%), while the total cost rate of products is the greatest at this point (615.75 \$/hr). Also, the minimum value for the total

Table 11.2 Optimization constraints and their rationales.

Constraint	Reason
$GTIT < 1550$ K	Material temperature limit
$P_2/P_1 < 22$	Commercial availability
$\eta_{AC} < 0.9$	Commercial availability
$\eta_{GT} < 0.9$	Commercial availability
$P_{HP} < 40$ bar	Commercial availability
$P_{LP} < 5.5$ bar	Commercial availability
$10\,°C < PP_{HP} < 22\,°C$	Heat transfer limit
$12\,°C < PP_{LP} < 22\,°C$	Heat transfer limit
$\eta_{ST} < 0.9$	Commercial availability
$\eta_p < 0.9$	Commercial availability
$2\,°C < T_{EVP} < 6\,°C$	Cooling load limitation
8 kPa $< P_{cond} < 10$ kPa	Thermal efficiency limit
500 kPa $< P_{ORC} < 750$ kPa	ORC commercial availability
180 kPa $< P_{ex,ORC} < 250$ kPa	ORC commercial availability
20 kPa $< P_{EVP,ORC} < 35$ kPa	Cooling load limitation

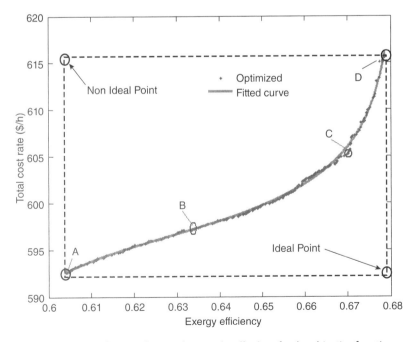

Figure 11.7 Pareto frontier showing best trade-off values for the objective functions.

Table 11.3 Optimized values for design parameters of the system based on multi-objective optimization.

Design parameter	A	B	C	D
η_{AC}	0.88	0.88	0.88	0.87
η_{GT}	0.90	0.89	0.90	0.90
r_{AC}	14.90	14.90	14.90	14.97
$GTIT$ (K)	1498	1495	1499	1496
P_{LP} (bar)	2.01	2.00	4.90	4.90
P_{HP} (bar)	12.29	23.40	29.90	29.90
PP_{HP} (°C)	14.98	14.90	14.90	4.46
PP_{LP} (°C)	14.92	14.90	14.80	14.95
T_{EVP} (°C)	5.00	1.10	2.31	2.10
η_{FWP}	0.84	0.84	0.75	0.87
η_{ST}	0.77	0.78	0.87	0.88
P_{cond} (kPa)	10.00	9.86	8.04	8.10
P_{ORC} (kPa)	718	689	503	506
$P_{ex,ORC}$ (kPa)	249	248	246	249
$P_{EVP,ORC}$ (kPa)	24.84	34.90	21.20	27.60

cost rate of product occurs at design point A and is about 592.6 $/hr. Design point A is the optimal situation when total cost rate of product is the sole objective function, while design point D is the optimum point when exergy efficiency is the sole objective function. In multi-objective optimization, a process of decision-making for selection of the final optimal solution from the available solutions is required. This process is usually performed with the aid of a hypothetical point in Figure 11.7 (the ideal point), at which both objectives have their optimal values independent of the other objectives. It is clear that it is impossible to have both objectives at their optimum point simultaneously and, as shown in Figure 11.7, the ideal point is not a solution located on the Pareto frontier. The closest point of the Pareto frontier to the ideal point might be considered as a desirable final solution. Nevertheless, in this case, the Pareto optimum frontier exhibits weak equilibrium, that is, a small change in exergy efficiency from varying the operating parameters causes a large variation in the total cost rate of product. Therefore, the ideal point cannot be utilized for decision-making in this problem. In selection of the final optimum point, it is desired to achieve a better magnitude for each objective than its initial value for the base case problem. Note that in multi-objective optimization and the Pareto solution, each point can be utilized as the optimized point. Therefore, the selection of the optimum solution depends on the preferences and criteria of the decision maker, suggesting that each may select a different point as the optimum solution depending on his/her needs. Table 11.3 shows all the design parameters for points A–D. The results of optimum exergy efficiency and total cost rate for all points evaluated over 300 generations are shown in Figure 11.8.

As shown in Figure 11.7, the optimized values for exergy efficiency on the Pareto frontier range between 60% and 68%. To provide a good relation between exergy efficiency and total cost rate, a curve is fitted on the optimized points obtained from the

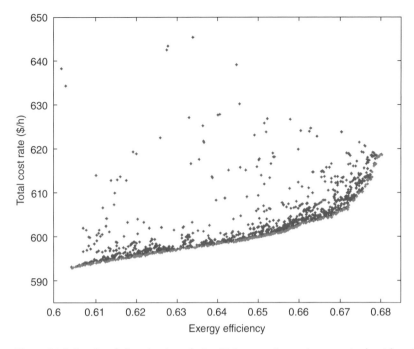

Figure 11.8 Results of all evaluations during 300 generations using genetic algorithm. A clear approximation of the Pareto frontier is visible in the lower part of the figure.

evolutionary algorithm. This fitted curve is shown in Figure 11.7. The expression for this fitted curve is given as follows:

$$\dot{C}_{total} = \frac{692.4\psi^3 - 2284\psi^2 + 1429\psi - 129.7}{\psi^5 + 51.48\psi^4 - 136.1\psi^3 + 130.7\psi^2 - 55.98\psi + 9.27} \tag{11.90}$$

To study the variation of thermodynamic characteristics, four points (A to D) on the Pareto frontier are considered. Table 11.4 shows the total cost rate of the system, the total exergy destruction, the system efficiency, the heating and cooling loads of the system, and the CO_2 emission of the system.

From point A to point D in this table, both the total cost rates of the system and the exergy efficiencies increase. As already stated, point A is preferred when total cost rate is a single objective function and design point D when exergy efficiency is a single objective function. Design point C has better results for both objective functions. Other

Table 11.4 Thermodynamic characteristics of four points on the Pareto frontier.

Point	$\dot{W}_{net}$ (kW)	ψ	$\dot{Ex}_{D,tot}$ (kW)	$\dot{Q}_{cooling}$ (kW)	$\dot{Q}_{heating}$ (kW)	$\dot{C}_{tot}$ ($/h)	CO_2 emission (kg/kWh)	$\dot{m}_{H_2}$ (kg/h)	$\dot{m}_{DWH}$ (kg/h)
A	10 304	0.60	14 911	929.95	4858	592.61	136.94	0.71	2983
B	10 817	0.63	14 437	915.10	5207	597.44	130.13	0.71	2938
C	11 393	0.67	13 909	904.34	6625	605.31	116.46	1.25	2981
D	11 451	0.68	13 845	930.35	6833	615.75	114.78	1.29	3064

thermodynamic properties correctly confirm this trend. For instance, from point B to C, the total exergy destruction rate decreases when the exergy efficiency increases.

11.3 Biomass Based Multigeneration Energy System

Renewable energy is a source of energy that comes from natural resources such as sunlight, wind, rain, tides, waves, geothermal heat, and biomass. These are naturally replenished when used. Biomass, as a renewable energy source, is biological material from living, or recently living, organisms [18]. Comprehensively, biomass comprises all the living matter present on earth and, as an energy source, biomass can either be used directly or converted into other energy products such as biofuels [18]. Currently, biomass resources are mainly used in the production of heating, cooling, and electricity. Direct combustion of biomass with coal is the most common method of conversion and provides the greatest potential for large scale utilization of biomass energy in the near term [19]. Other thermochemical conversion technologies such as gasification and pyrolysis are technically feasible and potentially efficient, compared to combustion, for power generation. However, these technologies either lack maturity and reliability or are not economically viable for large scale utilization [14]. Biomass based cogeneration systems have been studied over many years by numerous researchers for various industries (e.g., sugar, rice, palm oil, paper, and wood) as a means of waste disposal and energy recovery [20].

Figure 11.9 illustrates an integrated multigeneration system containing a biomass combustor, an ORC cycle to produce electricity, a double-effect absorption chiller for cooling, a heat exchanger for heating, a proton exchange membrane (PEM) electrolyzer to produce hydrogen, a domestic water heater to produce hot water, and a reverse osmosis (RO) desalination unit to produce fresh water. Pine sawdust is used as the biomass fuel and it is burned in a biomass combustor. The heat from the biomass combustor is input to the ORC cycle. The waste heat from the ORC is utilized to produce steam in the heating process via the heat exchanger, and to produce cooling using a double-effect absorption chiller. To have an efficient ORC, its working fluid should have a high critical temperature so that the waste heat can be used efficiently [5]. A typical organic fluid used in ORCs is n-octane, which has a relatively high critical temperature (569 K) [21]. This organic fluid is selected here as the working fluid of the ORC. The ORC cycle produces electricity, part of which is used for residential applications depending on the electricity needs of the building, and the remainder of which drives a PEM electrolyzer for hydrogen production and RO desalination to produce fresh water. The hydrogen and fresh water are stored in a hydrogen tank and fresh water tank respectively. Since the flue gases leaving the ORC evaporator still have energy, they are utilized to produce hot water in a domestic water heater.

As shown in Figure 11.9, biomass enters the combustor at point 30 and air enters at point 29. Hot flue gases leave the biomass combustor at point 31 and then enter a cyclone to remove ash. Hot flue gases without ash enter an ORC evaporator to produce steam at point 27 to rotate the ORC turbine blades and produce shaft work. The high pressure and temperature vapor at point 27 is expanded through the turbine to generate electrical power, and the extracted vapor from the turbine enters the heat exchanger for the heating process. Saturated vapor exits the heating process unit at point 24 and is input to the generator of the double-effect absorption system to provide the

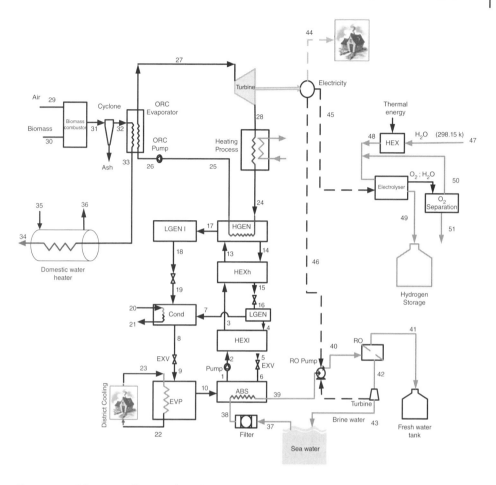

Figure 11.9 Schematic of biomass based a multigeneration energy system for the provision of heating, cooling, electricity, hydrogen, fresh water, and hot water.

cooling load of the system. Saturated liquid leaves the absorption generator and enters the ORC pump at point 25. The ORC pump increases the pressure of ORC working fluid, and high pressure ORC fluid enters the ORC evaporator at point 26. The flue gases exiting the ORC evaporator at point 33 are utilized in a domestic water heater. Water enters the domestic water heater at point 35 and exits at a higher temperature at point 36. Reverse osmosis (RO) desalination is used to produce fresh water (see lower right portion of Figure 11.9). Sea water at point 37 enters a filter to remove dissolved species and then passes through the absorber of the double-effect absorption chiller to increase the temperature to improve the efficiency of the OR desalination unit. A high pressure RO pump is used to increase the pressure of the water. High pressure sea water leaves the RO pump and enters the RO unit at point 40. Fresh water is produced at point 41 and stored in a fresh water tank for later use while high pressure brine enters a hydraulic turbine where it expands, generating electricity. Finally, low pressure brine exits the RO unit and is returned to the sea. The cooling load of the system is provided by a double-effect absorption chiller. A weak LiBr solution at point *a* leaves the pump at point 2 then passes through a high temperature heat exchanger where its temperature increases. In the high temperature generator, water is removed from the solution and the strong solution returns to the absorber after passing through the high and low temperature heat exchangers. On the other side, vapor leaves the high temperature generator at point 17 and enters the low temperature generator. The refrigerant steam produced by the low pressure generator is condensed by cooling water and then enters the expansion valve at point 8 where its pressure is reduced before it enters the evaporator at point 9. This low pressure vapor exits the evaporator as a saturated vapor at point 10 and enters the absorber. The absorption heat is removed by the seawater entering the absorber at point 38 to improve the efficiency of the RO desalination unit.

11.3.1 Thermodynamic Analysis

The thermodynamic modeling of the biomass based multigeneration system shown in Figure 11.9 is divided into four subsystems: (1) biomass combustor, (2) organic Rankine cycle and domestic water heater, (3) double-effect absorption chiller and proton exchange membrane (PEM) electrolyzer, and (4) reverse osmosis desalination unit. We determine the temperature profile in the multigeneration plant, input and output specific enthalpies, exergy flow rates, environmental impacts, exergy destruction rates, and exergy efficiencies. The relevant energy rate balances and governing equations for the main sections of the multigeneration plant in Figure 11.9 are described in the following subsections.

11.3.1.1 Biomass Combustion

As shown in Figure 11.9, biomass enters the combustor at point 30 and air enters at point 29. The composition of the biomass considered in this study (pine sawdust) is described in Table 11.5. A general chemical equation for biomass combustion with air assuming complete combustion is:

$$C_xH_yO_z + \omega\, H_2O + \lambda(O_2 + 3.76\, N_2) \rightarrow a\, CO_2 + b H_2O + c\, N_2 \tag{11.91}$$

where ω is the moisture content in the biomass fuel. The molar mass flow rate of the biomass can be expressed as

$$\dot{n}_{C_xH_yO_z} = \frac{\dot{m}_{biomass}}{M_{C_xH_yO_z}} \tag{11.92}$$

Table 11.5 Composition of pine sawdust biomass.

Composition quantity	Value (%)
Moisture content in biomass (by weight)	10
Elemental analysis (dry basis by weight)	
Carbon (C)	50.54
Hydrogen (H)	7.08
Oxygen (O)	41.11
Sulfur (S)	0.57

Here, $M_{C_xH_yO_z}$ is the molar mass of the biomass. The coefficients on the right hand side of Equation 11.91 are determined with element balances:

$$a = x \tag{11.93}$$

$$b = \frac{y + 2\omega}{2} \tag{11.94}$$

$$c = \frac{79}{21}\lambda \tag{11.95}$$

where

$$\lambda = \frac{2a + b - \omega - Z}{2} \tag{11.96}$$

To calculate the flue gas temperature leaving the combustor, we write an energy balance for a control volume around the biomass combustor, as follows:

$$\overline{h}_{C_xH_yO_z,30} + \omega\overline{h}_{H_2O,29} + \lambda\overline{h}_{O_2,29} + 3.76\lambda\overline{h}_{N_2,29} = a\overline{h}_{CO_2,31} + b\overline{h}_{H_2O,31} + c\overline{h}_{N_2,31} \tag{11.97}$$

Here, $\overline{h}_{C_xH_yO_z}$ is defined as [106]:

$$\overline{h}_{C_xH_yO_z,30} = x\overline{h}_{CO_2,30} + \left(\frac{y}{2}\right)\overline{h}_{H_2O(l),29} + \overline{LHV}_{biomass}M_{C_xH_yO_z} \tag{11.98}$$

For pure and dry biomass fuels, nitrogen and sulfur are usually negligible. Then, the lower heating value can be expressed for dry biomass with a chemical formula of CH_aO_b as follows:

$$\overline{LHV}_{dry} = \frac{400\,000 + 100\,600y - \frac{b}{1+0.5a}(117\,600 + 100\,600a)}{12 + a + 16b} \tag{11.99}$$

The lower heating value for biomass with moisture is expressible as [106]:

$$\overline{LHV}_{moist} = [1 - \mu_m - H_u]\overline{LHV}_{dry} - 2500H_u \tag{11.100}$$

where μ_m and H_u respectively denote the mineral matter content and moisture content of the biomass. Once the temperatures at points 29 and 30 are determined, Equation 11.97 can be solved for the temperature at point 31.

11.3.1.2 ORC Cycle
The ORC cycle considered here has four main components as follows:

Evaporator To determine the temperatures and specific enthalpies for flows interacting with the ORC evaporator, the following energy rate balance equation for the evaporator can be used:

$$\dot{m}_{32}h_{32} + \dot{m}_{26}h_{26} = \dot{m}_{33}h_{33} + \dot{m}_1h_1 \tag{11.101}$$

Considering a pinch point temperature in the evaporator T_{pp}, the following expression can be used to calculate the gas temperature leaving the evaporator, which is an important parameter for hot water production:

$$T_{pp} = T_{33} - T_{26} \tag{11.102}$$

ORC Turbine An energy rate balance for the ORC turbine and condenser yields the following relation:

$$\dot{m}_{27}h_{27} = \dot{W}_T + \dot{m}_{28}h_{28} \tag{11.103}$$

Also,

$$\eta_{ORC,T} = \frac{\dot{W}_{ORC,act}}{\dot{W}_{ORC,is}} \tag{11.104}$$

where h_{27} and h_{28} are the inlet and outlet specific enthalpies and $\dot{W}_{ORC,act}$ and $\dot{W}_{ORC,is}$ are actual and isentropic turbine power outputs.

ORC Condenser An energy rate balance equation for the condenser can be written as

$$\dot{m}_{28}h_{28} = \dot{m}_{24}h_{24} + \dot{Q}_{Cond} \tag{11.105}$$

ORC Pump The ORC pump work rate can be expressed using an energy rate balance for a control volume around the ORC pump as follows:

$$\dot{W}_{ORC,pump} = \dot{m}_{25}(h_{26} - h_{25}) \tag{11.106}$$

11.3.1.3 Domestic Water Heater

The hot gases leaving the evaporator enter the water heater and heat domestic hot water to 60°C. Water enters this heater at a pressure of 2 bar and a temperature of 20°C. An energy rate balance for this component follows:

$$\dot{m}_{33}C_{pg}(T_{33} - T_{34}) = \dot{m}_{DWH}(h_{36} - h_{35}) \tag{11.107}$$

11.3.1.4 Double-effect Absorption Chiller

Absorption chillers can be used for air conditioning and cooling purposes. Compared to the more conventional vapor compression refrigeration systems, absorption refrigeration systems replace the electricity consumption associated with vapor compression by a thermally driven system. This is accomplished by making use of absorption and desorption processes that employ a suitable working pair (a refrigerant and an absorbent). LiBr–water is a common working fluid for absorption systems in various cooling applications, including use in multigeneration systems.

In this analysis, the LiBr–water mixture is heated in the generator as shown in Figure 11.9. Heat provided by saturated water vapor via the heating process unit (point 24) allows separation of the refrigerant (H_2O) from the absorbent (LiBr solution). To model the single effect LiBr–water absorption chiller system used in this multigeneration system, the principle of mass conservation and the first and second laws of thermodynamics are applied to each component. In our analysis, each component is considered as a control volume with inlet and outlet streams, and heat and work interactions are considered. Mass rate balances are applied for the total mass and each material of the working fluid solution.

Absorber A mass rate balance equation for absorber can be written as

$$\dot{m}_{10} = \dot{m}_1 + \dot{m}_6 \tag{11.108}$$

A concentration rate balance equation for absorber can be expressed as

$$\dot{m}_1 x_1 = \dot{m}_6 x_6 \tag{11.109}$$

An energy rate balance equation for the absorber is expressible as

$$\dot{m}_{10} h_{10} + \dot{m}_6 h_6 = \dot{m}_1 h_1 + \dot{Q}_{ABS} \tag{11.110}$$

where

$$\dot{Q}_{ABS} = \dot{m}_{38}(h_{39} - h_{38}) \tag{11.111}$$

Pump   The absorption pump work rate can be expressed using mass balance and an energy rate balance for a control volume around the absorption pump respectively as follows:

$$\dot{m}_1 = \dot{m}_2 \tag{11.112}$$

$$\dot{W}_{\text{pump}} = \dot{m}_2(h_2 - h_1) \tag{11.113}$$

HEX I   Mass rate balances equation for the first heat exchanger can be written as

$$\dot{m}_2 = \dot{m}_3 \tag{11.114}$$

$$\dot{m}_4 = \dot{m}_5 \tag{11.115}$$

Concentration rate balance equations can be expressed as

$$x_2 = x_3 \tag{11.116}$$

$$x_4 = x_5 \tag{11.117}$$

An energy rate balance equation can be written as

$$\dot{m}_2 h_2 + \dot{m}_4 h_4 = \dot{m}_3 h_3 + \dot{m}_5 h_5 \tag{11.118}$$

and the pinch point temperature *PP* can be determined with the following:

$$T_{14} = T_{24} - PP \tag{11.119}$$

Low Temperature Generator (LGEN)   To determine the temperature and specific enthalpy for the low temperature generator (LGEN), the following energy and mass rate balance equations can be used:

$$\dot{m}_{16} = \dot{m}_4 + \dot{m}_7 \tag{11.120}$$

$$\dot{m}_4 x_4 = \dot{m}_{16} x_{16} \tag{11.121}$$

$$\dot{m}_{17} h_{17} - \dot{m}_{18} h_{18} + \dot{m}_{16} h_{16} = \dot{m}_7 h_7 + \dot{m}_4 h_4 \tag{11.122}$$

HEX II   Mass rate balance equations for the second heat exchanger can be written as

$$\dot{m}_3 = \dot{m}_{13} \tag{11.123}$$

$$\dot{m}_{14} = \dot{m}_{15} \tag{11.124}$$

Concentration rate balance equations can be expressed as

$$x_3 = x_{13} \tag{11.125}$$

$$x_{14} = x_{15} \tag{11.126}$$

An energy rate balance equation can be written as

$$\dot{m}_3 h_3 + \dot{m}_{14} h_{14} = \dot{m}_{13} h_{13} + \dot{m}_{15} h_{15} \tag{11.127}$$

High Temperature Generator (HGEN)   To determine the temperature and specific enthalpy for the high temperature generator (HGEN), the following energy and mass rate balance equations for the HGEN can be used:

$$\dot{m}_{13} = \dot{m}_{14} + \dot{m}_{17} \tag{11.128}$$

$$\dot{m}_{14}x_{14} = \dot{m}_{13}x_{13} \tag{11.129}$$

$$\dot{m}_{24}(h_{24} - h_{25}) + \dot{m}_{13}h_{13} = \dot{m}_{17}h_{17} + \dot{m}_{14}h_{14} \tag{11.130}$$

$$h_{17} = h(P_{14}, x = 1) \tag{11.131}$$

First Low Temperature Generator (LGEN I)   Mass and energy rate balances for the first low temperature generator can be written as

$$\dot{m}_{17} = \dot{m}_{18} \tag{11.132}$$

$$\dot{Q}_{LGEN} = \dot{m}_{17}h_{17} - \dot{m}_{18}h_{18} \tag{11.133}$$

Expansion Valves   Mass rate balance equations for the expansion valves can be written as follows:

$$\dot{m}_5 = \dot{m}_6 \tag{11.134}$$

$$\dot{m}_{18} = \dot{m}_{19} \tag{11.135}$$

$$\dot{m}_8 = \dot{m}_9 \tag{11.136}$$

$$h_5 = h_6 \tag{11.137}$$

$$h_{18} = h_{19} \tag{11.138}$$

$$h_8 = h_9 \tag{11.139}$$

Condenser   Mass and energy rate balance equations for the condenser can be written as

$$\dot{m}_8 = \dot{m}_{19} + \dot{m}_7 \tag{11.140}$$

$$\dot{m}_{19}h_{19} + \dot{m}_7h_7 = \dot{m}_8h_8 + \dot{Q}_{cond} \tag{11.141}$$

Evaporator   Mass and energy rate balance equations for the evaporator can be written as

$$\dot{m}_9 = \dot{m}_{10} \tag{11.142}$$

$$\dot{Q}_{EVP} = \dot{m}_{10}h_{10} - \dot{m}_9h_9 \tag{11.143}$$

$$h_{10} = h(T_{EVP}, x = 1) \tag{11.144}$$

11.3.1.5   Reverse Osmosis (RO) Desalination Unit

A typical seawater reverse osmosis desalination plant consists of three main processes: seawater intake, pre-treatment, and the RO system. Here, the RO system is the main process in which the separation occurs. The RO system includes a high pressure pump, the membrane separation unit, and an energy recuperation system. The raw water is pressurized by a high pressure pump and it is then supplied to the membranes where the seawater desalination occurs. In this analysis, a standard RO unit based on a typical seawater RO plant consisting of a single RO stage of *b* trains is considered, according to Salcedo *et al.* [107]. The rejected brine is pressurized at the outlet of the RO stage and then passes through a hydro-turbine in order to recover part of the energy consumed by the high pressure pump.

In order to model the RO desalination unit, energy rate balances are used. The net work input rate is expressible as follows:

$$\dot{W}_{net} = b_n(\dot{W}_{pump} - \dot{W}_{turbine}) \tag{11.145}$$

Here, b_n is the number of trains (7 in our analysis) and $\dot{W}_{pump}$ and $\dot{W}_{turbine}$ are RO pump required work rate and hydro-turbine work generation rate, which are expressible as follows:

$$\dot{W}_{pump} = \frac{\Delta P \, \dot{m}_{39}}{\eta_{pump} \, \rho_{39}} \tag{11.146}$$

$$\dot{W}_{turbine} = \frac{\Delta P \, \dot{m}_{42} \, \eta_{turbine}}{\rho_{42}} \tag{11.147}$$

where ΔP is the transmembrane pressure, and η_{pump} and $\eta_{turbine}$ are RO pump and hydro-turbine isentropic efficiencies respectively. The target fresh water mass flow rate, $\dot{m}_{41}$, is determined from the electricity driving the RO unit and the recovery ratio RR, which is one of the technical characteristics of the membrane, as follows:

$$\dot{m}_{41} = \frac{\dot{m}_{37}}{RR} \tag{11.148}$$

The transmembrane pressure can be expressed by the following equation:

$$\Delta P = J_w \, k_m + \Delta \pi \tag{11.149}$$

where k_m is the membrane permeability resistance, which has a value of $8.03 \times 10^{-11} \frac{m^2 s}{kgPa}$, and J_w is the volumetric permeate flow rate, expressed as

$$J_w = \frac{\dot{m}_{41}}{\rho_{41} \cdot n \, A_{mem}} \tag{11.150}$$

Here, n is the total number of membranes, which is 600 in this analysis, ρ_{41} is the density at point 41 and A_{mem} is the membrane area. In Equation 11.149, $\Delta \pi$ is the transmembrane osmotic pressure, which can be expressed as follows:

$$\Delta \pi = 805.1 \times 10^5 C_w R \tag{11.151}$$

Here, C_w is the membrane wall concentration, which can be expressed as

$$C_w = \frac{e^{\left(\frac{J_w}{K_{mass}}\right)} x_{39}}{e^{\left(\frac{J_w}{K_{mass}}\right)} (1 - R) + R} \tag{11.152}$$

where R denotes the membrane rejection coefficient, which has a value of 0.9975 in this analysis based on Salcedo $et\ al.$ [107]. Also, K_{mass} is the mass transfer coefficient, expressed as follows:

$$K_{mass} = 0.04 \, Re^{0.75} \, Sc^{0.33} \frac{D_s}{d} \tag{11.153}$$

where D_s is the diffusivity and d is the feed channel thickness. These parameters have respective values of $1.45 \times 7 \frac{m^2}{s}$ and 0.71 mm in this analysis.

In Equation 11.153, the Reynolds number is determined as

$$Re_{39} = \frac{\dot{m}_{39}}{N_{ch} \, L_W \mu_{39} \cdot N_P} \tag{11.154}$$

where N_{ch} and N_p represent the number of feed channels and the number of pressure vessels respectively, μ_{39} is the dynamic viscosity of the water, and L_W is the membrane width. Also, in Equation 11.153, Sc is the Schmidt number, defined as

$$Sc = \frac{\mu_{39}}{\rho_{39}\, D_s} \tag{11.155}$$

11.3.2 Exergy Analysis of the System

The exergy rate of each flow is calculated at all state points and the changes in exergy are determined for each major component. Exergy destruction rate expressions for all components in this multigeneration system (Figure 11.9) are listed in Table 11.6.

The exergy efficiency, defined as the product exergy output divided by the exergy input [47], can be expressed for the ORC power generation unit, the CHP unit and the multigeneration system, respectively, as follows:

$$\psi_{ORC} = \frac{\dot{W}_{net,ORC}}{\dot{Ex}_{biomass}} \tag{11.156}$$

$$\psi_{CHP} = \frac{\dot{W}_{net,ORC} + \dot{Ex}_{heating}}{\dot{Ex}_{biomass}} \tag{11.157}$$

$$\psi_{multi} = \frac{\dot{W}_{net,ORC} + \dot{Ex}_{heating} + \dot{Ex}_{cooling} + \dot{Ex}_{H_2} + \dot{Ex}_{36} + \dot{Ex}_{41}}{\dot{Ex}_{biomass}} \tag{11.158}$$

Table 11.6 Expressions for exergy destruction rates for components of the system.

Component	Exergy destruction rate expression
Combustor	$\dot{Ex}_{D,comb} = \dot{Ex}_{29} + \dot{Ex}_{30} - \dot{Ex}_{31}$
ORC evaporator	$\dot{Ex}_{D,evp} = \dot{Ex}_{32} + \dot{Ex}_{26} - \dot{Ex}_{27} - \dot{Ex}_{33}$
ORC turbine	$\dot{Ex}_{D,T} = \dot{Ex}_{27} - \dot{W}_T - \dot{Ex}_4$
Heating process	$\dot{Ex}_{D,heat} = \dot{Ex}_{28} - \dot{Ex}_Q - \dot{Ex}_{24}$
ORC pump	$\dot{Ex}_{D,P} = \dot{Ex}_{25} + \dot{W}_P - \dot{Ex}_{26}$
Absorption condenser	$\dot{Ex}_{D,cond} = \dot{Ex}_{19} + \dot{Ex}_{20} - \dot{Ex}_8 - \dot{Ex}_{21}$
Expansion valves	$\dot{Ex}_{D,exv} = \dot{Ex}_{18} - \dot{Ex}_{19} + \dot{Ex}_8 - \dot{Ex}_9 + \dot{Ex}_{15} - \dot{Ex}_{16} + \dot{Ex}_5 - \dot{Ex}_6$
Absorption evaporator	$\dot{Ex}_{D,evp} = \dot{Ex}_9 + \dot{Ex}_{21} - \dot{Ex}_{23} - \dot{Ex}_{10}$
Absorber	$\dot{Ex}_{D,abs} = \dot{Ex}_{10} + \dot{Ex}_6 + \dot{Ex}_{38} - \dot{Ex}_1 - \dot{Ex}_{39}$
Absorption pump	$\dot{Ex}_{D,P} = \dot{Ex}_1 + \dot{W}_P - \dot{Ex}_2$
Absorption heat exchanger I	$\dot{Ex}_{D,hexl} = \dot{Ex}_2 + \dot{Ex}_4 - \dot{Ex}_3 - \dot{Ex}_5$
Absorption heat exchanger h	$\dot{Ex}_{D,hexh} = \dot{Ex}_3 + \dot{Ex}_{14} - \dot{Ex}_{13} - \dot{Ex}_{15}$
High temperature absorption generator	$\dot{Ex}_{D,genh} = \dot{Ex}_{24} + \dot{Ex}_{13} - \dot{Ex}_{14} - \dot{Ex}_{17} - \dot{Ex}_{25}$
Low temperature absorption generator	$\dot{Ex}_{D,genl} = \dot{Ex}_{17} - \dot{Ex}_{18} - \dot{Ex}_Q$
PEM electrolyzer	$\dot{Ex}_{D,PEM} = \dot{Ex}_{22} + \dot{W}_{PEM} - \dot{Ex}_{24} - \dot{Ex}_{23}$
Domestic hot water heater	$\dot{Ex}_{D,DWH} = \dot{Ex}_{33} + \dot{Ex}_{35} - \dot{Ex}_{36} - \dot{Ex}_{34}$
RO pump	$\dot{Ex}_{D,RO\ pump} = \dot{Ex}_{39} + \dot{W}_P - \dot{Ex}_{40}$
RO desalination unit	$\dot{Ex}_{D,RO\ desalination} = \dot{Ex}_{40} - \dot{Ex}_{41} - \dot{Ex}_{42}$
RO hydraulic turbine	$\dot{Ex}_{D,h\ turbine} = \dot{Ex}_{42} - \dot{W}_T - \dot{Ex}_{43}$

where

$$\dot{Ex}_{\text{heating}} = \dot{Q}_{\text{cond}} \left(1 - \frac{T_0}{T_{\text{cond}}} \right) \tag{11.159}$$

$$\dot{Ex}_{\text{cooling}} = \dot{Q}_{\text{cooling}} \left(\frac{T_0 - T_{EVP}}{T_{EVP}} \right) \tag{11.160}$$

$$\dot{Ex}_{H_2} = \dot{m}_{H_2} ex_{H_2} \tag{11.161}$$

$$\dot{Ex}_{35} = \dot{m}_{35}(h_{35} - h_0) - T_0(s_{35} - s_0) \tag{11.162}$$

Also, $\dot{Ex}_{\text{biomass}}$ is the exergy flow rate of biomass, defined as [24]:

$$\dot{Ex}_{\text{biomass}} = \dot{m}_{\text{biomass}} \beta LHV_{\text{moist}} \tag{11.163}$$

Here, β is defined as

$$\beta = \frac{1.0414 + 0.0177 \left(\frac{H}{C} \right) - 0.3328 \left(\frac{O}{C} \right) \left\{ 1 + 0.0537 \left(\frac{H}{C} \right) \right\}}{1 - 0.4021 \left(\frac{O}{C} \right)} \tag{11.164}$$

where LHV_f denotes the lower heating value of the biomass. For pure and dry biomass fuels, nitrogen and sulfur are usually negligible and the lower heating value can be expressed for a biomass with a chemical formula of CH_aO_b as follows:

$$\overline{LHV}_{\text{dry}} = \frac{400\,000 + 100\,600y - \frac{b}{1+0.5a}(117\,600 + 100\,600a)}{12 + a + 16b} \tag{11.165}$$

The lower heating value for biomass with moisture is expressible as [106]:

$$\overline{LHV}_{\text{moist}} = [1 - \mu_m - H_u]\overline{LHV}_{\text{dry}} - 2500H_u \tag{11.166}$$

where μ_m and H_u respectively denote mineral matter content and moisture content in the biomass. Also, $\dot{Q}_{\text{heating}}$ and $\dot{Q}_{\text{cooling,chiller}}$ denote the heating load of the multigeneration system and the double-effect absorption chiller cooling load, while the last three terms in the numerator denote the energy values of the hydrogen, hot water, and fresh water products. It can be seen from these expressions that the energy efficiency of the multigeneration system must exceed that for power generation cycle.

11.3.3 Economic Analysis of the System

In order to perform an economic analysis and optimization of the multigeneration system in Figure 11.9, we first need to determine the purchase cost of each component used in the system as a function of the main design parameters. In this section, the cost function of each component and some economic parameters are defined and explained. Some parts of the biomass based multigeneration system shown in Figure 11.9 are similar to those in the multigeneration system analyzed earlier in this chapter. Here, the cost functions of the components that differ from those in the system examined earlier are provided.

11.3.3.1 Biomass Combustor and Evaporator

The cost of the biomass combustor and evaporator can be expressed as a function of the flue gas energy exiting the combustor, and the main pressure and temperature of the

ORC cycle [79]:

$$Z_{com}(\$) = 740(H_B)^{0.8} \exp\left(\frac{0.01P_{27} - 2}{14.29}\right) \exp\left(\frac{T_{27} - 350}{446}\right) \tag{11.167}$$

where

$$H_B = \dot{m}_g h_{31} \tag{11.168}$$

Here, $\dot{m}_g$ is the flue gas mass flow rate in kg/s.

11.3.3.2 Heating Process Unit

Several cost functions are available for the heating process considered for this biomass based multigeneration system. The cost function of the heating process here is defined as [103]:

$$Z_{heating}(\$) = 5714\dot{m}_{28} \tag{11.169}$$

11.3.3.3 Reverse Osmosis (RO) Desalination Unit

In this biomass based multigeneration system, a RO desalination unit is applied as previously noted. Its cost can be expressed as follows [108]:

$$Z_{RO}(\$) = 0.98m^3 \tag{11.170}$$

where m is the fresh water mass in kg.

11.3.4 Multi-objective Optimization

A multi-objective optimization method based on an evolutionary algorithm is applied to the multigeneration system for heating, cooling, electricity, hot water, fresh water, and hydrogen to determine the best design parameters for the system. Objective functions, design parameters and constraints, and the overall optimization procedure are described in this section.

11.3.4.1 Definition of Objectives

Two objective functions are considered here for multi-objective optimization: exergy efficiency (to be maximized) and total cost rate of product (to be minimized). The cost of pollution damage is assumed to be added directly to the required expenditures, making the second objective function the sum of thermodynamic and environmental objectives. Consequently, the objective functions in this analysis can be expressed as follows:

Exergy Efficiency

$$\Psi_{multi} = \frac{\dot{W}_{net,ORC} + \dot{Ex}_{heating} + \dot{Ex}_{cooling} + \dot{Ex}_{H_2} + \dot{Ex}_{36} + \dot{Ex}_{41}}{\dot{Ex}_{biomass}} \tag{11.171}$$

Total Cost Rate

$$\dot{C}_{tot} = \sum_k \dot{Z}_k + \dot{C}_f + \dot{C}_{env} \tag{11.172}$$

where the cost rates of environmental impact and fuel are expressed as

$$\dot{C}_{env} = C_{CO_2}\dot{m}_{CO_2} \tag{11.173}$$

$$\dot{C}_f = c_f\dot{m}_f \, LHV \tag{11.174}$$

Table 11.7 Optimization constraints and their rationales.

Constraint	Reason
$0.2 \text{ kg/s} < \dot{m}_{\text{biomass}} < 0.4 \text{ kg/s}$	Biomass fuel limitation
$T_{\text{in,pump}} < 115 \text{ K}$	Material temperature limit
$1500 \text{ kPa} < P_{\text{main}} < 3000 \text{ kPa}$	Commercial availability
$320°\text{C} < T_{\text{main}} < 400°\text{C}$	Commercial availability
$10°\text{C} < PP < 35°\text{C}$	Heat transfer limit
$\eta_T < 0.9$	Commercial availability
$\eta_p < 0.9$	Commercial availability
$2°\text{C} < T_{EVP} < 6°\text{C}$	Cooling load limitation

Here, $\dot{Z}_K$ is the purchase cost rate of component and the summation is for all components. More details about equipment purchase costs can be found elsewhere [64]. The purchase cost of each component in this multigeneration system was described in section 11.3.3. Also C_f is the fuel cost which is taken to be 0.01 \$/kWh in this analysis. In this analysis, we express the environmental impact as the total cost rate of pollution damage (\$/s) due to CO_2 emissions by multiplying the respective flow rates by the corresponding unit damage cost (C_{CO2}), which is taken to be 0.024 \$/kg [64]. The cost of pollution damage is assumed here to be added directly to other system costs.

11.3.4.2 Decision Variables

The following decision variables (design parameters) are selected for this analysis: biomass flow rate ($\dot{m}_{\text{biomass}}$), ORC pump inlet temperature ($T_{\text{in,pump}}$), ORC evaporator pinch point temperature (PP) difference, ORC turbine inlet pressure (P_{main}), ORC turbine inlet temperature (T_{main}), ORC turbine isentropic efficiency (η_T), pump isentropic efficiency (η_p) and absorption chiller evaporator temperature (T_{EVP}). Decision variables may be varied within a reasonable range in the optimization procedure, and constraints for this based on earlier reports are listed in Table 11.7.

11.3.5 Optimization Results

The results of the optimization are described. The genetic algorithm optimization is performed for 250 generations, using a search population size of $M = 100$ individuals, a crossover probability $p_c = 0.9$, a gene mutation probability $p_m = 0.035$, and a controlled elitism value $c = 0.55$. Figure 11.10 shows the Pareto frontier solution for the multi-objective optimization of the multigeneration system for the objective functions in Equations 11.171 and 11.172. It can be seen in this figure that the total cost rate of products increases moderately as the total exergy efficiency of the cycle increases to about 32%. Increasing the total exergy efficiency from 32% to 34% increases the cost rate of product significantly.

The results for the optimum exergy efficiency and total cost rate for all points evaluated over 300 generations are shown in Figure 11.11. The Pareto optimal curve (best rank) is clearly visible in the lower part of the figure (solid line) which is separately shown in Figure 11.10. There, the maximum exergy efficiency exists at design point C (33.5%), while the total cost rate of products is the greatest at this point (874.62 \$/hr). But, the

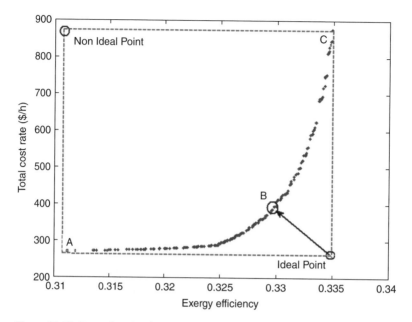

Figure 11.10 Pareto frontier showing best trade-off values for the objective functions.

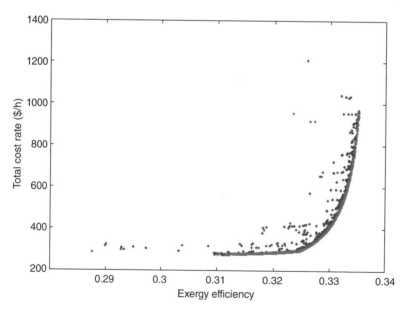

Figure 11.11 Results of all evaluations during 300 generations using a genetic algorithm. An approximation of the Pareto frontier is visible in the lower part of the figure.

minimum value for the total cost rate of product occurs at design point A which is about 271.84 \$/hr. Design point A is the optimal situation when total cost rate of product is the sole objective function, while design point C is the optimum point when exergy efficiency is the sole objective function. In multi-objective optimization, a process of decision-making for selection of the final optimal solution from the available solutions is required. The process of decision-making is usually performed using Figure 11.10 with the aid of a hypothetical point (the ideal point), at which both objectives have their optimal values independent of the other objectives. It is clear that it is impossible to have both objectives at their optimum point simultaneously and, as shown in Figure 11.10, the ideal point is not a solution located on the Pareto frontier. The closest point of the Pareto frontier to the ideal point might be considered as a desirable final solution.

In this case, the Pareto optimum frontier exhibits weak equilibrium, that is, a small change in exergy efficiency due to varying the operating parameters causes a large variation in the total cost rate of product. Therefore, the ideal point cannot be utilized for decision-making in this problem. In selection of the final optimum point, it is desired to achieve a better magnitude for each objective than its initial value for the base case problem. Since each point can be utilized as the optimized point in the Pareto solution to a multi-objective optimization, the selection of the optimum solution depends on the preferences and criteria of the decision maker who may select a different point as for the optimum solution depending on needs. Table 11.8 shows all the design parameters for points A–C.

As shown in Figure 11.10, the optimized values for exergy efficiency on the Pareto frontier range between 31% and 34%. To provide a good relation between exergy efficiency and total cost rate, a curve is fitted on the optimized points obtained from the evolutionary algorithm. This fitted curve is shown in Figure 11.11, and an expression for it is:

$$\dot{C}_{total} = \frac{360\psi^3 + 194.8\psi^2 - 253.4\psi + 49.76}{\psi^3 - 57.46\psi^2 - 28.92\psi - 3.25} \times 100 \tag{11.175}$$

This is allowable when the exergy efficiency varies between 29% and 34%. To study the variation of thermodynamic characteristics, three points (A to C) on the Pareto frontier are considered. Table 11.9 shows for the system the total cost rate, the total

Table 11.8 Optimized values for design parameters of the system based on multi-objective optimization.

Design parameter	Point on Pareto frontier		
	A	B	C
$\dot{m}_{biomass}$ (kg/s)	0.20	0.20	0.21
$T_{in,pump}$ (°C)	115	114	114
P_{main} (kPa)	1500	2049	3818
T_{main} (°C)	320	400	399
PP (°C)	12.0	10.3	10.6
η_T	0.80	0.90	0.9
η_p	0.82	0.84	0.85
T_{EVP} (°C)	5.2	7.0	6.9

Table 11.9 Thermodynamic characteristics of three points on the Pareto frontier.

Point	$\dot{W}_{net}$ (kW)	ψ	$\dot{Ex}_{D,tot}$ (kW)	$\dot{Q}_{cooling}$ (kW)	$\dot{Q}_{heating}$ (kW)	$\dot{C}_{tot}$ ($/h)	CO_2 emission (kg/kWh)	$\dot{m}_{H_2}$ (kg/h)	$\dot{m}_{DWH}$ (kg/s)	$\dot{m}_{fresh}$ (kg/s)
A	278.35	0.31	3749	2000	1487	271.8	346.8	1.19	0.52	1.08
B	307.10	0.32	3473	1543	1741	362.0	364.0	1.30	0.53	1.19
C	351.10	0.33	3477	1614	1644	874.0	361.7	1.50	0.52	1.40

exergy destruction rate, the exergy efficiency, the heating and cooling loads, and the CO_2 emission. From point A to point C in this table, both the total cost rates of the system and the exergy efficiencies increase. As already stated, point A is preferred when total cost rate is a single objective function and design point C when exergy efficiency is a single objective function. Design point B provides better results for both objective functions. Other thermodynamic properties correctly confirm this trend. For instance, from point B to C, the total exergy destruction rate decreases when the exergy efficiency increases.

To better understand the variations of all design parameters, the scattered distribution of the design parameters are shown in Figures 11.12 and 11.13. The results show that ORC pump inlet temperature (Figure 11.12b) and absorption chiller evaporator temperature (Figure 11.13d) tend to become as high as possible. This observation means that an increase in these parameters leads to better optimization results. For example, an increase in these design parameters leads to improvement for both objective functions in multi-objective optimization. We also observe that the ORC turbine inlet pressure (Figure 11.12c), the ORC turbine inlet temperature (Figure 11.12d), the evaporator pinch point temperature difference (Figure 11.13a), ORC turbine isentropic efficiency (Figure 11.13b), and the ORC pump isentropic efficiency (Figure 11.13c) have scattered distributions in their allowable domains, suggesting that these parameters have important effects on the trade-off between exergy efficiency and total cost rate. Design parameters selected with their maximum values indicate that they do not exhibit a conflict between two objective functions, indicating that increasing those design parameters leads to an improvement to both objective functions.

To better understand the multi-objective optimization results, a sensitivity analysis is performed. The effects of varying each of the design parameters for points A–C on both objective functions are investigated. Figure 11.14 shows the effects of biomass flow rate on system exergy efficiency and total cost rate. An increase in biomass flow rate is seen to have a negative effect on both objective functions. An increase in biomass flow rate leads to a decrease in system exergy efficiency as the denominator of Equation 11.171 increases. Also, an increase in biomass flow rate parameter increases the total cost rate of the system because the cost associated with the fuel increases as the biomass mass flow rate increases.

Figure 11.15 shows the variation of ORC turbine inlet pressure on both objective functions. As shown in this figure, an increase in turbine inlet pressure results in an increase in both objective functions. When the turbine inlet pressure rises, the exergy efficiency of the system increases, mainly due to an increase in cooling load and net power output of the system.

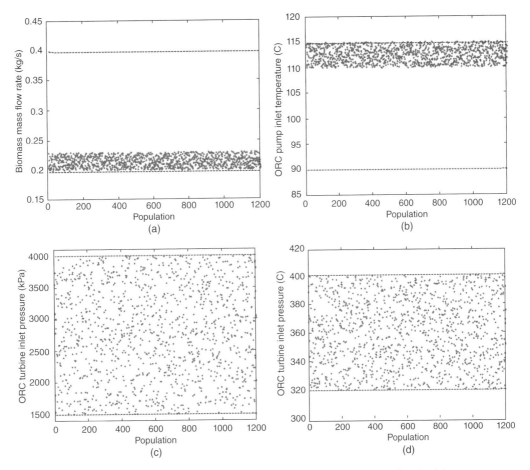

Figure 11.12 Scattered distribution of decision variables with population in Pareto frontier: (a) biomass flow rate, (b) ORC pump inlet temperature, (c) ORC turbine inlet pressure, (d) ORC turbine inlet temperature.

An increase in ORC turbine inlet pressure reduces the heating load of the system while an increase in this pressure has a positive effect on the cooling load of the system. An energy balance for a control volume around the ORC evaporator shows that, when the energy input from biomass is constant, a reduction in turbine inlet specific enthalpy increases the ORC mass flow rate. Since inlet and outlet specific enthalpies of the generator in the absorption cycle are constant, an increase in the ORC mass flow rate leads to an increase in energy input rate to the absorption system, which increases its cooling load.

For the heating load, an increase in ORC turbine inlet pressure while holding other design parameters fixed decreases the turbine inlet specific enthalpy and, since the turbine outlet specific enthalpy is a function of the turbine inlet specific enthalpy and turbine isentropic efficiency, this a corresponding decrease in turbine outlet enthalpy (h_{28}), which is the inlet energy for the heating process unit. Although the ORC mass flow rate increases, as already discussed, the reduction in specific enthalpy of the heating process

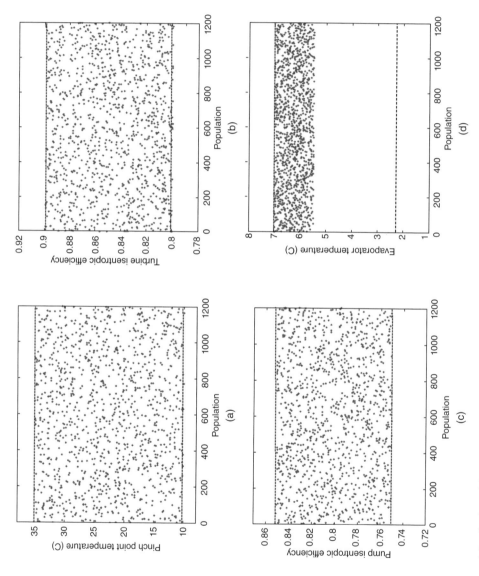

Figure 11.13 Scattered distribution of decision variables with population in Pareto frontier: (a) Pinch point temperature, (b) ORC turbine isentropic efficiency, (c) ORC pump isentropic efficiency, (d) absorption chiller evaporator temperature.

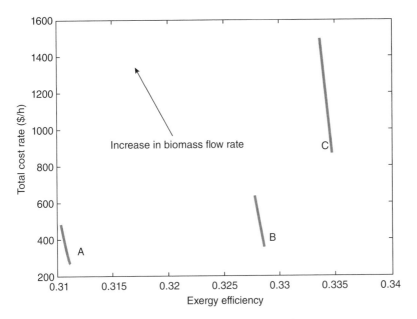

Figure 11.14 Effects of biomass flow rate on both objective functions.

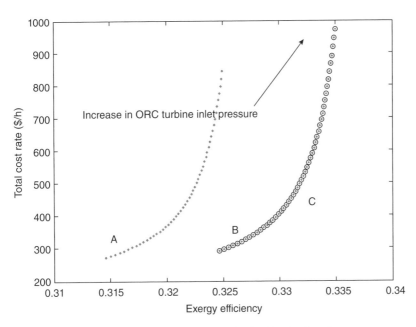

Figure 11.15 Effects of turbine inlet pressure on both objective functions.

dominates. In addition, an increase in ORC turbine inlet pressure results in increase in the ORC mass flow rate entering the turbine, which increases the turbine output work rate. Therefore, the combination of these effects leads to an increase in the system exergy efficiency. Figure 11.15 also shows that an increase in turbine inlet pressure results in an increase in total cost rate of the system. This is due to an increase in turbine purchase

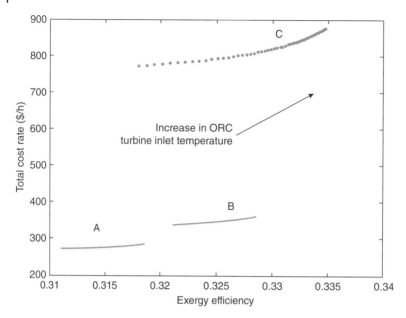

Figure 11.16 Effects of turbine inlet pressure on both objective functions.

cost as the inlet pressure increases. As a result, an increase in ORC turbine inlet pressure has negative and positive effects on the objective functions. This is why the scatter distribution for the ORC turbine inlet pressure is widely scattered.

Figure 11.16 shows the variation of turbine inlet temperature on both objective functions. An increase in turbine inlet temperature is seen to increase the exergy efficiency of the system for points *A–C* on the Pareto curve. Increasing the turbine inlet temperature leads to an increase in turbine inlet specific enthalpy, when other parameters are kept fixed. This increase leads to a rise in the turbine work rate which results in an increase in the exergy efficiency, according to Equation 11.171. However, an increase in turbine inlet temperature raises the total cost of the system, due to an increase in turbine purchase cost. Since an increase in this parameter has positive and negative effects on both objective functions, the variation of this design parameter within its allowable range exhibits a scattered distribution as shown in Figure 11.12c.

Figure 11.17 shows the effect of evaporator pinch point temperature on both objective functions. It is seen that an increase in pinch point temperature reduces the system exergy efficiency. This is due to the fact that the higher the pinch point temperature becomes, the lower is the energy being utilized in the evaporator, which leads to a reduction of ORC turbine power output. But, an increase in pinch point temperature, while other design parameters are held fixed, reduces the heat transfer area of the evaporator, explaining why the total cost rate of the system decreases.

Since an increase in pinch point temperature has positive and negative effects on both objective functions, the variation of this design parameter within its allowable range exhibits a scattered distribution as shown in Figure 11.16a.

Figure 11.18 illustrates the effect of turbine isentropic efficiency (η_T) on both objective functions. An increase in η_T results in an increase in system exergy efficiency and in total cost rate of the system. Increasing the turbine isentropic efficiency results in an

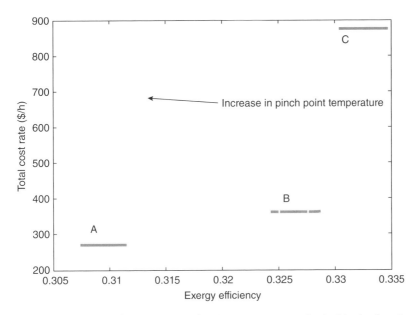

Figure 11.17 Effects of evaporator pinch point temperature on both objective functions.

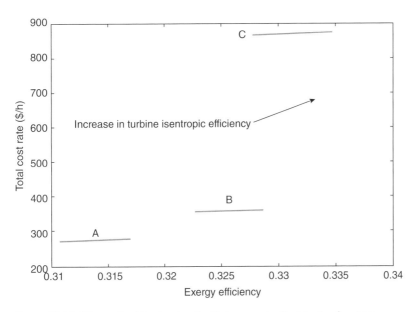

Figure 11.18 Effects of turbine isentropic efficiency on both objective functions.

increase in the steam turbine power output, which directly leads to an increase in the exergy efficiency. It is also seen that increasing the turbine isentropic efficiency leads to an increase in steam turbine purchase and maintenance costs. Since an increase in the turbine isentropic efficiency has positive and negative effects on both objective functions, its variation within its allowable range exhibits a scattered distribution as shown in Figure 11.13b.

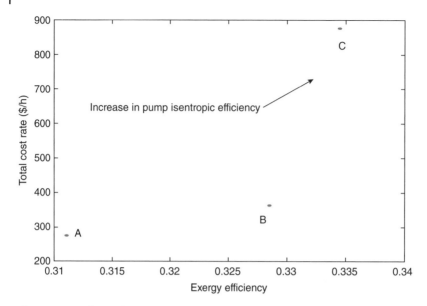

Figure 11.19 Effects of pump isentropic efficiency on both objective functions.

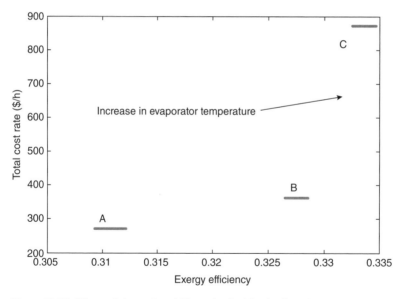

Figure 11.20 Effects of absorption chiller on both objective functions.

Two remaining components are now considered. Figure 11.19 shows that pump isentropic efficiency does not have a significant effect on both objective functions, as its purchase cost is small. Figure 11.20 shows the variation of the objective function by changing the absorption chiller evaporator temperature. An increase in evaporator temperature results in an increase in cooling load of the absorption chiller and at a same time increase the cost of the chiller.

11.4   Concluding Remarks

This chapter discusses the concept of multigeneration and introduces the advantages of multigeneration energy systems for producing multiple product outputs. Novel multigeneration energy systems, ranging from non-renewable to renewable based systems for various locations are considered, and optimization is applied to determine the optimal design parameters. The benefits of using exergy methods are emphasized. The process of defining of objective functions subject to reasonable constraints to ensure reliable results is covered. An evolutionary algorithm based optimization is applied to each multigeneration system considered and the optimal design parameters are carefully determined. To enhance understanding of the design criteria, a sensitivity analysis is conducted to investigate how each objective function varies in response to small changes in selected design parameters.

References

1 Dincer, I. (2000) Renewable energy and sustainable development: a crucial review. *Renewable and Sustainable Energy Reviews* **4**:157–175.

2 Dincer, I. and Zamfirescu, C. (2012) Renewable-energy-based multigeneration systems. *International Journal of Energy Research* **36**(15):1403–1415.

3 Ahmadi, P., Rosen, M. A. and Dincer, I. (2012) Multi-objective exergy-based optimization of a polygeneration energy system using an evolutionary algorithm. *Energy* **46**:21–31.

4 Khaliq, A., Kumar, R. and Dincer, I. (2009) Performance analysis of an industrial waste heat-based trigeneration system. *International Journal of Energy Research* **33**:737–744.

5 Ahmadi, P., Dincer, I. and Rosen, M. A. (2012) Exergo-environmental analysis of an integrated organic Rankine cycle for trigeneration. *Energy Conversion and Management* **64**:447–453.

6 Farshi, L. G., Mahmoudi, S. S., Rosen, M. A. and Yari, M. (2012) Use of low grade heat sources in combined ejector–double effect absorption refrigeration systems. *Proceedings of the Institution of Mechanical Engineers, Part, A.: Journal of Power and Energy* **226**:607–622.

7 Wang, J., Dai, Y. and Sun, Z. (2009) A theoretical study on a novel combined power and ejector refrigeration cycle. *International Journal of Refrigeration* **32**:1186–1194.

8 Ahmadi, P., Dincer, I. and Rosen, M. A. (2012) Energy and exergy analyses of hydrogen production via solar-boosted ocean thermal energy conversion and PEM electrolysis. *International Journal of Hydrogen Energy* **38**:1795–1805.

9 Ni, M., Leung, M. K. and Leung, D. Y. (2008) Energy and exergy analysis of hydrogen production by a proton exchange membrane (PEM) electrolyzer plant. *Energy Conversion and Management* **49**:2748–2756.

10 Dincer, I. and Rosen, M. A. (2012) *Exergy: Energy, Environment and Sustainable Development*, Second edition, Elsevier.

11 Roosen, P., Uhlenbruck, S. and Lucas, K. (2003) Pareto optimization of a combined cycle power system as a decision support tool for trading off investment vs. operating costs. *International Journal of Thermal Sciences* **42**:553–560.

12 Peters, M. S., Timmerhaus, K. D., West, R. E., Timmerhaus, K. and West, R. (1968) *Plant Design and Economics for Chemical Engineers*, McGraw-Hill, New York.

13 Mabrouk, A. A., Nafey, A. and Fath, H. (2007) Thermoeconomic analysis of some existing desalination processes. *Desalination* **205**:354–373.

14 Lian, Z., Chua, K. and Chou, S. (2010) A thermoeconomic analysis of biomass energy for trigeneration. *Applied Energy* **87**:84–95.

15 Genç, G., Çelik, M. and Serdar Genç, M. (2012) Cost analysis of wind-electrolyzer-fuel cell system for energy demand in Pınarbaşı-Kayseri. *International Journal of Hydrogen Energy* **37**(**17**): 12,158–12,166.

16 Bejan, A., Tsatsaronis, G. and Moran, M. (1995) *Thermal Design and Optimization*, John Wiley & Sons, Inc., New York.

17 Ahmadi, P., Dincer, I. and Rosen, M. A. (2011) Exergy, exergoeconomic and environmental analyses and evolutionary algorithm based multi-objective optimization of combined cycle power plants. *Energy* **36**:5886–5898.

18 Cohce, M., Dincer, I. and Rosen, M. (2011) Energy and exergy analyses of a biomass-based hydrogen production system. *Bioresource Technology* **102**: 8466–8474.

19 Hughes, E. E. and Tillman, D. A. (1998) Biomass cofiring: status and prospects 1996. *Fuel Processing Technology* **54**:127–142.

20 Mujeebu, M., Jayaraj, S., Ashok, S., Abdullah, M. and Khalil, M. (2009) Feasibility study of cogeneration in a plywood industry with power export to grid. *Applied Energy* **86**:657–662.

21 Ahmadi, P., Dincer, I. and Rosen, M. A. (2013) Development and assessment of an integrated biomass-based multi-generation energy system. *Energy* **56**(**1**):155–166.

Study Questions/Problems

1 Consider the solar based multigeneration system shown in Figure 11.21.

I) Carry out the following solution procedure:

a. Define a problem statement.

b. Show the cycle on a *T-s* diagram, and include saturation lines.

c. Make assumptions and approximations.

d. Identify relevant physical laws.

e. List relevant properties.

f. Write all rate balance equations (mass, energy, entropy, exergy).

g. Perform necessary calculations using EES software.

h. Determine which type of solar collector will be appropriate for this power plant and include solar data in the calculations.

i. Conduct a performance evaluation using energy and exergy efficiencies and exergy destruction rates.

j. Discuss and interpret the results, providing reasoning and verification.

k. Discuss associated performance, environmental, and sustainability issues.

l. Make recommendations for enhanced performance, environmental stewardship, and sustainability.

II) Perform a parametric study to show how system performance is affected by varying the operating conditions (e.g., *T*, *P*, *m*).

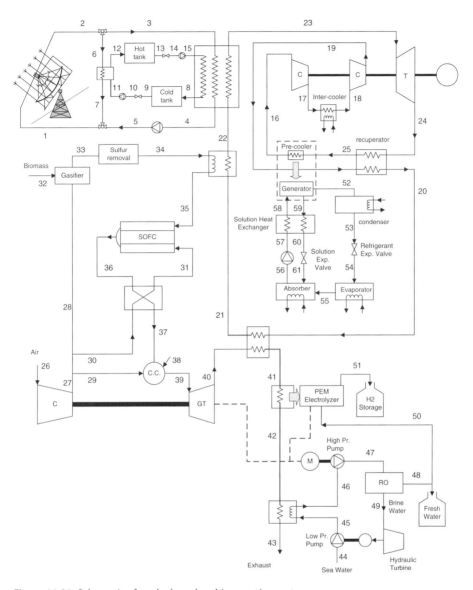

Figure 11.21 Schematic of a solar based multigeneration system.

III) Apply a single objective optimization to determine the maximum exergy efficiency of the system and discuss the results.

2 A solar based multigeneration energy system is shown in Figure 11.22. Apply mass, energy, entropy, and exergy rate balance equations to determine the exergy destruction rate and exergy efficiency of the system and find the optimal exergy efficiency of the system. Make reasonable assumptions.

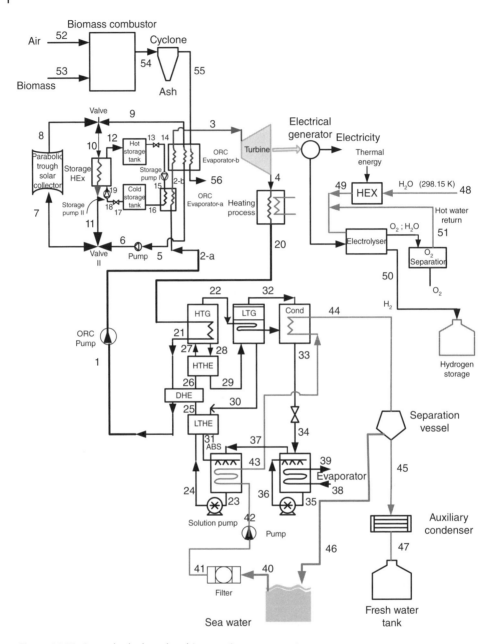

Figure 11.22 A novel solar based multigeneration energy system.

Index

Optimization of Energy Systems, First Edition. Ibrahim Dincer, Marc A. Rosen, and Pouria Ahmadi.
© 2017 John Wiley & Sons Ltd. Published 2017 by John Wiley & Sons Ltd.